CW00521144

Guidebook on

# Molecular Modeling in Drug Design

# Guidebook on Molecular Modeling in Drug Design

*Edited by*

**N. CLAUDE COHEN**

Ciba-Geigy Limited

Pharmaceuticals Division

CH-4002 Basel, Switzerland

**ACADEMIC PRESS**

San Diego New York Boston

London Sydney Tokyo Toronto

*Front cover photograph*: Ray traced figure of van der Waals representation of the peptide. See Chapter 2, Figure 14.

This book is printed on acid-free paper. ♾



Academic Press, Inc.
A Division of Harcourt Brace & Company
525 B Street, Suite 1900, San Diego, California 92101-4495

*United Kingdom Edition published by*
Academic Press Limited
24-28 Oval Road, London NW1 7DX

Library of Congress Cataloging-in-Publication Data

Guidebook on molecular modeling in drug design / edited by N. Claude Cohen.
p. cm.
Includes bibliographical references and index.
ISBN 0-12-178245-X (alk. paper)
1. Drugs--Design. 2. Molecules--Models. I. Cohen, N. Claude.
RS420.G85 1995 95-21396
CIP

PRINTED IN THE UNITED STATES OF AMERICA
96 97 98 99 00 01 BB 9 8 7 6 5 4 3 2 1

# Contents

## 2 Molecular Graphics and Modeling: Tools of the Trade

RODERICK E. HUBBARD



## 3 Molecular Modeling of Small Molecules

TAMARA GUND

## 4 Computer-Assisted New Lead Design

**AKIKO ITAI, MIHO YAMADA MIZUTANI, YOSHIHIKO NISHIBATA, AND NUBUO TOMIOKA**

## 5 Experimental Techniques and Data Banks

**JOHN P. PRIESTLE AND C. GREGORY PARIS**



## 6 Computer-Assisted Drug Discovery

**PETER GUND, GERALD MAGGIORA, AND JAMES P. SNYDER**

## 7 Modeling Drug–Receptor Interactions

**KONRAD F. KOEHLER, SHASHIDHAR N. RAO, AND JAMES P. SNYDER**

# Contributors

*Numbers in parentheses indicate the pages on which the authors' contributions begin.*

**N. Claude Cohen** (1) Ciba-Geigy Limited, Pharmaceuticals Division, CH-4002 Basel, Switzerland

**Peter Gund** (219) Molecular Simulations Incorporated, Burlington, Massachusetts 01803

**Tamara Gund** (55) New Jersey Institute of Technology, Newark College of Engineering, Biomedical Engineering Program, Department of Chemistry, Newark, New Jersey 07102

**Roderick E. Hubbard** (19) Department of Chemistry, University of York, Heslington, York Y01 5DD, United Kingdom

**Akiko Itai** (93) Laboratory of Medicinal Molecular Design, Faculty of Pharmaceutical Sciences, The University of Tokyo, Bunkyo-ku, Tokyo 113, Japan

**Konrad F. Koehler** (235) Department of Medicinal Chemistry, Istituto di Ricerche di Biologia Molecolare (IRBM), 00040 Pomezia, Roma, Italy

**Gerald Maggiora** (219) Upjohn Laboratories, Kalamazoo, Michigan 49001

**Yoshihiko Nishibata** (93) Laboratory of Medicinal Molecular Design, Faculty of Pharmaceutical Sciences, The University of Tokyo, Bunkyo-ku, Tokyo 113, Japan

**C. Gregory Paris** (139) Research Division, Ciba-Geigy Pharmaceuticals, Summit, New Jersey 07901

**John P. Priestle** (139) Ciba-Geigy Limited, Pharmaceuticals Division, CH-4002 Basel, Switzerland

**Shashidhar N. Rao** (235) Drug Design Section, Searle Research and Development, Skokie, Illinois 60025

**James P. Snyder** (219, 235) Emerson Center for Scientific Computation, Department of Chemistry, Emory University, Atlanta, Georgia 30322

**J. P. Tollenaere** (337) Department of Theoretical Medicinal Chemistry, Janssen Pharmaceutica, B-2340 Beerse, Belgium

**Nobuo Tomioka** (93) Laboratory of Medicinal Molecular Design, Faculty of Pharmaceutical Sciences, The University of Tokyo, Bunkyo-ku, Tokyo 113, Japan

**Miho Yamada Mizutani** (93) Laboratory of Medicinal Molecular Design, Faculty of Pharmaceutical Sciences, The University of Tokyo, Bunkyo-ku, Tokyo 113, Japan

# Preface

Molecular modeling has become a well-established discipline in pharmaceutical research. It has created unprecedented opportunities for assisting the medicinal chemist in the rational design of new therapeutic agents. This book is intended to be a guide for advanced students and chemists who are entering the field of molecular modeling. It elucidates the important role this research area is assuming in the understanding of the three-dimensional (3-D) aspects of drug–receptor interactions at the molecular level.

Advances in computer hardware and software and in theoretical medicinal chemistry have brought high-performance computing and graphic tools within the reach of most academic and industrial laboratories, thus facilitating the development of useful approaches to rational drug design. This book provides the reader with the basic background necessary to approach such systems. The discipline has various aspects, i.e., computer software and hardware, structural and quantum chemistry, structure–activity relationships, force-field simulations, superimposition techniques, 3-D database searching, etc., which we have presented in a unified way.

Our central aim was to provide a comprehensive overview of the strategies currently used in computer-assisted drug design, such as pharmacophore-based and structure-based drug design.

The reader of this book will find not only the fundamental concepts (the molecular basis of drug action, molecular simulations, molecular mimicry, etc.), but also ways of applying them to real problems. Although it is not necessary to read the book strictly in chapter order, this may be preferred as there is some progression in technicalities as the subject is developed.

The first chapter places the field in perspective, whereas Chapter 2 presents the molecular modelist's "panoply" of modeling hardware and software equipment. Chapter 2 covers the areas of computer graphic basic operations that are common to all molecular modeling applications. More specialized aspects such as molecular mechanics and dynamics, the 3-D representation of various molecular properties, and methods for deriving bioactive conformations are introduced in Chapter 3. Manual docking has been used extensively in a number of projects and has contributed to the creation of sophisticated automated treatments, as presented in Chapter 4. This chapter shows how tailor-made molecules can be identified either by construction methods or by extraction from a 3-D database. The most important experimental techniques for obtaining relevant 3-D information on small molecules and proteins, namely X-ray crystallography and nuclear magnetic resonance spectroscopy, are discussed in the first part of Chapter 5, whereas the second part presents 3-D databases and current techniques used in database searching approaches. Chapter 6 gives an overview of the current practice of computer-aided drug discovery and development. It includes useful tables with some examples of successful use of molecular modeling for drug discovery, and discusses the accuracy of current computational methods. In addition, organizational considerations concerning the reporting relationships of molecular modeling groups in the pharmaceutical industry are presented. Chapter 7, on drug–receptor interactions, shows how it is possible to take full advantage of the knowledge of the 3-D structure of the target protein in the design of drugs rationally conceived on the basis of their complementarity with the target macromolecule. The book closes with a glossary of more than 100 terms currently used in the field.

More and more medicinal chemistry publications contain substantial molecular modeling analyses. Students and chemists often encounter difficulties in learning the principles of modeling. Available review articles are not sufficient and are either too specialized or too general. Likewise, available books in the field generally are those of scientific proceedings and appear as mere collections of chapters cataloging what has been presented. The intent of this book is to fill this gap by providing advanced students and chemists with the very information they need to learn the fundamental concepts of molecular modeling, and enabling them to understand and apply these concepts in their current research.

*N. Claude Cohen*

# 1

# The Molecular Modeling Perspective in Drug Design

**N. CLAUDE COHEN**
Ciba-Geigy Limited
Pharmaceuticals Division
CH-4002 Basel, Switzerland

## I. DEFINITION OF MOLECULAR MODELING

Medicinal chemists today are facing many complicated challenges. The most demanding and perhaps the most rewarding one is the rational design of new therapeutic agents for treating human disease. For many years the strategy for discovering new drugs consisted of taking a lead structure and developing a chemical program for finding analog molecules exhibiting the desired biological properties. Generally found by chance observation or by random screening, initial lead compounds also encompassed, in the last decade, the natural ligand of the system concerned. The process involved several trial and error cycles patiently developed and analyzed by medicinal chemists utilizing their experience and chemical intuition to ultimately select a candidate analog for further development. The entire process is laborious, expensive, and, perhaps when looked at today, conceptually inelegant. However, the undeniable fact is that this process has provided most of the existing medications that are used today for indications ranging from the treatment of minor pain to life-threatening diseases. The traditional methods of drug discovery are now being supplemented by more direct approaches made possible by the understanding of the molecular processes involved in the underlying disease. In this perspective, the starting point in drug design is the molecular target (receptor, enzyme) in the body instead of the existence of an already known lead structure.

The scientific concepts underlying this approach have been understood for generations, but their practical application was beyond the reach of existing technology. The existence of receptors and the lock-and-key concepts currently considered in drug design were formulated by P. Ehrlich (1909) and E. Fischer (1894). It was only in the seventies that it became possible to understand some of the subtleties of the mechanisms involved in life processes. Pure samples of protein targets were isolated and X-ray crystallography revealed their molecular architecture. It then became possible to learn how precisely three-dimensional (3-D) structures control the regulation of life processes. In order to further such progress, a rational approach to drug discovery has emerged in the pharmaceutical industry and has contributed to the rapid development of molecular modeling as a full discipline.

The concepts used in 3-D drug design are quite simple. New molecules are conceived either on the basis of similarities with known reference structures or on the basis of their complementarity with the 3-D structure of known active sites. Molecular interactions are regulated by subtle recognition and discrimination processes whereby the 3-D features and the binding energies play an important role. Molecular modeling is a discipline that contributes to the understanding of these processes in a qualitative and sometimes quantitative way. It not only presents means for analyzing the details of the molecular machinery involved in a known system and understanding the way the biological system functions, but it also provides the necessary tools for predicting the potential possibilities of prototype candidate molecules. Molecular modeling can be simply considered as a range of computerized techniques based on theoretical chemistry methods and experimental data that can be used either to analyze molecules and molecular systems or to predict molecular and biological properties.

The techniques currently available provide extensive insight into the precise molecular features that are responsible for the regulation of biological processes: molecular geometries, atomic and molecular electronic aspects, and hydrophobic forces. All these structural characteristics are of primary importance in the understanding of structure–activity relationships and in rational drug design.

The field has grown rapidly since the 1980s. A number of spectacular advances have been made in molecular biology and in experimental and theoretical structural chemistry as well as in computer technologies. They all constitute important elements of the molecular modeling framework. The discipline is now fully recognized and integrated in the research process. In the past the emergence of this new discipline had occasionally encountered some opposition here and there. Nowadays, the science is mature and there is a growing number of success stories that continuously expand the armory of drug research.

## II. THE FIRST GENERATION OF RATIONAL APPROACHES IN DRUG DESIGN

Rational drug design is based on the principle that the biological properties of molecules are related to their actual structural features. What has changed along the years is the way molecules are perceived and defined. In the early 1970s, medicinal chemists considered molecules as mere topological two-dimensional (2-D) entities with associated chemical and physicochemical properties. Quantitative structure–activity relationships (QSAR) concepts began to be considered and became very popular. It was implemented in computers and constituted the first generation of computer-aided rational approaches in drug design. This discipline was promoted by Hansch and his group (Fujita, 1990). It was based on the determination of mathematical equations expressing the biological activities in terms of molecular parameters such as log P (the partition coefficient), steric substituent constants (Es), molar refractivity (MR) (Fujita, 1990). This has been expanded to the use of structural indexes obtained by quantum mechanical treatments (i.e., HOMO and LUMO energies, total dipole moments, charge and hybridization moments, molecular polarizability, Mulliken electronegativity, and frontier orbital indices).

Such methods have dominated the area of medicinal chemistry since the 1980s. Looking back now, one can notice that they have proven to be useful only for the optimization of a given series. Being constructed on a fixed (therefore invariant) 2-D formula, these approaches were unable to go beyond the 2-D frame of the structure considered. This has been a great limitation for those interested in lead finding. It is worth noticing that, in general, most of the parameters defined in QSAR approaches are conceptually relevant and not very different from those currently used today in molecular modeling (e.g., steric parameters, electronic indexes, hydrogen bonding capabilities, and hydrophobicities). However, most of these properties have not been well represented by the simplistic numerical parameters considered to represent these features: the interactions between a ligand and a protein require much more detailed information than the ones included in substituent indexes characterizing the molecular properties. The second generation has shown that consideration of the full detailed properties in 3-D is necessary in allowing the subtle stereochemical features to be appreciated.

In fact, the increasing interest elicited by molecular modeling in the mid-1980s was a direct consequence of the limitations that were found in the QSAR approaches by those attempting to find lead compounds. Not only were QSAR methods ill suited for that, they were also used more often for retrospective analyses than for predictive undertakings.

## III. MOLECULAR MODELING: THE SECOND GENERATION

### A. Conceptual Frame and Methodology of Molecular Modeling

In molecular modeling the perspective is no longer restricted, as it was in the past, to the design of closely related analogs of known active compounds. Molecular modeling has opened the way to the discovery of lead structures by a rational approach, and its central role in rational drug design has become fully apparent.

The acceptance by medicinal chemists of molecular modeling was favored by the fact that the structure–activity correlations are represented by 3-D visualizations of molecular structures and not by mathematical equations. The latter do not exactly correspond to the natural way of representing chemical systems. The 3-D representations have improved the perception of the chemists and have contributed to expanding their current chemical intuition. Moreover, looking at a drug in 3-D actually does give the impression of knowing everything about it, including its biological properties. Not so far from this intuition, molecular modeling perceives biological function as being embedded in the 3-D structure of the molecules. From that point of view, the "lock and key" complementarity between a drug and its biological receptor suggested in the early 1900s is literally considered.

Along these lines, the biological activity of drugs is to be recognized in their actual 3-D molecular features. Computer-aided molecular design (CAMD) is expected to contribute to the discovery of "intelligent" molecules conceived on the basis of precise three-dimensional stereochemical considerations.

"Direct" and "indirect" designs are the two major modeling strategies currently used in the conception of new drugs. In the first approach the three-dimensional features of a known receptor site are directly considered whereas in the latter the design is based on the comparative analysis of the structural features of known active and inactive molecules that are interpreted in terms of their complementarity with a hypothetical receptor site model (Fig. 1).

### B. The Field Currently Covered

Molecular modeling has widened the horizons of pharmaceutical research by providing tools for finding new leads. The fields currently covered by this discipline include:

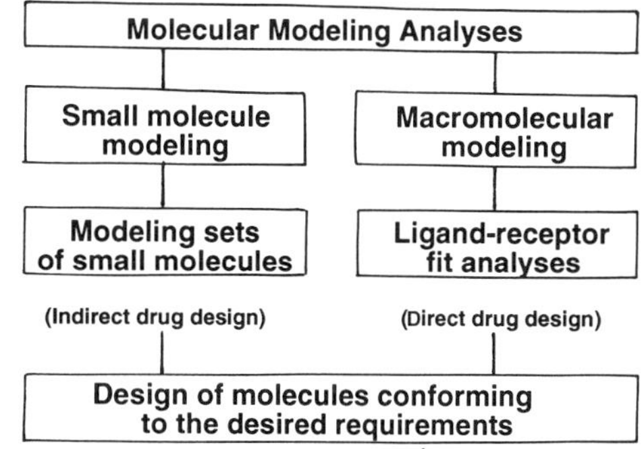


**FIGURE 1** Conceptual frame in computer-aided drug design.

- Direct drug design: the three-dimensional features of the receptor site (i.e., known X-ray structure or 3-D model of a receptor) are directly considered for the design of lead structures.
- Indirect drug design: the analysis is based on the comparison of the stereochemical and physicochemical features of a set of known active/inactive molecules; lead structures are designed on the basis of the pharmacophore model obtained by such analyses.
- Database searches: lead compounds are identified from searches using databases defined in 3-D. The input query describes the pharmacophore; it consists of a set of molecular fragments together with their relative location in 3-D and additional structural constraints (geometrical or chemical).
- Three-dimensional automated drug design: new lead compounds are generated by the computer on the basis of a "growing" procedure inside the active site of a protein whose 3-D structure is known or by a computerized treatment by assembling a set of pharmacophoric fragments defined in 3-D.
- Molecular mimicry: lead molecules are conceived as mimics of a known reference compound as, for example, the design of mimics of selected peptide ligands.

Each of the areas just mentioned has been developed in the different chapters of this book.

## C. Importance of the "Bioactive Conformation"

The existence of an experimental structure for the protein–ligand complex allows one to use this information to design new molecules. The knowledge of the bioactive conformation of the ligand as it binds to the receptor or enzyme is of great utility in designing new mimic molecules that are potent and specific.

Experience shows that the bioactive conformation of a molecule is not necessarily the one found in solution or in the crystal. Rather it is the specific conformation of the lowest energy in the context of the receptor. Knowing how the ligand is oriented in the binding pocket is also extremely valuable in designing new structures. The chemist tries to design a better analog of the existing prototype structure on which he is working or a completely new chemical structure. Typical questions include: What are the possibilities for increasing potency? Where can hydrophobic groups be added so that they penetrate well into hydrophobic pockets? Which hydrogen bond can be created between the ligand and the protein to improve selectivity? Where can polar groups be added in the prototype molecule that will strengthen the binding energy of the compound? Will this steric contact be favorable to the binding of the molecule? Can another binding mode be considered for the compound? Will this side chain of the protein be able to move? Is this modification of the prototype candidate molecule energetically favorable for its internal energy? Is this polar group "happy" in this subpocket? By providing answers to such questions the molecular modeling approach allows full utilization of the structural information and capitalizes on what is known about the mechanism of action of the protein–ligand complex.

## IV. MOLECULAR MIMICRY AND STRUCTURAL SIMILARITIES

### A. Molecular Mimicry

Molecular mimicry is an activity of central importance in drug research. Very often the molecules that are created are conceived as mimics of substances known to interact with the biological system considered (i.e., hormones, peptides, and transmitters). For example, peptidomimetic molecules are conceived to mimic the structure of an endogenous peptide and are converted into a regular organic molecule. The reason is that very often peptide molecules cannot be developed as drugs. In general, peptide molecules are:

- biologically unstable
- poorly absorbed
- rapidly metabolized.

A nonpeptide molecule should permit one to overcome these drawbacks. It is expected that the synthetic molecules may provide the structural diversity necessary to allow the molecule to be optimized for:

- specificity
- oral bioavailability
- pharmacokinetic properties

Molecular mimicry can be considered from various points of view and can cover many molecular properties. Actually, this is not really a new approach in drug research; it has been inherent in most attempts aimed at the design of drugs in a rational way. New compounds were, and continue to be, conceived on the basis of analogies with the structures of known active reference molecules. What has improved is the way the molecular structures are characterized.

Molecular modeling has become very popular because it gives access to the very structural information necessary for the understanding of the biological properties.

### B. Structural Similarities and Superimposition Techniques

The visualization in 3-D of molecular properties (i.e., steric aspects, electrostatic potentials, and hydrophobicity) is one step toward the determination of similarities and differences between active and inactive compounds. The easiest way to reveal 3-D structural features common to a series of compounds is the use of superimposition procedures. This technique has been extensively used since the early seventies. For example, the analyses made by Horn and Snyder (1971) proposing the dopamine hypothesis of neuroleptic action provide a good illustration of this method: the X-ray structures of chlorpromazine and dopamine are perfectly superimposable (Fig. 2).

The superimpositions are obvious when topological analogs are considered, but they are not obvious when chemically unrelated structures are compared. Extensive analyses were made in a variety of therapeutic areas showing how different molecules can elicit similar biological activities when they share common stereochemical features in 3-D (see Chapter 3 and references cited therein). This has led to the development of 3-D pharmacophoric models that are necessary for the rational design of new lead compounds.

The structures are usually aligned on the basis of their atomic positions, but very often the alignment can be very subjective. The best way to reduce subjectivity is to compare the molecules in 3-D in the actual space of their physicochemical properties [sometimes designated as four-dimensional (4-D) representations] (Fig. 3).

As far as computerized superimposition techniques are concerned, this perspective is relatively new and this book reviews the progress that has been made in this area. New developments in this field are in progress and should provide a powerful means for obtaining good structural pharmacophore models.

## V. RATIONAL DRUG DESIGN AND CHEMICAL INTUITION

### A. An Important Key and the Role of the Molecular Modelist

In pharmaceutical research the fundamental aspiration is to convert knowledge, intelligence, and imagination into practical results. In this multidisciplinary process the molecular modelist has become an important ele-

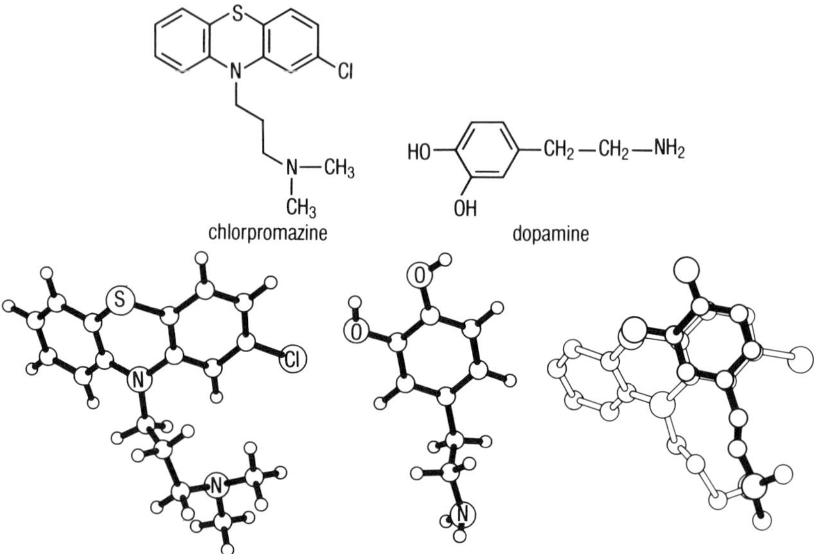


**FIGURE 2** Common 3-D features between dopamine and chlorpromazine. The superposition suggests that antipsychotic activity may be controlled by modulating central dopaminergic activity. [Adapted from Horn and Snyder (1971), with permission.]

ment (Cohen *et al.*, 1991; Snyder, 1991). As already mentioned, molecular modeling in the past was not always accepted by the traditional medicinal chemist. It was very often considered an unproven and less efficient approach than those that were currently in use. Now the approach is much more widely understood and molecular modeling is accepted as a full discipline by the new generation. Even the term "rational drug design" describing this approach does not induce emotional opposition any more. This, of course, does not imply that all other approaches were irrational! Nevertheless, it is

**Structural similarities**

**molecules are superimposed in terms of**

| | |
|---|---|
| **2D** | **topology** |
| **3D** | **atomic positions** |
| **Beyond** | **spatial arrangement of physical and chemical properties** |

**FIGURE 3** Common features in 2-D, 3-D, and beyond.

desirable to replace whenever possible the expression "rational" by terms such as "structure-based," "mechanism-based," "direct," or "indirect" drug design.

Beyond the emotional ambiguities carried by the terminology, an important issue that needs to be solved concerns the seeming inconsistency between "intuition" and "rational." The combination of these two aspects is actually the very driving force of any modern scientific innovation. Both dimensions are combined within a single person, and whereas one aspect might be more dominant than the other, they are never dissociated. The same should occur with teams. When the two aspects (intuition and rationalism) are not combined, the result falls short of the goal. How admirable it is when a team succeeds in integrating this dialectic and reaches a steady state that allows the combination of intuitive feelings and current knowledge with the special perception that they are on the right track for making outstanding breakthroughs!

In drug design, chemical intuition is an important element because in this field one has to be willing to make predictions based on incomplete information and models (Snyder, 1991). It also plays an important role in helping to unify disorderly observations, to promote the consideration of novel hypotheses, and to compensate for the lack of detailed information. All of this is needed for making progress and to realize achievements.

## B. Limitations of Chemical Intuition

Progress in structural chemistry and in molecular modeling has considerably contributed to improve the perception of the chemists by expanding their current chemical intuition. Chemical intuition is a good complement and develops together with knowledge. Chemical intuition and knowledge both expand continuously and each one redefines new limits to the other. The interactions occurring in this process constitute the dynamics of creation in research. One example is presented here to illustrate some of their relationships along with some of their limits.

The example concerns the inhibition of cytochrome P450. This class of enzymes is a widely occurring class of heme proteins that catalyze the hydroxylation of aliphatic and aromatic molecules. Camphor is the natural substrate of the bacterial cytochrome P450cam and is stereospecifically hydroxylated in the 5-*exo* position. It was found that phenylimidazoles, such as phenyl-1, phenyl-2, or phenyl-4 (Fig. 4), inhibit this hydroxylation process (Griffin and Peterson, 1972; Poulos *et al.*, 1987). From the structural point of view, a simple rationale could be derived to explain these observations; it seems reasonable to assume that the three inhibitors bind to the enzyme in a way where all the phenyl rings are aligned (Fig. 5).

This superimposition has the advantage of showing that the three molecules have almost identical shapes and can therefore occupy the same hypo-

camphor

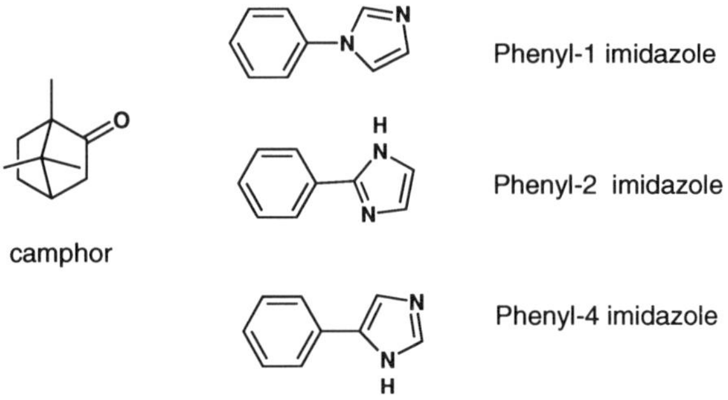

Phenyl-1 imidazole

Phenyl-2 imidazole

Phenyl-4 imidazole

**FIGURE 4** Camphor and phenylimidazole structures (see text).

thetical subpocket in the enzyme catalytic site. This should correspond to the true solution if the binding is driven by consideration of the geometrical features. If this is not the case and if, in particular, the electronic aspects play an important role, the model considered should be discarded because it does not distinguish between nitrogen and carbon atoms in the superimposition of the molecules. In an alternative solution satisfying the alignment

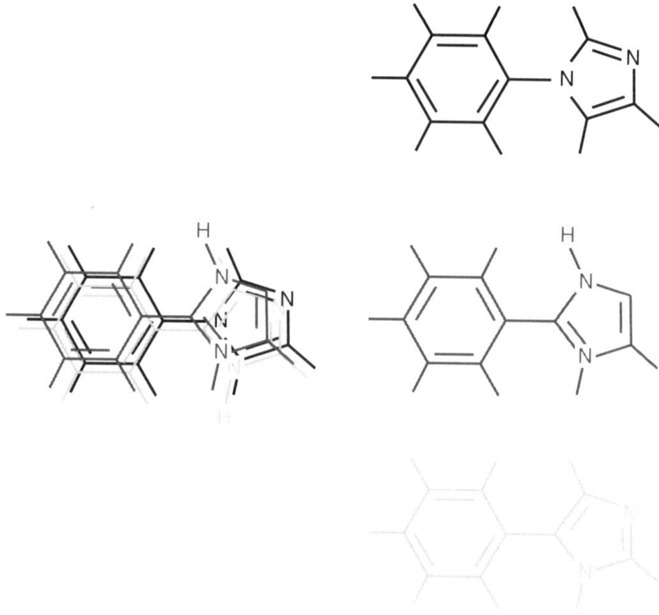

**FIGURE 5** Looking for 3-D features common to phenylimidazoles: superposition of the phenyl rings.

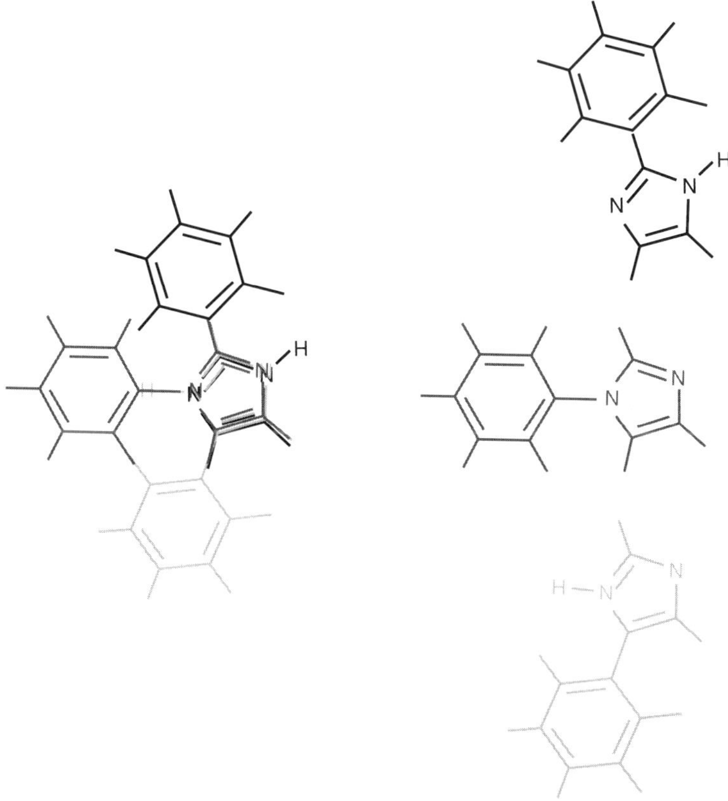


**FIGURE 6** Looking for 3-D features common to phenylimidazoles: superposition of the imidazole rings.

of electronic features, one should consider superimposing the molecules in such a way that the nitrogen atoms are optimally superimposed. This corresponds to the representation shown in Fig. 6. The new solution is far from being perfect: the 3-D diversity of the positions occupied by the phenyl rings does not provide convincing evidence of the validity of such a possibility.

At this point the important question is what are the determinant factors? Are they steric, electronic, or are both involved? In the absence of additional information for discriminating between the two alternative models, one would perhaps tend to favor, in the first place, the steric model of Fig. 5.

The correct answer is actually known: the X-ray structures of the complexes with the substrate camphor and with the three phenylimidazole inhibitors with cytochrome P450cam from *Pseudomonas putida* have been resolved (Poulos *et al.*, 1987; Poulos and Howard, 1987). Therefore one

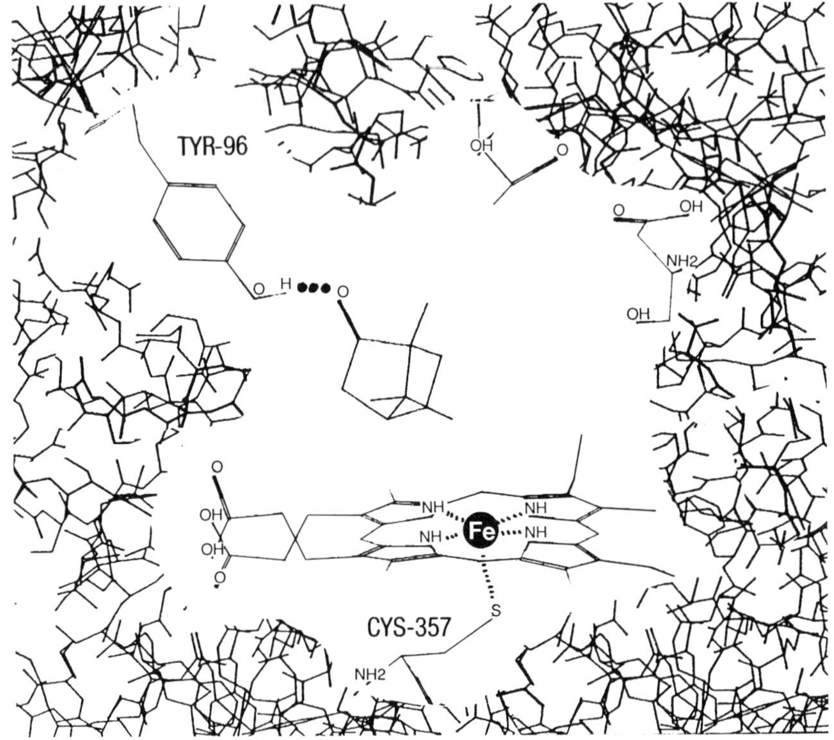


**FIGURE 7** Binding mode of camphor in the catalytic site of cytochrome P450cam. [Adapted from Poulos and Howard (1987) with permission. Copyright by the American Chemical Society.]

can look at the exact binding modes observed for the different ligands. The camphor molecule binds with a strong hydrogen bond to the hydroxyl group of a Tyr-96 residue of the heme protein (Fig. 7). This explains why, in the natural pathway, the 5-*exo* position is stereospecifically hydroxylated, being located just on top of the heme iron.

It is interesting to see how the different phenylimidazoles bind to the protein. The first complex with phenyl-2-imidazole shows a situation where the imidazole ring is anchored to the protein via the Tyr-96 residue already mentioned earlier, with the phenyl fragment pointing toward the heme (Fig. 8).

In the case of the phenyl-1- and phenyl-4-imidazoles, the situation is very different. A chelation is observed between the iron of the heme and a nitrogen atom of the imidazole ring (Fig. 9).

It becomes possible to visualize the superimposition of the three molecules in their observed binding modes. Figure 10 shows clearly that the

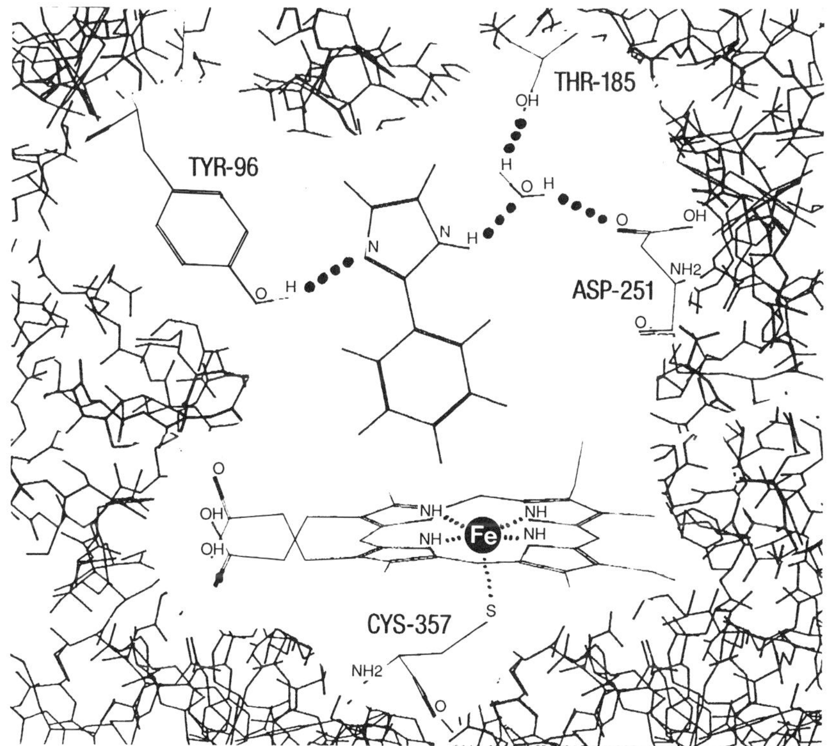


**FIGURE 8** Binding of phenyl-2-imidazole with cytochrome P450cam. [Adapted from Poulos and Howard (1987) with permission. Copyright by the American Chemical Society.]

correct answer to the question posed earlier corresponds to a situation where neither the phenyl groups nor the imidazoles are aligned.

Useful achievements have been made in this project; in particular, very potent inhibitors have been designed showing a double anchorage to both the iron of the heme and the tyrosine residue. This example illustrates that although chemical intuition has often proven to be very useful in compensating for the lack of knowledge, there are situations where a minimum of indispensable data is necessary, in the absence of which the reality is masked and an approach could be counterproductive. In the example discussed earlier, the knowledge that the biological activities of the compounds involve the binding with a heme protein is of high informational content. From this information one could already exclude the possibility of a common binding mode for the three ligands: whereas in phenyl-1- and phenyl-2-imidazole the substitution patterns allow a complexation of the heme group by the imidazole moiety, this is impossible in the phenyl-2 congener due to a steric clash between the phenyl and the porphyrin rings.

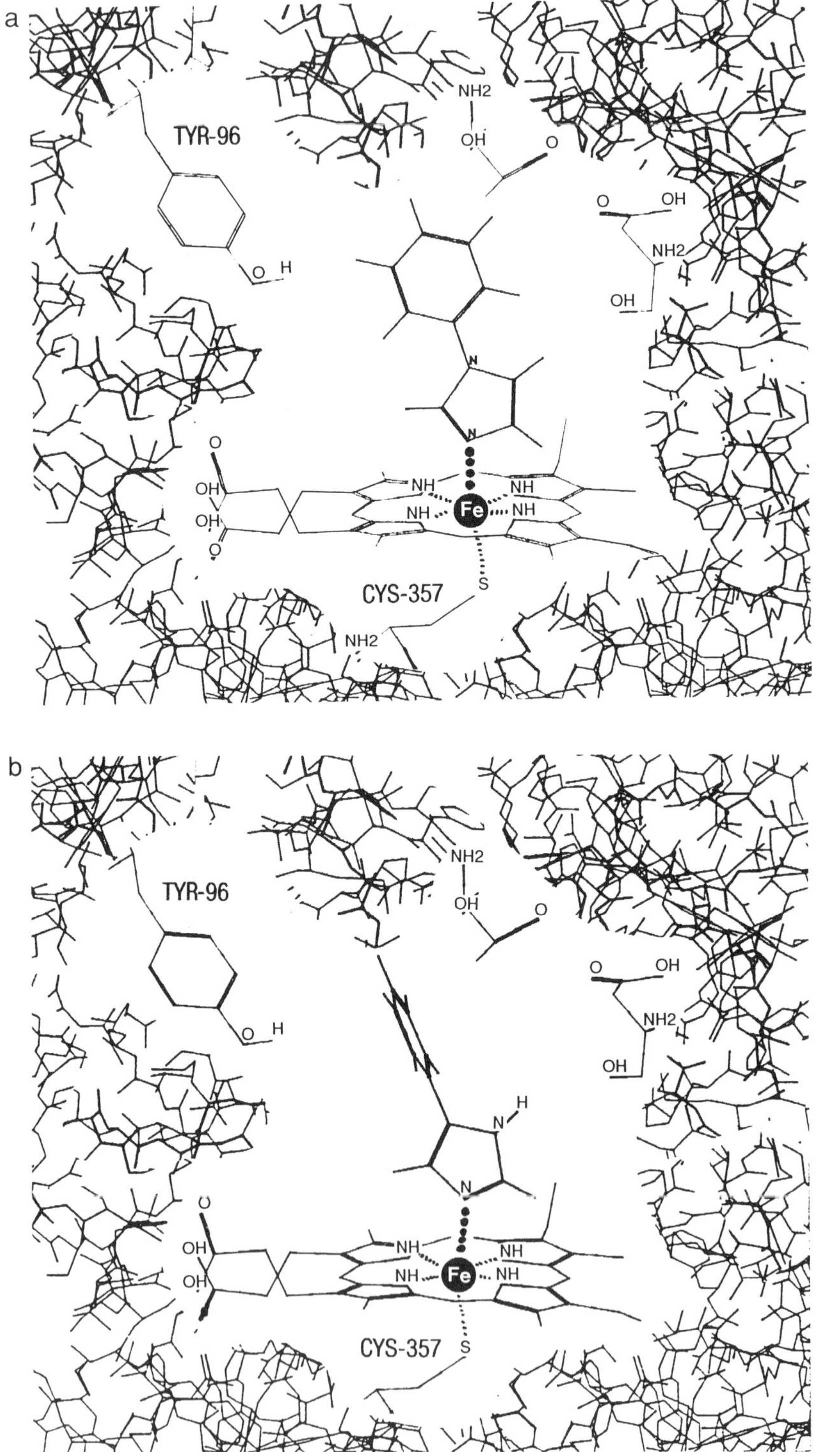
a
TYR-96
NH2
OH
O
O
OH
NH2
OH
O H
N
N
O
OH
OH
O
NH
NH
Fe
NH
NH
CYS-357
S
NH2
b
TYR-96
NH2
OH
O
O
OH
NH2
OH
O H
N H
N
O
OH
OH
NH
NH
Fe
NH
NH
CYS-357
S

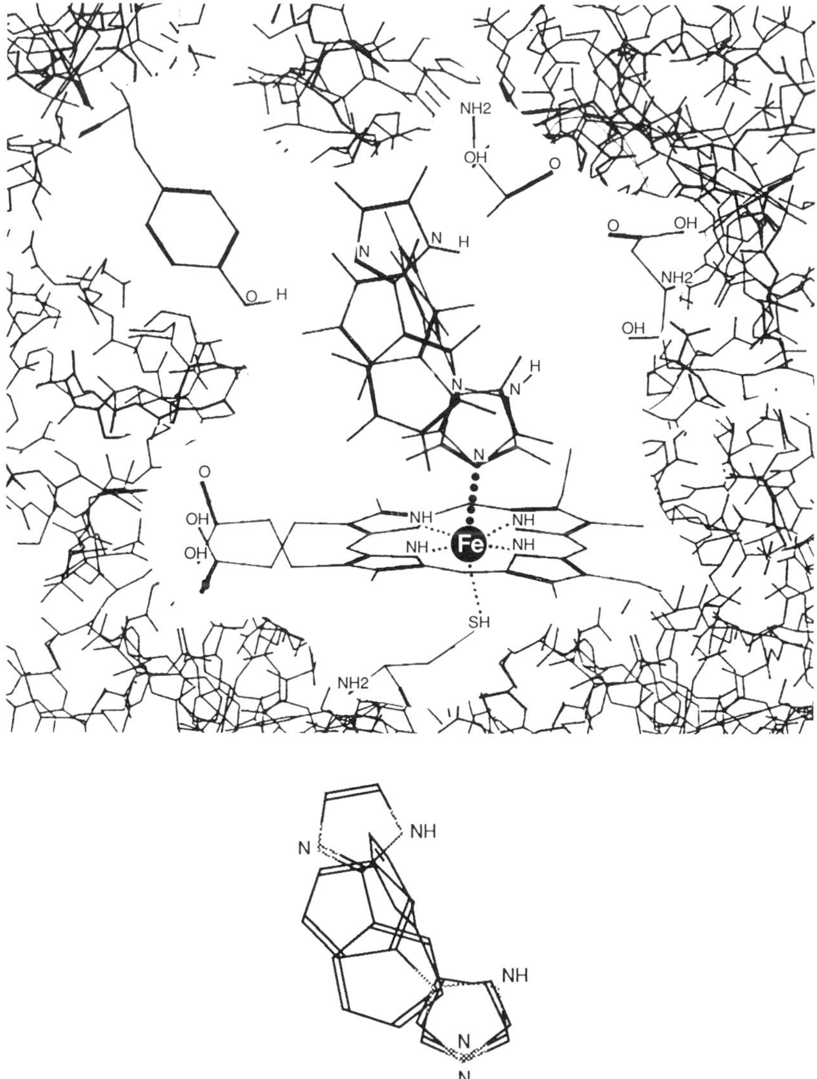


**FIGURE 10** Binding of phenylimidazoles on cytochrome P450cam: their "common structural features"! (see text).

---

**FIGURE 9** Binding modes of phenyl-1- (a) and phenyl-4- (b) imidazoles with cytochrome P450cam. [Adapted from Poulos and Howard (1987) with permission. Copyright by the American Chemical Society.]

This example suggests the following:

- optimal conditions for direct molecular modeling consist of having some knowledge about the macromolecular target or, even better, an X-ray structure of a complex between a ligand and the protein.
- in indirect modeling, good models can be derived through the design of appropriate experiments to resolve the observed contradictions.
- optimal conditions in a project exist when working hypotheses and experimental undertakings are combined.

## VI. MAJOR MILESTONES AND FUTURE PERSPECTIVES

The field of molecular modeling has advanced since the 1980s. The achievements made in computer technology, theoretical structural chemistry, biotechnology, protein crystallography, and high field NMR have permitted the development of molecular modeling. The visualization of complex three-dimensional structures with their associated properties such as geometries, electrostatic, and hydrophobicity provide useful means in understanding some of the driving forces operating in life processes. Such information allows one to rationalize structure–activity relationships and has proven to reduce the role of empiricism in the design of new prototype drugs. It is probably just an esthetic hope to envisage that a new drug will be the pure result of this approach. Nevertheless, it is now evident that the technique has already contributed to the realization of outstanding achievements in a number of research programs. It is rare that the announcement of the development of new drugs obtained through such rationales are published in mainstream international financial journals, but examples exist, such as the announcement in the *The Wall Street Journal* (issue of January 1994) of the development of a new class of AIDS drugs by the DuPont Merck Pharmaceutical Co.; others will certainly follow.

As far as methodology is concerned, the limitation that currently exists concerns, in the first place, the calculation of the free energy of binding of ligands to a protein enzyme or receptor. The exact calculation of this quantity is crucial for structure-based drug design because the interaction of the ligand with the macromolecular protein (which eventually leads to inhibition of the enzyme or antagonism of the receptor) is generally directly related to these energies. When it is possible to perform this calculation in a reliable manner, direct drug design will require much less iteration in the discovery cycle and will significantly shorten the drug development cycle. However, new progress in this direction requires taking into consideration complex desolvation and solvation processes that occur during the binding process. New advances in this area will occur only if substantial breakthroughs are also made in computer technology in order to have acceptable computing times for such treatments.

The development of automatized methods as well as that of databanks for computer new lead design has exploded in the early nineties and is changing the now classical panoply of the molecular modelist of the new century. The very ambitious "peptidomimetic dream" of medicinal chemists for converting the structure of a peptide into a regular organic molecule is far from being solved. New examples are continuously keeping up with the set of known pairs of peptides and mimics; however, no general rules have yet emerged and it is too early to talk of a "method."

## VII. CONCLUSION

Molecular modeling has become a new discipline in pharmaceutical research that has grown very rapidly since the 1980s and has contributed to the discovery of new lead compounds. The evolution of this discipline is progressively unfolding, leading to a new era of intelligent approaches in drug design. A new generation of useful therapeutic agents is expected to emerge from this unprecedented effort.

## REFERENCES


Cohen, N. C., Furet, P., Sele, A., Tintelnot-Blomley, M., and Bach, L. H. (1991). New professionals in drug research: The molecular modelists. *Drug News Prospect.* **4**, 459–467.

Ehrlich, P. (1909). Uber den jetzigen Standder Chemotherapie. *Chem. Ber.* **42**, 17.

Fischer, E. (1894). Einfluss der Configuration auf die Wirkung der Enzyme. *Chem. Ber.* **27**, 2985.

Fujita, T. (1990). The extrathermodynamic approach to drug design. *In* "Comprehensive Medicinal Chemistry" (C. Hansch, P. G. Sammes, and J. B. Taylor, eds.), Vol. 4, pp. 497–560. Pergamon, Elmsford, NY.

Grivvin, B. W., and Peterson, J. A. (1972). Camphor binding by *Pseudomonas putida* cytochrome P-450: Kinetics and thermodynamics of the reaction. *Biochemistry* **11**, 4740–4746.

Horn, A. S., and Snyder, S. H. (1971). Chlorpromazine and dopamine: Conformational similarities that correlate with the antischizophrenic activity of phenothiazine drugs. *Proc. Natl. Acad. Sci. U.S.A.* **68**, 2325–2328.

Poulos, T. L., Finzel, B. C., and Howard, A. J. (1987). High-resolution crystal structure of cytochrome P-450cam. *J. Mol. Biol.* **195**, 687–700.

Poulos, T. L., and Howard, A. J. (1987). Crystal structures of metyrapone- and phenylimidazole-inhibited complexes of cytochrome P-450cam. *Biochemistry* **26**, 8165–8174.

Snyder, J. P. (1991). Computer-assisted drug design. I. Conditions in the 1980s. *Med. Res. Rev.* **11**(6), 64–662.

# 2

# Molecular Graphics and Modeling: Tools of the Trade

**RODERICK E. HUBBARD**
Department of Chemistry
University of York
Heslington, York
United Kingdom

## I. INTRODUCTION

The goal of molecular modeling is to understand the fundamental relationship between the chemical and physical properties of a molecule, its chemical structure, and the three-dimensional (3-D) structure that the molecule adopts. From this understanding comes the ability to relate these properties to the chemical, catalytic, or biological function of the molecules and, most importantly, the ability to rationally design molecules of altered function. This link between chemical structure and function is of central importance to many scientific endeavors, be it molecular biology, protein science, drug design, chemical catalysis, or materials and polymer science.

The computer-based techniques of molecular graphics and modeling initially developed quite separately into two distinct scientific disciplines. The first was in the pharmaceutical industry, and the often elusive search for an understanding between structure and function in drug design. This has driven the development of a diverse set of techniques for receptor modeling, structure–activity relationships, molecular similarity, and 3-D database handling. The second area of initial development came from the protein crystallographic community, which requires tools for fitting and optimization of molecular structure into experimental electron density maps, and the subsequent analysis of protein structure and function. Since 1988, these two areas have come together with increasing success in the area of structure-based ligand design. At the same time, there has been a substantial expansion in the application of many of these tools and techniques in most areas of the natural sciences, including polymer and materials modeling.

Molecular graphics and modeling approaches to scientific problems are particularly exciting because of the phenomenal range and diversity of techniques that can be applied. Practitioners can swing from consideration of gene organization and biological function, through crystal structures or NMR data, synthetic organic chemistry, molecular mechanics, and molecular orbital techniques in pursuit of an understanding of the function of a molecule. Together this provides a tremendously exciting and daily challenging working environment.

Alongside these scientific developments has been a dramatic evolution in computer graphics hardware and molecular modeling software and, not insignificantly, a dramatic fall in the cost of such systems. This has led to the availability of systems with the capability for displaying and probing structure effectively in most laboratories. This chapter discusses the components of a modern molecular graphics and modeling system, and the tools with which molecular modelers tackle the extraordinary breadth of tasks that are presented to them. One of the features of the development of molecular graphics and modeling techniques is that many of them were developed in different laboratories at about the same time. Because these developments were generally in response to a particular scientific need, there are very few publications that have dealt solely with these new methods. For these reasons this discussion is necessarily rather generic and descriptive and contains very few references.

## II. DEVELOPMENT OF MOLECULAR GRAPHICS HARDWARE

The first major developments in molecular graphics came in the mid-1960s when Project MAC at MIT produced the first multiple access computer, a prototype for the development of modern computing. The computer included a high performance oscilloscope on which programs could draw vectors very rapidly and a closely coupled "trackball" through which the user could interact with the representation on the screen. Using this equipment, Levinthal and his team developed the first molecular graphics system and his article in *Scientific American* (Levinthal, 1966) remains a classic in the field. In this paper, he described their achievements and laid the foundations for many of the features that characterize modern day molecular graphics systems. It became possible to produce a vector representation of the bonds in a molecule and to rotate it in real time. The representation could be of the whole molecule or a reduced representation such as an $\alpha$ carbon backbone.

Because the computer holds the atomic coordinates of the molecule, it became possible to interrogate the structure and to use a computational model to perform crude energy calculations on the molecule and its interaction with other molecules.

Since that time it became possible to discern four main phases in the development, spread, and application of molecular graphics systems. Although, for many of us, the computer hardware is the least important aspect of what we do, these four phases are defined by the developments of computer technology: unless the relevant technology is available, the software and, most importantly, the techniques cannot flourish.

The first phase lasted until 1980 or so. A small number of academic laboratories interested in protein structure and function gained access to early high performance graphics systems produced by Evans and Sutherland and by Vector General that were linked to predominately DEC PDP minicomputers. This provided the embryo of software and technique development.

The second phase came about primarily through the development of effective interactive computers, and it was the DEC–VAX series of computers that dominated laboratory molecular computing until the late 1980s. Although this equipment was expensive ($300–500,000), it was justifiable by large research groups.

As mentioned in the introduction, the scientific imperative for provision of such expensive equipment came from two areas: protein crystallography and drug design. For the protein crystallographers the main advantage was to assist in the initial fitting of a model of a protein molecule into an electron density map, and the manual model building required in subsequent refinement of the model against the electron density. Before graphics systems became available, this required many man months of painstaking and tedious work with plastic models, ruler, and pencil. The graphics systems allowed the electron density to be represented as a "chicken-wire" grid that could be manipulated in real time to give a full 3-D perception of the shape of the molecule (see, for example, Jones, 1978, 1982).

The principle underlying the early days of computational chemistry in the pharmaceutical and chemical industries was that there is a fundamental link between the structure of a molecule and its function or activity. It was this elusive relationship that sponsored the growth of molecular modeling in the pharmaceutical industry throughout the 1980s. As the structure of very few receptor target molecules was known, the main emphasis was on comparing the structure, physical, and chemical properties of a series of active molecules. From this, the goal was to derive some structural rationale for activity, which could then be developed in other molecules to provide more potent binding. A molecular graphics system was a powerful focus for the computational and graphics techniques needed to explore these structural relationships (see, for example, Humblet and Marshall, 1981).

It was during the early 1980s that the high performance systems made the transition from black and white to color. It is fascinating looking back to remember that very strong arguments were needed to make the case for color molecular graphics. Many eminent scientists were strongly of the view that color was not necessary. It very quickly became clear, however, that color was one of the most important attributes of a molecular graphics system for illustrating and dissecting molecular structure and properties, and today many of the structure-related publications include color figures as a central part of their editorial policy.

During this second phase of development, two distinct types of graphics device were discernible. The first were machines such as the Evans and Sutherland and Megatek which through specialized hardware and programming techniques tried to provide very high performance, real-time rotation of many thousands of vectors. These were necessarily single user systems and contributed to the centralized facility attitude to molecular graphics which still pervades the discipline. The second category of graphics devices were color graphics terminals such as the Sigmex/Lundy where most of the graphics was driven by the host computer, giving slower but much less expensive facilities. With these devices it was possible to have a number of terminals sharing the central, expensive computer resources.

The third phase in the development of molecular graphics systems lasted from about 1986 to 1991 and was characterized by the increasing spread of the techniques of molecular graphics and modeling to the wider physical science community. Although the development and successful demonstration of the techniques played a major role in this expansion, the availability of a new generation of powerful and relatively inexpensive raster workstations based on RISC technology was crucial. During this period, the price of the hardware required for an effective molecular modeling system fell from about $150,000 in 1985 to around $30,000 in early 1991. This made the hardware accessible to any laboratory dedicated to studying molecular properties that could be interpretable in structural terms. The rapid rate of technical change brought its own problems. Although equipment becomes obsolete within a few years, the move from essentially VAX-dominated systems to UNIX workstations took some considerable time to complete, both in user acceptance of the new operating systems but also for the software providers.

Finally, we are now in the fourth phase of computer development. During the summer of 1991, Silicon Graphics introduced a desktop workstation that provided a complete environment for molecular graphics and modeling for around $10,000. In addition, we have recently seen the arrival of computers from a number of manufacturers which provide computational performance that would have seemed incredible just a year or two ago. The main effect of this has been to broaden the acceptance of molecular modeling as an additional tool for the molecular scientist.

The phenomenal pace of development of computer technology is summarized in Fig. 1. It should be emphasized that the examples and prices chosen are illustrations only and are based rather loosely on my own memory of the equipment obtained in York since 1985. Figure 1a summarizes how the price of representative modeling workstation hardware has changed since 1985. In 1985, a VAX 11/750 with Evans and Sutherland PS300 cost around $300,000 for a reasonably configured system. The performance of the VAX computer was just less than 1 MIP. By 1987 a Silicon Graphics 4D/70GT cost around $150,000 with a performance of about 10 MIPS. From 1989 on, the rate of change increased dramatically, and an acceptably configured workstation of around 100 MIPS with memory, disk, and graphics to support a full molecular modeling system costs around $10,000. Figure 1b highlights how the ratio of cost per MIP of computing power has fallen so dramatically in these systems over the years.

Interestingly, the price of the lowest cost workstations has remained relatively stable since 1991, but with significant improvements in their compute and graphics performance. In many ways, this base cost of around $7,000 appears to represent the lowest price the manufacturers can provide a full workstation and still make a profit.

The other illustration (Fig. 1c) is the cost of disk space on a standard workstation expressed as cost per Mbyte. A similar pattern is seen as for computer performance, although there is not such a dramatic change. The price of memory for computers has shown similar changes. Overall, the change in computer technology has been dramatic. It is quite staggering to realize that modern workstations are 100 times more powerful than a laboratory minicomputer was in 1985, at approaching one-fiftieth of the price. The effect of this has been to see the spread of molecular graphics and modeling devices both in number and in the different scientific areas in which they are used.

The pace of all this change in technology means that consideration of hardware is much less important than it was a few years ago. Most workstations are adequate in computer performance, and many now support the level of graphics necessary for effective molecular modeling. It is the availability of software that dominates the choice of equipment, and any comparative benchmarking of equipment should always be in terms of how quickly your particular application will run on a machine instead of any quoted performance figures from the manufacturer.

The next section discusses the various attributes of hardware, emphasizing the features that are important for molecular graphics and modeling.

## III. HARDWARE FOR GRAPHICS AND MODELING

This section discusses the various classes of computers and graphics that make up the modern environment for molecular graphics and modeling.

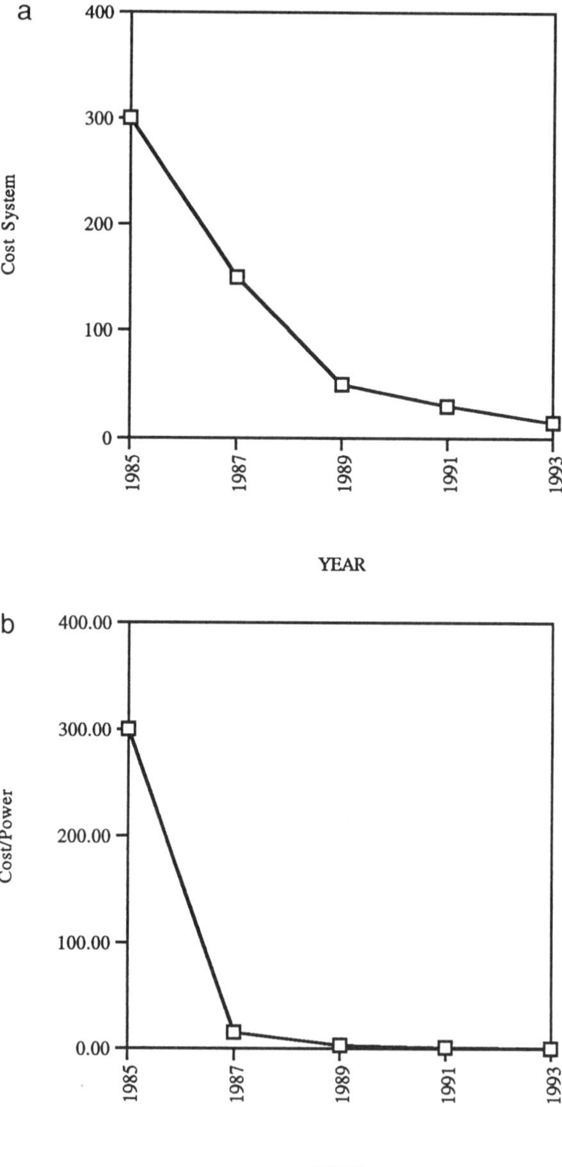


**FIGURE 1** Plots illustrating the dramatic fall in price and the increase in performance of computing equipment used for molecular graphics and modeling since 1985. (a) Cost of a representative molecular graphics and modeling system in pounds. (b) Cost of this system per MIP of computing power. (c) Cost of a Gbyte of disk.

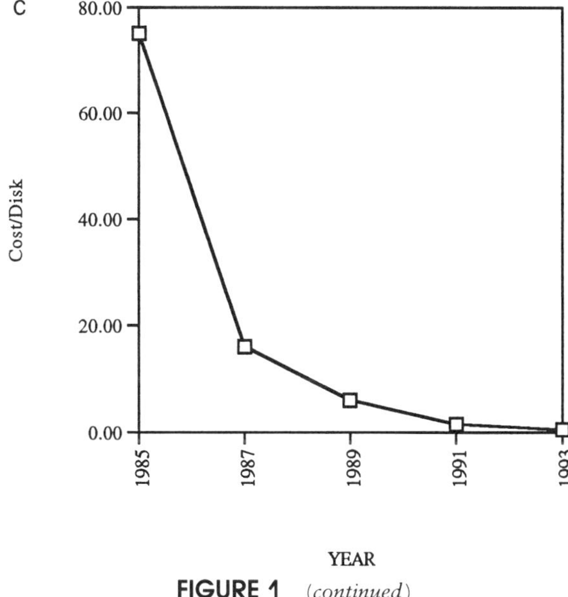


**FIGURE 1** *(continued)*

I have only named particular equipment to be illustrative; one of the clearest lessons of the past decade is the increasing speed with which computing and graphics equipment becomes obsolete. So this discussion mainly focuses on describing the essential characteristics of different classes of computing, trying to draw out the main considerations and attributes.

Although the boundaries are a little tenuous, three distinct classes of computers can be identified: central departmental compute and file servers, high performance graphics workstations, and personal workstations. In addition to these three categories, the various services such as networking, printing, and plotting essential for providing an efficient working environment for the molecular modeler will be considered. This section finishes with a discussion of how the various facilities can best be integrated to ensure the most effective use.

## A. Central Servers

When the number of computer users gets above just a few users, then it becomes necessary to define and maintain some computer resource that can act as a central filestore for users' data and shared databases. In addition, some types of calculation also require a central compute server with a large memory or particularly fast processors on which large calculations can be run. There has always been the need for this type of facility. In the mid-

1980s the requirements were met by the central VAX computer, which had the very important property of being able to share available computer time between different processes.

A key development was the efficient provision of virtual mèmory. What this means is that the various jobs that the computer is operating on do not all have to reside completely in memory for the computer to be able to share its time between them. Instead, the computer can operate a system of paging and swapping operations to extend the physical memory across onto disk-based virtual memory. In this way, the interactive requirements of the users running small calculations or driving graphics devices were satisfied, while at the same time larger calculations were present in the background where they could soak up any unused computer time.

This type of broadly based compute and file server is the central business of the current major computer manufacturers. Currently, the molecular modeling community is using equipment from a variety of manufacturers, such as Digital, IBM, Sun, Hewlett-Packard, and Silicon Graphics. New models are being introduced monthly, and in general the manufacturers are leapfrogging each other in performance; currently most have machines with single processors in the 100- to 150-MIPS range. In addition, there is a bewildering variety of different speeds of memory, architectures of the main infrastructure or backplane of the computer, disk speeds, and technologies, some of which are discussed next.

Fortunately, the current generation of machines predominantly run the UNIX operating system, making the transfer of software between machines relatively straightforward. There is also a relatively broad concensus on the way in which systems can work with each other, in areas such as how data are held on disk and how systems can share file systems and batch queues. However, there are a number of different flavors of the UNIX operating system that can make integration of many different systems rather difficult. Although there have been attempts at standardization, some manufacturers have a strategy of developing their own variant of UNIX that can lock the user into their systems. In addition, major problems can also arise as the manufacturers introduce new versions of their own operating system. These can be as costly as rendering older models obsolete, requiring more memory or giving such inconsistencies that all the software has to be recompiled or relinked.

Computer speed is discussed first. In general, all calculations relevant to the molecular modeler go faster on computers that have higher ratings, but there can be quite large discrepancies. The various benchmark figures quoted for the speed of a particular machine need to be treated carefully; wherever possible the comparison should be made by running a range of computer programs that are representative of how the computer will be used. A series of modern UNIX computers can have the same quoted performance on a single calculation, but will respond quite differently to being used as a shared, central resource put under pressure from many users and calculations. For example, when assessing a large central computer as the central user and compute area for a laboratory of some 30 scientists, it is essential to check how the computer performs under a simulation of the loading the computer will have to bear. Program scripts exist which simulate a number of interactive users, editing and listing files. These can be run at the same time as a representative set of interactive and background jobs are being run which compete for computer time and memory. An important measurement that can then be made is how much longer a particular calculation takes when the system is stretched in this way, using virtual memory on disk, against how long it takes to run the calculation when the computer is empty.

In general, the hardware infrastructure required to support heavy computational load costs more money; thus although the very large computers costing many hundreds of thousands of dollars are only a few times faster than a cheaper workstation, they are able to cope with much higher loading. There are also specialized calculations that may need an exotic disk or memory architectures for efficient calculation, e.g., the very large quantum mechanics calculations where the construction of very large basis sets and integrals for computing orbitals can place significant demands on the computer requiring large memory or disk space.

In summary, the only way to compare the true performance and thus suitability of a computer for a particular purpose is to benchmark under conditions that reflect the applications and loading that they will actually have to carry.

In the past, there have been many new computer architectures developed to provide the speed for our calculations that we all seek. Although many of the more specialized machines have now disappeared, it is still relevant to provide a brief overview of the major differences among serial, vector, and parallel computers. Perhaps the best way to do this is to use the analogy of a production line, such as assembling a car, or to draw on one of the largest activities in York, the confectionery industry and packing chocolates into a box. The task is to fill a number of boxes with four different chocolates and then to wrap the chocolate box and attach a label. A maximum of six people are available.

The serial case is the most analogous to the general processor in standard workstations. There is only one person who places each of the four chocolates in turn into the box, then wraps the box and adds a label. They then move onto the next box. Each box takes six cycles of time. The person is capable of a number of different tasks and the amount of time taken to complete the filling of the box is constant no matter how many boxes are filled.

In the vector or production line case, the boxes go past six different operators. The first operator spends all their time placing the first chocolate in the box, the second operator the second chocolate, and so on. The fifth and sixth operators, respectively, wrap the box and add a label. The vector case is analogous to specialized pieces of hardware such as produced by Cray and Convex and is how many of the very high performance graphics systems achieve such high speeds of rendering. In vector computation, the maximum throughput of boxes of chocolates is achieved when the production line gets full, i.e., all the different operators are performing their different functions at the same time. A major benefit arises in optimizing the performance of each operator for the particular task they have to perform. The disadvantage is that the operators or production line is rather inflexible. Changing the number of chocolates would require restructuring of the line, and it is rather inefficient for only a few boxes.

For the parallel case, there are a number of different ways in which the operations could be carried out. The six operators could all work on the boxes independently but in synchronization, like six serial operators. Alternately, each operator is doing the same operation at the same time, but there is some redundancy. As a box is filled, four of the operators put the different chocolates into the box in one cycle, and then one operator can wrap it up and attach the label. This means that the operators need to be flexible and perform many different tasks, but there is not the lag in starting up and finishing the production line of the vector case. There is, however, redundancy in that at different times some of the operators are not fully occupied.

In the nonserial cases there are real problems if there are any dependencies among different tasks, e.g., rejecting a product halfway through because one particular component is not working properly. For the serial worker, this does not affect their productivity very seriously; they are always busy. For the production line, the workers downstream will be idle for some time.

There are a number of variants on the basic architectures. The most successful is the development of super-scalar processors which are becoming the standard in modern workstations; in our analogy this can be thought of as using two hands at once. This is helpful sometimes (e.g., two chocolates can be put in a box at once), but not at others (as in wrapping the box).

To provide more computer power, some manufacturers offer multiple processor machines, whereas various systems are available for others that can coordinate the individual processors in computers across a network to all work together on the same computation. Again, this is only of value for very large computations where the time for a single job to complete is important. In general, the parallelization of computer programs to run across a number of processors can rarely be achieved with 100% efficiency.

### B. High Performance Graphics Systems

The real-time manipulation of representations of large macromolecular structures requires specialized graphics hardware to be able to draw the many tens of thousands of objects with the depth cueing, stereo, and high line quality necessary for full three-dimensional perception of the molecules. At the same time, the manipulation, interrogation, and modeling of these molecules require a reasonable level of computing power, memory, and disk space. To provide all these facilities requires specialist hardware; in most laboratories this is provided as a centralized service where the most powerful graphics machines are shared among a number of users.

Up until 1988, Evans and Sutherland was the only company that produced laboratory workstations with the performance necessary for these tasks. In particular this was the only device capable of drawing the many vectors necessary as they are still the main technique for representing large molecular systems in real time. These workstations can draw in excess of half a million vectors per second, with good stereo and high quality lines.

This class of workstations has undergone a discrete change in that the high performance graphics can be provided at a moderate cost on personal workstations. Specialist high performance graphics devices such as the Reality Engine from Silicon Graphics are, however, available. These offer great improvements in the performance of raster graphics hardware such that a large number of shaded polygons can be drawn rapidly. This means that cartoon-like representations of protein secondary structure or shaded molecular surfaces are now possible as real-time displays.

### C. Personal Workstations

This class of machine has been developed considerably. The current Indy machine from Silicon Graphics costs well below $10,000 and for this price will provide the computing power (100 MIPS), memory (32 Mbytes +), disk space (500 Mbytes +), and graphics performance necessary for most molecular modeling applications. Although the quality of the graphics can be reduced (the cheapest option is of poorer quality because the machine has fewer bit planes and achieves shading by dithering), it is an impressive machine on which all but the most ambitious modeling projects can be carried out. At the same time, the performance and facilities available within microcomputers have grown quite rapidly. In many ways these two classes of computing are coming together, particularly as the microcomputers gain graphics enhancement boards, run UNIX, or can be used as X terminals to a central server.

As this is the class of machines on which most molecular modeling is carried out, then it is useful to go through the requirements and characteristics of the various components of the hardware.

## IV. COMPONENTS OF MOLECULAR GRAPHICS HARDWARE

The main requirement is to hold a model of the molecule in the computer and to be able to interact with it. This means a visual representation. Most of the current generations of workstations have adequate graphics for most molecular modeling applications. The rapid increase in computer power means that the graphics can be produced in software reasonably quickly, without needing the specialist graphics hardware of previous generations. The computer power needed for manipulation and analysis of structures is also adequate on most workstations; however, in a large laboratory, the machine can be networked to larger central servers, allowing more intensive calculations (such as molecular dynamics simulations or molecular orbital calculations) to be done elsewhere.

### A. Memory/Disk

Computer memory continues to get less expensive and 32 Mbytes is usually adequate. The need for computer memory on the personal workstation has increased steadily over the years, both because the operating and windowing systems require more and because the scope and size of molecular modeling software have increased. The fundamental requirement for disk space is sufficient to hold all the operating system and swapping space. Five hundred Mbytes is a minimum and 1 Gbyte is becoming the standard as the price of disks falls. The balance between local disk space and central file servers is discussed later. In the late 1980s, there was a temporary move to diskless workstations in which all the operating system and swap space was provided by a central file server over the network. This did not work very well, as there were too many bottlenecks, and with the fall in disk prices, this strategy has now been abandoned.

### B. Communications

The close coupling of machines on the same site has developed slowly since the late 1980s. Ethernet remains the normal means, although a much faster protocol/hardware known as FDDI has been introduced. FDDI is currently at a cost where it is mainly reserved for connecting major systems together, not personal workstations. Both approaches are fast enough to allow significant sharing of disk and other resources between workstations. More recent software developments even share processing between machines linked by ethernet.

Perhaps the most significant development in communications is the establishment of a worldwide network, Internet, which means that a user can interact with computers and users all over the world. The significance of this development has relied not only on the installation of national pathways to link all the sites together, but also on the rapid evolution of a software culture. It is now possible to access databases and software all over the world, essentially interactively. This is changing quite dramatically the way in which the academic community works—with shared software, central servers, and information bureaux.

## C. Data Security

It is vital on any computer system to have backup strategies in place; that is, a means and schedule for making a copy of all the disks on the system to a tape medium of some sort. The reasons are obvious: hardware failures or occasional accidental deletion of files can be disastrous. Currently, three different devices are used for backing up data: a $\frac{1}{4}''$ tape, a VIDEO8 tape, or a 4-mm DAT tape. The tape drives are either found on the personal workstations or in a larger working group attached to particular central machines, with backup occurring across the network. The different tape names reflect the physical nature of the device. Beware that different machines can have different formats for writing the tapes; byte reversal can be dealt with in software, but some variants of these devices have a built-in compression mechanism that increases the amount of data which can be stored on a tape but which makes it impossible to read on another machine. Software updates and databases can be provided on CD-ROM drives; as yet writable CDs are rather expensive and not in general use.

## D. User Interaction

How the user interacts with a workstation is considered in this section. A typical workstation is shown in Fig. 2. The basic mode of operation has not changed since the mid-1980s: the user interacts with the programs running on the workstation through typing at the keyboard, or the 2-D mouse. The mouse is used to point at items on the screen, or by a combination of mouse movements and mouse buttons to interact with a molecule on the screen. There have been many attempts to improve the way in which the user interacts with the three-dimensional objects—in many ways the 2-D mouse is not ideal. Recent developments and interest in virtual reality systems and the entertainment application of computers have led to the development of a number of innovative devices for true three-dimensional interaction. These include the spaceball, in which a sphere can be grasped by the user and by both lateral and rolling hand movements allow the manipulation of the molecules. In addition, there are a number of hand-held devices which allow a much freer three-dimensional movement. It remains to be seen whether these devices will become standard options on a molecular modeling workstation.

## E. Organization

Finally, this section considers how a number of workstations could be organized in a larger grouping to ensure the most effective working. Although the exact division of different tasks will vary on the requirements of the group, it is possible to identify a number of different types of computers in an integrated environment such as is shown in Fig. 3. These different categories are:

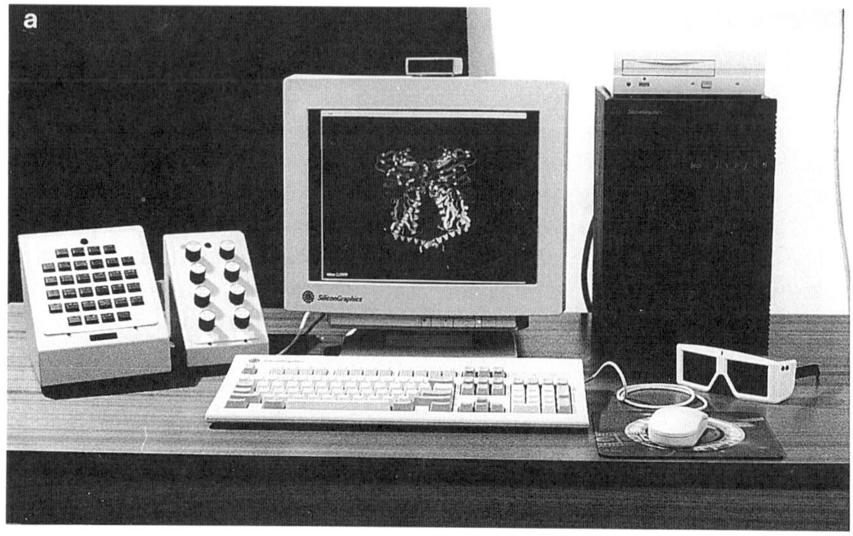


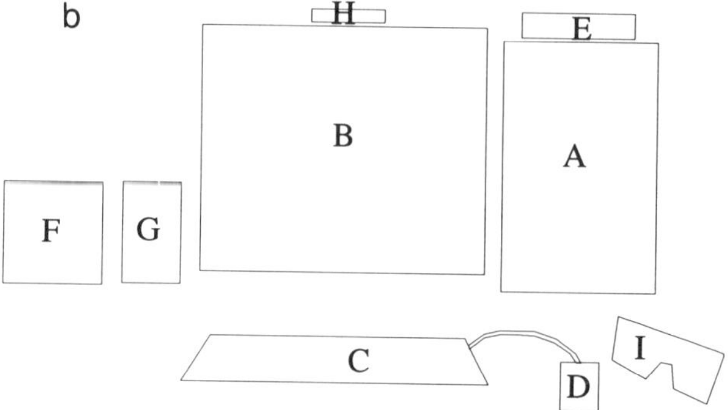


**FIGURE 2** A molecular graphics and modeling workstation. (a) A typical workstation. (b) Labeled schematic of the various components: A, main system box containing processor, memory, disks, and graphics hardware; B, screen; C, keyboard; D, mouse; E, CD drive; F, button box; G, dials box; H, infrared emitter used to drive the stereo glasses; and I, stereo glasses.

*Compute server.* The machine on which large calculations can be run in batch.

*File server.* The main machine for all the centralized databases and the machine on which the files of most users will reside. This can also maintain the queues for use of centralized printers and plotters for hardcopy output.

*General access/high performance workstations.* These are workstations for occasional users (whose files reside on the file server) and also the high performance graphics facilities shared by the group.

*Personal workstations.* In general, these will be of moderate performance and will be dedicated to individual workers who spend a considerable

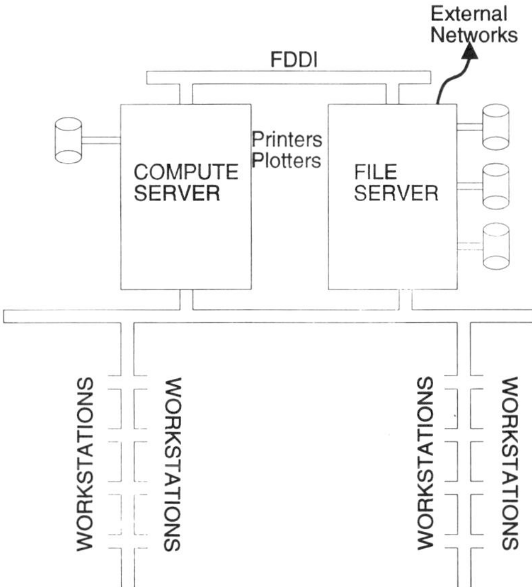


**FIGURE 3** A computer network for molecular graphics and modeling. See text for description of the various components.

amount of time working on the computer. They will have significant amounts of local disk space to accommodate the user.

The goals of this organization are to maximize the use of the most expensive computer resources, centralize specialist resources and facilities so they can be shared, managed, and operated effectively, and to keep network traffic to a minimum to ensure good performance and interactivity. On large computer installations (say more than 10 workstations) it is generally the organization and the performance of the network that causes more problems than any other feature.

### F. Visualization

The development of molecular graphics has had a profound effect on our ability to view, interrogate, and model molecular structure. The most important advantages are the ability to visualize and manipulate the three-dimensional structure of molecules and to provide rapid and detailed analyses of molecular properties, especially when closely coupled to molecular calculations.

All modern graphics workstations are based on raster graphics, and to understand some of the features and limitations of graphics displays, it is necessary to give a brief idea of how these work.

In the early 1980s, the main graphics devices were calligraphic or vector drawing systems (such as the Evans and Sutherland PS300) which produced graphics by causing the electron beam in a tube to directly draw lines on

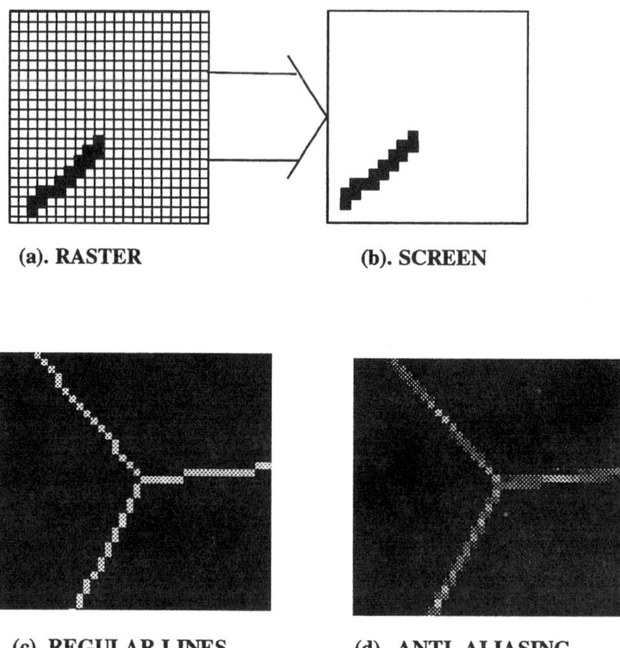


**FIGURE 4** How raster graphics works. The computer changes the values held at particular positions in the raster (a); this is then displayed on the screen (b). The line drawing in particular suffers from the granularity of the raster (c). The technique of antialiasing can help this (d).

the phosphors of the display screen. In raster graphics, the whole of a specially designated piece of memory (known sometimes as the frame buffer) is continuously read and displayed on the screen, usually at a frequency of 50 or 60 Hz. This is very similar to television and video technology. Figure 4 is a schematic of how this works.

The graphics memory is organized as a grid or raster of a particular dimension, typically 1280 in the horizontal direction and 1024 in the vertical, giving a total of 1,310,720 pieces of memory known as pixels. Graphic displays are produced by the computer by changing the value of each of these pixels. So to draw a vector, the value of each of the pixels along this line is changed.

In most graphics systems, the memory is organized as blocks in which a single bit is available for each pixel. The number of blocks in the system controls how many different colors are available for each of the pixels. These blocks of memory can be thought of as planes through the display memory, and are often called bit planes. Currently, systems are available with between 8 and 32 bit planes, although some of the more sophisticated devices have many hundreds of bit planes available.

The number of bit planes determines how big a number can be held at each of the pixel positions in memory. This varies from 256 for an 8-bit plane machine to many millions for a 24-bit plane machine. The way in which these numbers are translated to color on the screen depends on how the look up table is configured. Most of the current generation of 24-bit plane workstations use the notion of true color in which each 24-bit value stored in graphics memory is divided into 8 bits for red, 8 bits for green, and 8 bits for blue. Other systems still rely on the graphics program to allocate different colors for each possible pixel value.

Variations on this general rule arise on some 8-bit machines in which true color is emulated by a process known as dithering. Here, variations in color are achieved by using only 256 colors to achieve shading by modifying the color of surrounding pixels.

A similar process is used on some workstations to provide smooth edges to graphics primitives such as lines or polygons by a process known as antialiasing. A line drawn on a raster graphics device is often jagged at the edges because of the discrete elements of the raster. This effect is reduced on higher resolution devices but can still be intrusive when displaying vectors. Figure 4 contains expanded images taken from a workstation screen for lines drawn on a raster with and without antialiasing. The pixels surrounding a line are modified to smooth off the edge, giving the impression of a smooth line.

The advances in raster graphics have led to the development of a wide diversity of graphics hardware, with different levels of performance for different types of graphics and very different prices. This section discusses the various features of the graphics of a workstation that are important for a molecular modeling system. First let us consider the various primitives that are available. The most widely used is the vector, still the basic graphics operation for representing bonds between atoms. Antialiasing has been discussed; perhaps just as important is the ability to draw vectors of different thicknesses, particularly with the high resolution of raster graphics on most modern workstations. Line thickness is important to be able to see the vectors clearly and is a useful way of distinguishing between different molecules or representations; surprisingly, there are still some workstation graphics systems on which only single pixel thickness vectors are available. In general, it is still not possible to get antialiased vectors more than a pixel wide without real performance degradation.

The second basic primitive used extensively in modern molecular graphics systems is the ability to draw polygons, particularly triangular meshes. The mesh is important as it is relatively fast to render, and it is reasonably straightforward to fill or tile a surface using triangles. Figure 5 explains how a triangular mesh is generated and why it is so fast to draw. Once triangle A is defined by the vertices 1, 2, and 3, then triangle B only requires vertex 4, as the system remembers the positions of 2 and 3. In the same way,

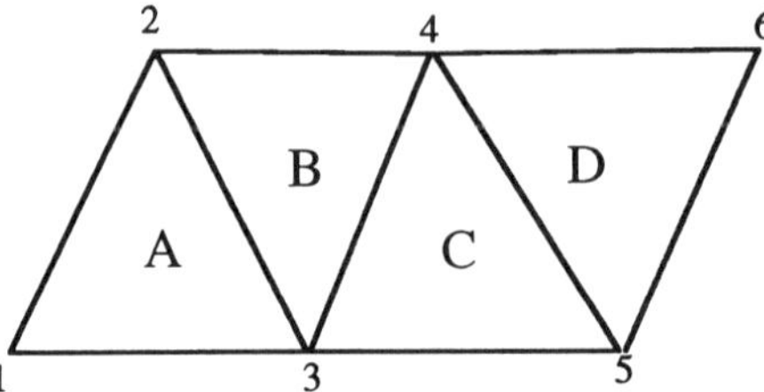


**FIGURE 5** How triangular meshes work in raster graphics systems. Triangle A is first defined by vertices 1, 2, and 3. Then only vertex 4 needs to be specified to define triangle B, vertex 5 for triangle C, and vertex 6 for triangle D.

triangle C in the mesh only requires vertex 5 and triangle D requires vertex 6. This reduces quite dramatically (approaching a factor of 3) the number of points or vertices that need to be transformed by the computer and hence allows the surface to be rendered more quickly. Some graphics workstations are optimized to deal with this type of primitive, e.g., the Reality Engine from Silicon Graphics.

The technique of shading and lighting of such graphical primitives as these meshes and polygons is a necessary enhancement to visualize the shape and depth that these surfaces represent. Transparency is also very useful to be able to see the details of molecular structure underneath the surface representation. This requires special hardware (known as $\alpha$-planes on some equipment) to produce the full effect. However, a reasonable effect can be achieved using a bit pattern mask for rendering polygons and meshes, which causes the surface to be drawn with some of the pixels blank, so that portions of the surface can be seen through. The well-established technique of depth cueing can be included as a form of shading. Here, objects that are further back in a scene are drawn slightly darker than those in the front.

An addition to the techniques of rendering these primitives is known as texture mapping. This is only feasible at present on specialist hardware but will probably become more widely available on lower performance workstations with time. In this technique, the surface of a primitive is rendered from a texture map, allowing various properties to be represented more rapidly. Section V, E of this chapter contains examples of how many of these different techniques can be used for representing structure.

The general trend in graphics performance has been interesting to watch. In terms of vector representation, there has been little change in the performance of graphics devices since the early 1980s. The Evans and Sutherland of the early 1980s could transform a similar number of vectors in real time to the high performance workstations of today. However, the Evans and Sutherland devices relied on expensive, specialized hardware to achieve the performance, whereas modern workstations are increasingly achieving graphics performance through software alone. The area in which there has been considerable development is in the performance in displaying raster primitives, with the potential this offers for more spectacular interactive graphics. As the molecular modeling software techniques develop to exploit these, so it is hoped that future generations of graphics hardware will satisfy the needs for interactive real-time manipulation of very complex shaded representations of large structures.

Because of the size of the molecules these are very demanding tasks, and the molecular graphics community continues to push at the limits of available technology. This is a totally 3-D application, and the favored graphics systems have incorporated such tricks as real-time rotation, depth cueing, and, in particular, stereo. Stereo perception has always been an important facility for molecular graphics and involves presenting a left and right eye view of the molecule to the appropriate eye of the user. The lack of stereo on many of the available graphics workstations has frustrated the molecular modeler for a long time, and alongside graphics performance the availability of suitable stereo has strongly influenced the hardware used by the molecular graphics community. Currently, Silicon Graphics is the system of choice. Stereo is provided by dividing the screen into a top and bottom half. The left eye view is drawn in the top half and the right eye view (usually rotated by 6° about the horizontal axis) is drawn in the bottom half. The monitor is then driven at double the frequency so that the screen presents the alternate eye views one after another. The user wears a pair of spectacles containing liquid crystal shutters, and an infrared emitter on the workstation synchronizes the visibility of the screen to each eye. This provides good stereo perception, crucial for visualizing the full three-dimensional nature of the molecule and essential for manipulating molecules effectively.

Real-time manipulation of a representation of molecular structure on a graphics screen can provide a clear three-dimensional image of the molecule. However, when modeling changes in the molecule where one part is moved against another, stereo is critical to maintain that three-dimensional perception. Examples include electron density fitting where the atoms are manipulated to be within the electron density map and docking experiments where the substrate is moved relative to an active site.

Unfortunately, there are very few other graphics applications that require stereo, and it has, with a few exceptions, proved difficult to persuade equipment manufacturers to make the required hardware and software modifications to support effective stereo. There is currently a lot of interest in virtual reality and this is producing a variety of different options for both visualizing graphics in stereo and some interesting devices for manipulation of objects within the computer (three-dimensional joysticks and force-feed back arms). At the current time, a number of pilot projects have demonstrated the potential of these techniques for molecular graphics and modeling. However, there needs to be significant improvement in technology before they are suitable for routine use by the molecular scientist.

## G. Hardcopy/Publication

One of the most visible impacts of molecular graphics and modeling techniques is in the production of presentation materials, be it for internal reports, scientific publications, or presentations at seminars and scientific meetings. The emergence of the Postscript format as a standard for 2-D graphics hardcopy has provided a useful standard around which a whole generation of printers has been developed. Black and white Postscript printers are now standard in most laboratories and cheap, reliable figure production is available to most users.

The production of color hardcopy remains more expensive and particularly difficult for obtaining exact reproduction of shaded images displayed on the workstation screen. At present, a number of technologies are available that use a mixture of colors to produce the full color picture. However, in general, the quality of shaded molecular graphics images is not good, the printers are not always reliable, and the cost of the consumable materials is rather high. The best medium for capturing molecular graphics information is still the 35-mm camera. Important requirements include using a long focal length lens and avoiding the appearance of straight lines, such as from the borders of the windows on the workstation, as they will appear distorted on the slide. The most effective technique is to use the standard workstation tools to capture parts of the image to disk and then compose a picture using presentation packages (such as showcase on the Silicon Graphics).

Molecular graphics is in three dimensions and the use of moving images to present results is particularly powerful. In the early days, 16-mm films were used, but these are very cumbersome and time-consuming to create, requiring production of each frame as a single-shot scene. Video boards, however, are now available for some workstations which allow the screen image to be captured directly onto a standard VHS recorder. Videos represent the only way to present some types of structural work and are particularly effective at lectures. A multitude of different options exist, ranging from driving a video display device directly from the workstation to creating a video which can itself then be played.

The video projection devices continue to evolve and fall into two main categories. The first are those with red, green, and blue lenses which project the appropriate images to form a composite colored image. These are the highest quality devices which can be driven directly from a workstation in a number of different ways, depending on the type of device. Some are capable of projecting the full resolution of current graphics systems directly; more commonly, a conversion device would need to be connected between the workstation and the projector to convert to standard video format. Alternatively, it is increasingly possible to obtain optional hardware for workstations which convert the image to standard video format which can then directly drive a standard projection device or be captured on video.

Perhaps the most exciting development in this type of projection device is the introduction of stereo capabilities. This is achieved either by liquid crystal shutter spectacles synchronizing the appearance of the left/right video images projected at high frequency or by projection of a stereo image through a liquid crystal shutter and polarizers onto a silvered projection screen which only requires the user to wear a pair of spectacles containing crossed polarizer lenses. These devices are extremely effective, but at present are rather expensive.

The second type of projection device uses a liquid crystal display to project standard video format signals using a conventional overhead projector. The quality of these images is not very good, but the device is very portable and quite acceptable for medium-sized lecture rooms. The major problem with video is the variation in video standards across the world, although it is possible to purchase devices that allow the recording of the video to be in a variety of formats. This complexity is compounded by the S-VHS format, which is very attractive because of the increased quality of color image that is recorded and projected, but does cause compatibility problems at different sites.

A very recent development is the incorporation of video into workstations themselves for communication across networks. It remains to be seen how this will develop, particularly in the case of the appropriate tools for constructing video and the capacity of the networks to handle the traffic.

## H. Standards

Finally, in this section discussing aspects of hardware for molecular graphics and modeling, it is appropriate to discuss standards. Over the years, a number of standards have emerged at all the levels and areas of computing. The problem for the smooth transition for the software and the user from one system to another has been the adherence to these standards by the manufacturers. Sometimes the enhancements introduced by a manufacturer are probably for commercial reasons. In other cases it has been vital that the manufacturers have ignored standards to ensure healthy development of new ideas that really work. For example, during the 1980s, there were strenuous efforts by various standards organizations to establish PHIGS as a standard in 3-D graphics. Silicon Graphics continued to very successfully develop their own graphics language known as GL, with the result that GL (or Open-GL) is becoming the standard in the graphics industry.

For the molecular modeling community, variances in operating system or compilers do not, in general, cause major problems for the computational code. However, the quite major changes and differences in 3-D graphics and even 2-D X11 graphics between manufacturers or revisions of software have caused serious problems.

## V. SOFTWARE COMPONENTS OF A MOLECULAR GRAPHICS AND MODELING SYSTEM

This section provides a brief outline of the features of modern software for molecular graphics and modeling. Many of the other contributions to this volume will deal with the detailed functionality and scientific merit of the various techniques. This chapter discusses the general features of the tools and techniques that can be defined as the core of any molecular modeling system, with particular reference to the demands this places on the workstation architecture and organization.

### A. User Interfaces

Essentially all modern software is presented to the user through a sometimes bewildering variety of user interface techniques. These include pull-down menus, slide right menus, dialogue boxes, control panels, file librarians, slide bars, icons, and much more. The purpose of all this is to use the features of a graphics workstation to make the software more accessible to the scientist and easy to use.

The representative screen shown in Fig. 6 demonstrates most of the types of the user interface technique. Much of the design of these interfaces is due to the influence of the Macintosh user interface which has had such an impact on computing. The main molecular visualization and interaction window is the central focus of the system. As the modeling session proceeds, messages appear in the textport which are both informational and report on the modeling process. The top level menu bar collects together a set of commands which can be seen on the pull down menu. Sub commands are then accessible through a pull right menu. At the right of the screen are a collection of commands as a palette. In general, these are interactive tools that can be used to affect the model. This palette can change with the application, reflecting the different tools required for different tasks.

At the bottom right of the screen are two control panels. These are continuously available slide bars and toggles which can act on how the molecular model is visualized. Finally, there is the dialogue box which, in response to a particular command, blocks and requires selection of a set of options to drive a particular calculation.

Although I have used the QUANTA (Molecular Simulations Inc.) program as an example, these types of user interface tools are available in essentially all molecular graphics and modeling packages. At the core, however, most of these programs are command driven; essential for batch processing and macroscripting of the modeling process, but also very desirable in providing shortcuts for the expert user who finds the continuous use of mouse and menu selection rather cumbersome.

## B. Constructing an Initial Model

There are two distinct methods for generating an initial model of a molecular system for a molecular modeling study. The model can either be derived from existing structural information held in databases or the chemical intelligence of the user and the molecular modeling system can be used to generate a model within the system.

A wide variety of techniques are available for generating an initial model. These include:

- entering a 1-D notation (such as SMILES) from which a rule-based or distance geometry-based approach can generate a 3-D structure.
- sketching in 2-D a diagram of the structure either generated free hand or by connecting fragments. The main requirements of the software are to maintain chirality, valency, or aromaticity as appropriate, to allow assignment of charge, generate an acceptable 3-D structure, and often assign atom types and parameters for force-field calculations (discussed later).
- building/editing in 3-D, in a way similar to the 2-D case; this can be sketching the molecule in 3-D or connecting together appropriate 3-D fragments to construct a molecule, with the addition of automatic refinement of bond distance, angles, and dihedrals as the building continues.

For all of these techniques, the requirements on computer hardware are fairly minimal. A good structure generation program will work just as adequately on a microcomputer as on a high performance workstation. The main criteria are the features of the software in how natural and straightforward it is for the chemist to build a molecule.

The second main category of techniques for generating an initial model is to search databases of known molecules related to the molecule under consideration and to generate an initial model from this. A wide range of databases need to be considered. There are databases of known experimentally determined structures such as the Brookhaven Data Bank for proteins (Bernstein *et al.*, 1977) and the Cambridge Crystallographic Data Centre's database for small molecule crystal structures (Allen *et al.*, 1979). The number of structures in these databases continues to grow at an impressive rate. These have been augmented by a number of small molecule databases in which the chemical information has been extended to a 3-D structure using techniques such as CONCORD. Examples here include the CAST 3-D database (available from Chemical Abstracts), the Available Chemicals Directory (available from MDL), the Maybridge directory of pharmaceuticals, or inhouse proprietary databases. The main requirements are large amounts of disk space to hold these databases, a flexible query language to allow searches to be defined in a natural way, and a browsing facility to help characterize the results of the search.

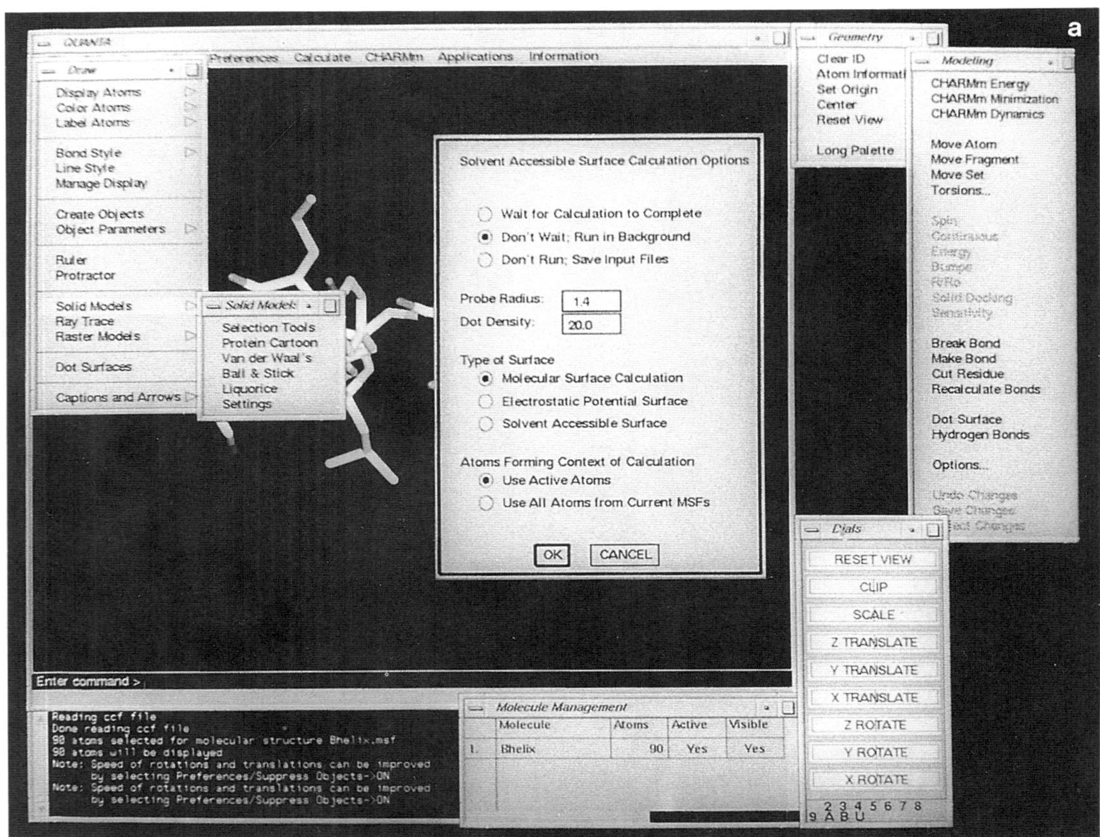
a
QUANTA
Preferences Calculate CHARMm Applications Information
Draw
Display Atoms
Color Atoms
Label Atoms
Bond Style
Line Style
Manage Display
Create Objects
Object Parameters
Ruler
Protractor
Solid Models
Ray Trace
Raster Models
Dot Surfaces
Captions and Arrows
Solid Models
Selection Tools
Protein Cartoon
Van der Waal's
Ball & Stick
Liquorice
Settings
Solvent Accessible Surface Calculation Options
Wait for Calculation to Complete
Don't Wait; Run in Background
Don't Run; Save Input Files
Probe Radius: 1.4
Dot Density: 20.0
Type of Surface
Molecular Surface Calculation
Electrostatic Potential Surface
Solvent Accessible Surface
Atoms Forming Context of Calculation
Use Active Atoms
Use All Atoms from Current MSFs
OK
CANCEL
Geometry
Clear ID
Set Origin
Center
Reset View
Long Palette
Modeling
CHARMm Energy
CHARMm Minimization
CHARMm Dynamics
Move Atom
Move Fragment
Move Set
Torsions...
Break Bond
Make Bond
Cut Residue
Recalculate Bonds
Dot Surface
Hydrogen Bonds
Options...
Dials
RESET VIEW
CLIP
SCALE
Z TRANSLATE
Y TRANSLATE
X TRANSLATE
Z ROTATE
Y ROTATE
X ROTATE
Molecule Management
Molecule Atoms Active Visible
1. Bhelix 90 Yes Yes
Enter command >
Reading ccf file
Done reading ccf file
90 atoms selected for molecular structure Bhelix.msf
90 atoms will be displayed
Note: Speed of rotations and translations can be improved
by selecting Preferences/Suppress Objects->ON
Note: Speed of rotations and translations can be improved
by selecting Preferences/Suppress Objects->ON

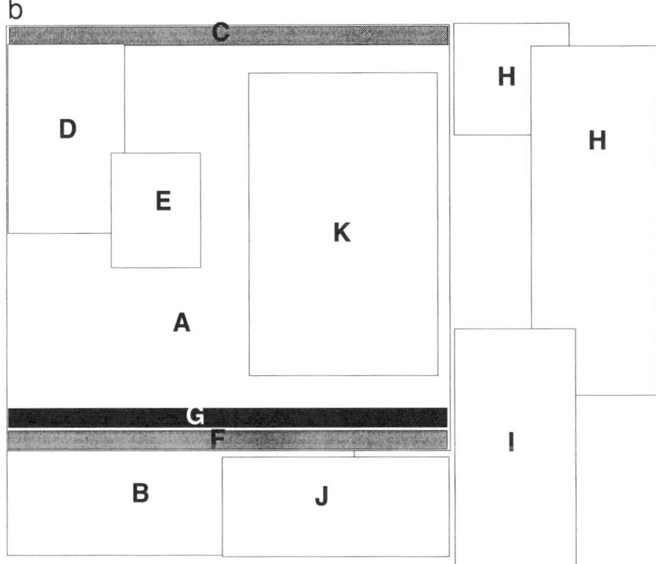


**FIGURE 6** User interfaces in molecular modeling systems. (a) A screen constructed to show all the user interface mechanisms in a typical molecular modeling package. (b) The various components are labeled as: A, the main molecule visualization window; B, textport; C, top menu bar; D, pull-down menu; E, pull right menu; F, message line; G, command input window; H, palettes of commands; I, control panel of pseudo-dials; J, control panel for molecule handling; and K, dialogue box.

## C. Refining the Model

Once an initial model of the molecular system has been constructed, it is necessary to explore the conformations available to the molecule. Sometimes it is the lowest energy structure of a molecule that is of interest. In other cases, the range of conformations available are necessary to represent the structure.

In theory, quantum mechanical calculations should provide a complete description of the energy of any particular conformation of a molecule. In practice this is not possible because of the number of atoms involved (and thus wave functions) and because of the inaccuracies from the approximations that must be made. For this reason, molecular mechanics is the main technique used to evaluate the energy of a molecule and from this search for the preferred conformations.

Molecular mechanics is essentially an approach to calculating the energy of a molecule using an empirical energy function which attempts to describe covalent bonds in terms of "springs" connecting atoms with van der Waals and electrostatic interactions between nonbonded atoms. A number of potential functions have been developed for use with polyatomic systems. In most of these a bond is described as a harmonic restoring force with penalties for deforming bond angles, a dihedral torsional potential to allow for hindered rotations of groups about a bond, and nonbonded interactions between separated atoms.

An expression of the following form is the most widely used and is the basis of many of the force fields widely in use. Some have additional terms defining additional interactions or constraints between groups of atoms.

$$
\begin{aligned}
V(r_1, r_2, \ldots, r_n) = & \sum_{\text{bonds}} \frac{1}{2} K_b(\mathrm{b} - \mathrm{b}_0)^2 + \sum_{\text{angles}} \frac{1}{2} K_\theta(\theta - \theta_0)^2 \\
& + \sum_{\text{torsions}} \frac{1}{2} K_\xi(\xi - \xi_0)^2 + \sum_{\text{dihedrals}} \frac{1}{2} K_\phi(1 + \cos(\mathrm{n}\phi - \delta))^2 \\
& + \sum_{\text{pairs}(i,j)} \left( \frac{\mathrm{C}_{12}}{r_{ij}^{12}} - \frac{\mathrm{C}_6}{r_{ij}^{6}} + \frac{\mathrm{q}_i \mathrm{q}_j}{4\pi \varepsilon_0 \varepsilon_r r_{ij}} \right),
\end{aligned}
$$

where the first term represents the covalent bond interaction (b) and the second term is for bond angles ($\theta$). Two terms are used for the dihedral angle: a harmonic term for torsions ($\zeta$) that do not undergo transitions (e.g., dihedral angles within aromatic rings) and a sinusoidal term for the other dihedral angles ($\phi$). The two terms that result from the interaction of pairs of atoms involve the Lennard–Jones (6–12) interaction and the coulombic effect between charge atoms. $C_{12}$ and $C_6$ are constants for a particular atom pair interaction, and $\varepsilon_0$ and $\varepsilon_r$ are the vacuum and relative permitivity, respectively.

In energy minimization, the coordinates of the system are changed so as to minimize the total energy of the system. Because of the number of degrees of freedom and the multiple minima nature of the potential energy surface, it is not possible to find the conformation corresponding to the global energy minimum except for quite small molecules. Generally, energy minimization finds a local minima close to the starting structure. In conformational analysis, the conformation of the molecule is varied while monitoring the energy of the system. If there are a limited number of flexible bonds in the molecule, then an exhaustive search of all possible conformations can be made. For more than about five to six bonds, then various strategies have to be adopted to make the search of conformational space more tractable. These include random generation of conformations and techniques such as Boltzmann Jump. The latter is a form of Monte Carlo sampling in which changes to the structure are always accepted if they result in a reduction in energy and with a Boltzmann probability if the energy of the system has increased. From these techniques a set of conformations can be achieved. Because conformational analysis usually works with a fairly coarse grid of dihedral angles (~10–30°), each step is usually followed by minimization to find the locally best structure.

Another technique used to explore the conformational space of molecules is molecular dynamics. The basic feature of molecular dynamics is the calculation of a trajectory of the molecule, i.e., a series of structures in which the system is moving under the influence of the forces acting on the atoms. These are calculated from the first derivative of the potential function with respect to the atom positions. By applying Newton's equations of motion, these forces can be used to calculate how the atomic positions change with time, giving rise to a dynamics trajectory, a series of conformations for the molecule which evolves with time. The general strategy is to calculate the forces on the atoms in the system and then allow the atoms to move under influence of this force, usually for a femtosecond. New forces are then calculated from these positions to be applied in the next step of dynamics and so on. In some methods, extra forces are considered to simulate the presence of solvent or some external heat bath (Langevin dynamics).

A critical aspect of using potential energy functions is providing appropriate parameters for the system under study. Many different parameter sets have been developed for a variety of different force fields. It is always important to check that the parameters and force field that are being used have been validated for the particular molecular system under study.

## D. Manipulating the Model

The essence of an interactive molecular modeling system is the ability for the scientist to explore the conformational variability and properties of molecules and how they may interact with each other. Sometimes these relationships can be explored automatically using algorithms based on the empirical force field conformational search techniques discussed earlier. In most molecular modeling systems, however, there is a need for some basic modeling tools, such as:

- moving molecules or fragments of molecules about relative to each other while monitoring bad contacts or energy of interaction
- specification of flexible bonds and rotating around these torsion angles with monitoring of contacts and energy
- solid docking procedures which prevent the manual modeling from moving the molecules so close to each other that they penetrate
- semiautomated docking procedures (such as least-squares fitting or distance geometry) which suggest ways in which two molecules can satisfy certain restraints.

All of these tools place strenuous demands on the molecular modeling software design and implementation. But in general, all of the techniques are essentially interactive and respond in real time on all the current generation of workstations.

## E. Visualization

This section briefly reviews and illustrates the different techniques available for display of molecular structure and properties. As an example, Figs. 7 to 15 show different representations of a small piece of protein structure, with the molecule in the same orientation in each figure to show the value of the various graphics renderings.

The simplest and by far the most widely used graphical method for representing molecular structure and properties is vector graphics where the bonds in the molecules are represented as lines joining the atoms. An important feature is the ability to rotate, scale, and translate with full three-dimensional clipping and depth cueing. Graphics hardware is now optimized so that on most 3-D workstations vector representations of quite large molecular structures can be manipulated in real time.

Molecular scientists find interactive vector graphics a very powerful tool for studying molecules; the representation of the bonds within a molecule as vectors not only reveals the essential chemistry of the molecule but also allows the complete detail of the structure to be appreciated. Figure 7 shows a vector representation of the molecule; even with such a small number of atoms it is difficult to get a real appreciation of the depth of the molecule. Figure 8 shows the same representation, this time with depth cueing in

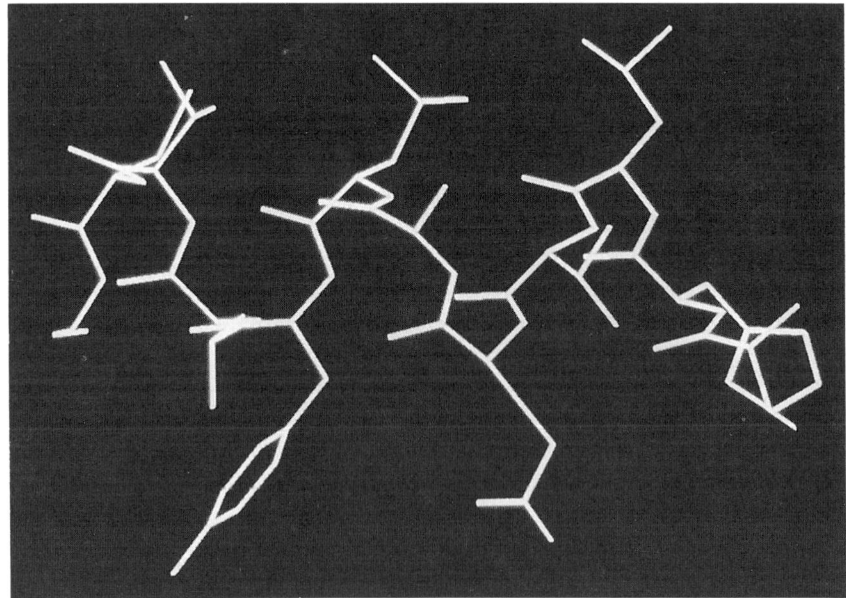

**FIGURE 7** The peptide with no depth cueing.

which the objects further away are drawn progressively darker. This provides a greater feeling of depth to the representation.

Advances in raster graphics now allow more detailed representations of the structure to be rendered in real time. For example, Fig. 9 shows a ball and stick representation in which the atoms are drawn as shaded spheres of varying radius to reflect the element type, connected by cylinders. When the radius of the atoms is uniform and the same as the radius of the bond cylinders, then the licorice representation of Fig. 10 results. These representations still allow the underlying chemical attributes of the molecule to be viewed. Shaded spheres at van der Waals radius (as in Fig. 11) give the impression of the overall shape and packing of the atoms in the molecule, but the details of the chemical structure are difficult to discern. In general, this latter type of representation is only useful in showing the overall organization of the molecule, such as in the distribution of polar atoms over a molecular surface, or in showing the packing together of molecules in a complex.

The molecular surface is a fundamental aspect of a structure as it is through the complementarity of shape and chemistry of the surface that molecules interact with each other. A variety of different representations of surfaces have developed. The most important for many years was that developed by Connolly (Langridge *et al.*, 1981) in which the molecular

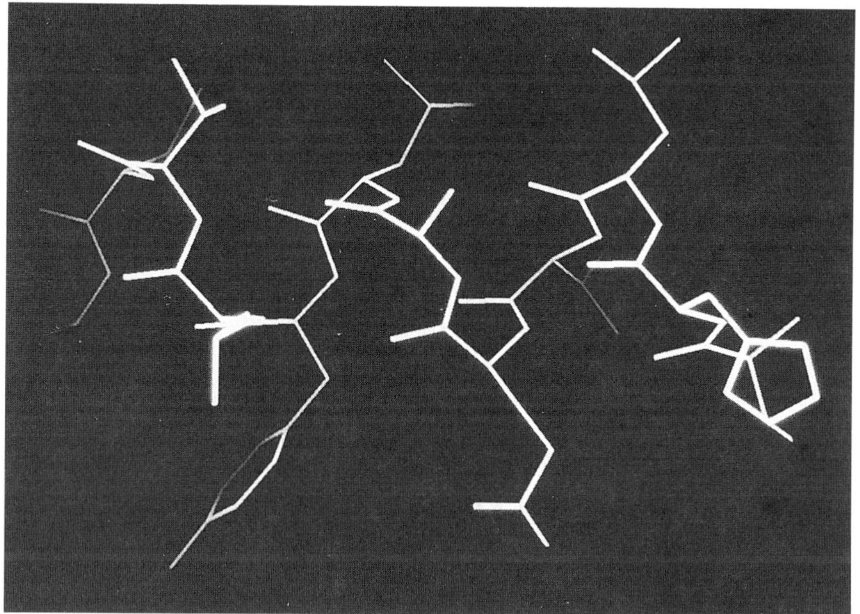

**FIGURE 8** The peptide with depth cueing.

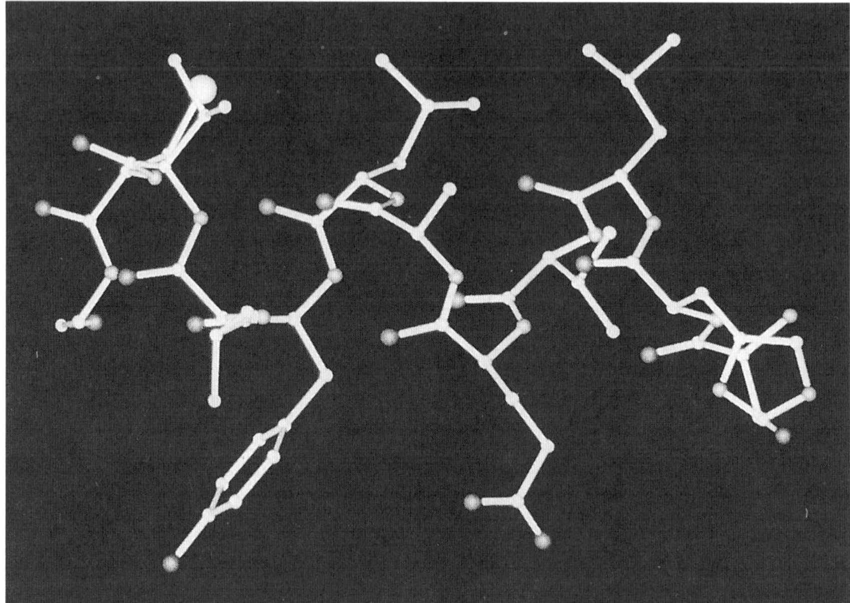

**FIGURE 9** Ball and stick representation of the peptide. The different radii represent different atom types.

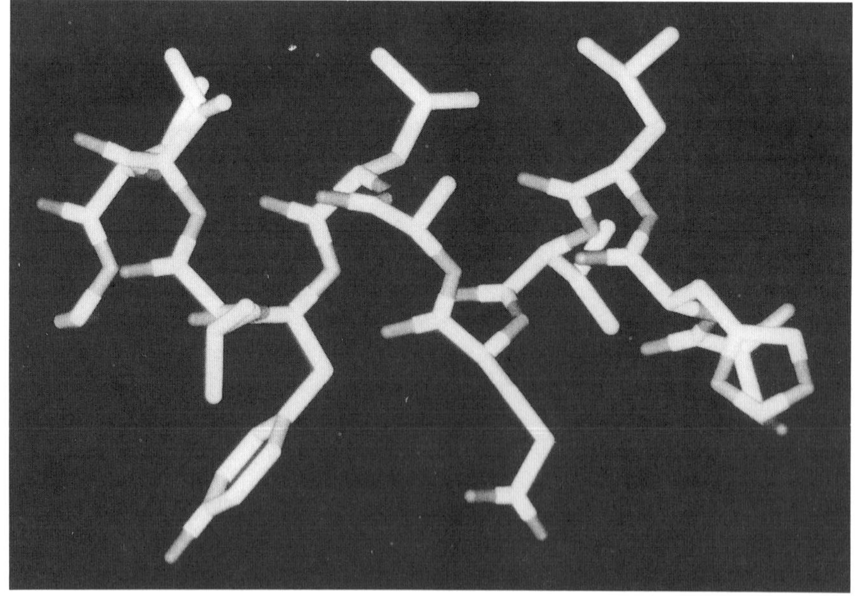

**FIGURE 10** Licorice representation of the peptide.

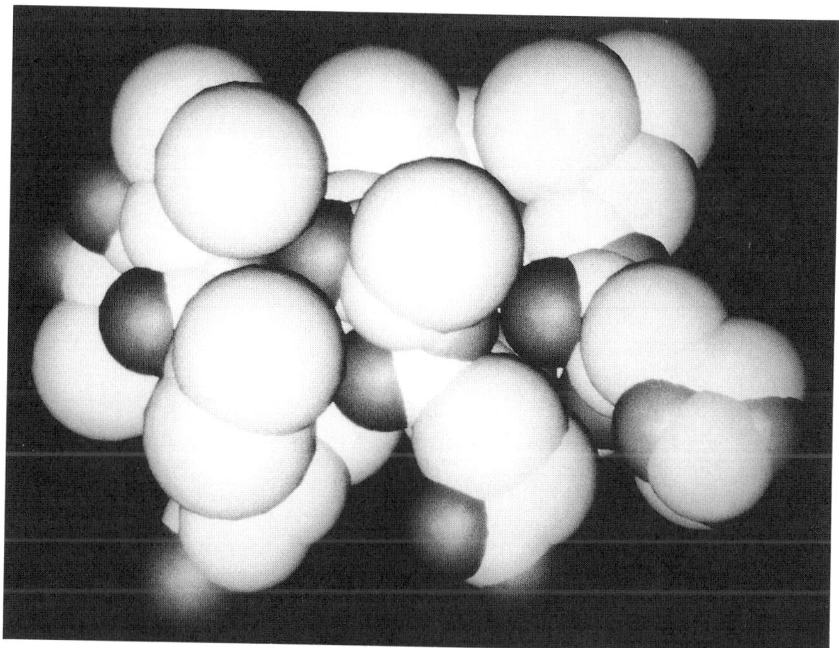

**FIGURE 11** Shaded van der Waals spheres representation of the peptide.

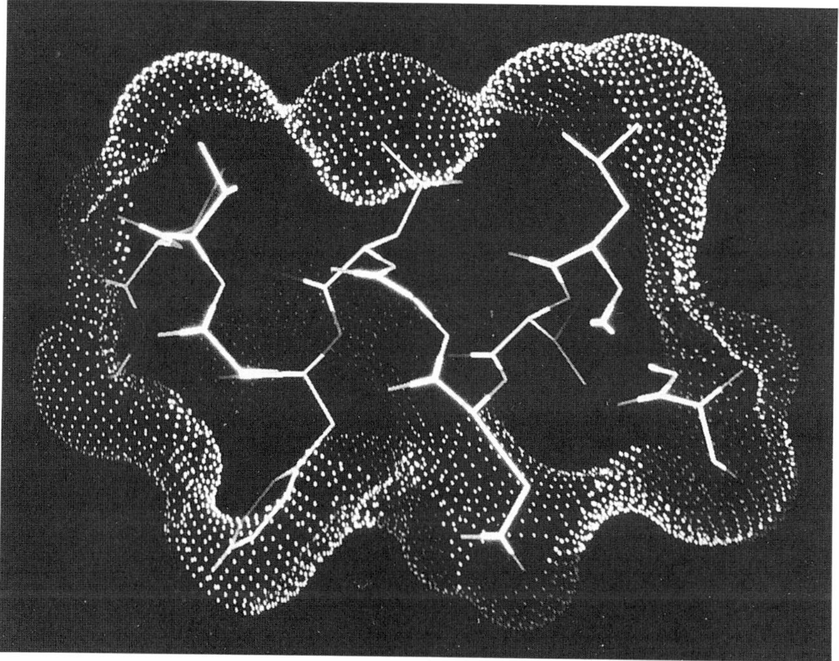

**FIGURE 12** Connolly surface on the peptide.

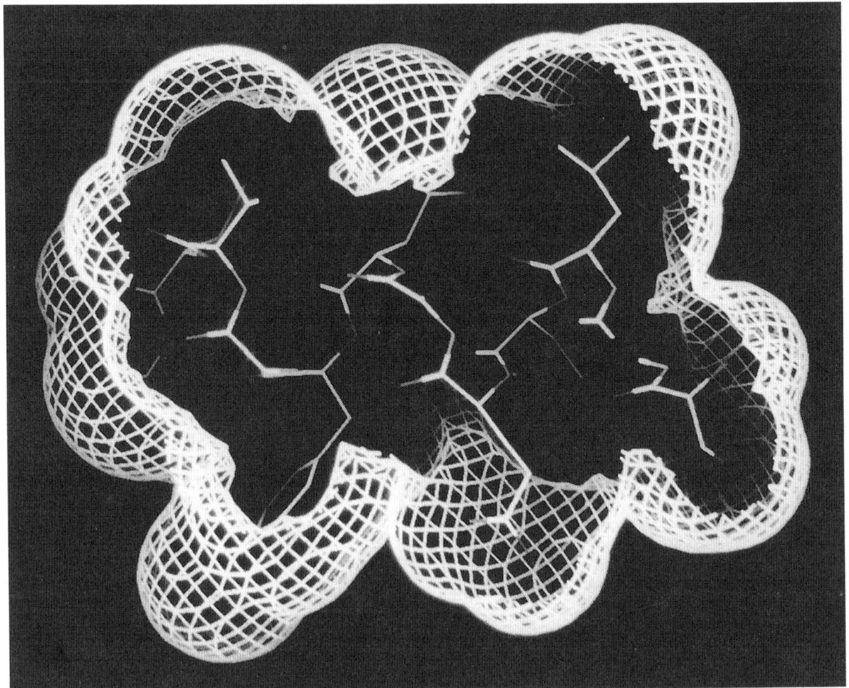

**FIGURE 13** Grid surface on the peptide.

surface is defined by the surface in contact with a probe sphere as the sphere "rolls" over the surface of the molecule. Alternatively, the extended solvent accessible surface can be calculated in which the surface is traced out by the center of the probe sphere as it rolls over the molecule. In both cases, the resulting surface is displayed as a continuous envelope of dots. Figure 12 shows the molecular surface for our peptide molecule. This dot representation is particularly powerful as it produces a smooth surface that can be dissected and viewed interactively on most graphics devices. At the same time, the underlying molecular architecture can be analyzed with the surface.

An alternative representation of the surface with similar properties is the production of a molecular envelope as a contoured grid of lines, as in Fig. 13. This type of representation developed from the early use of molecular graphics by crystallographers to view electron density maps (Jones, 1978, 1982). Here, the three-dimensional grid of electron density values were contoured in three dimensions to produce the chicken-wire representations of density. In a similar way, many molecular properties can be computed on a three-dimensional grid and displayed in this way. In the case of Fig. 13, this was a simple calculation in which a three-dimensional grid was

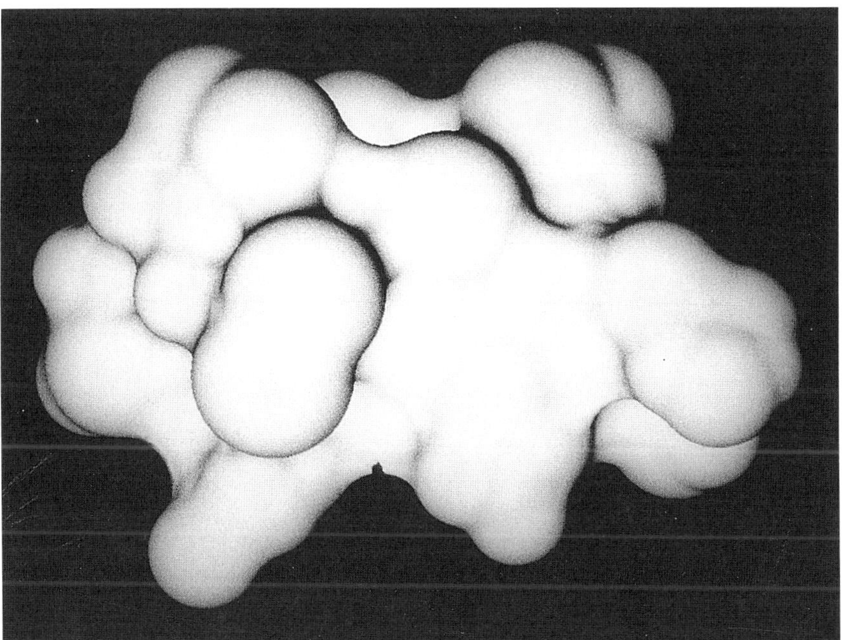

**FIGURE 14** Shaded molecular surface representation of the peptide.

placed over the molecule and the grid points were assigned a value related to their distance from an atom. Contouring this produces a molecular surface.

More recently, the increasing power of a raster graphics system allows more complex images to be viewed interactively and this has led to the development of many techniques for representing solid molecular surfaces. Figure 14 shows an example for our peptide produced by triangulating the dots produced in the Connolly surface.

Computer graphics developments in the early 1980s led to a technique known as ray tracing which can give added realism to a graphics image by calculating shadows and complex lighting and shading. Figure 15 shows such a figure for our peptide. The main advantage is the increased three-dimensional perception from the casting of shadows.

Although the figures presented in this chapter are necessarily monochrome, the use of color is a vital component of any molecular visualization. Color is extremely important for two main areas. The first is to aid in the presentation and interpretation of complex and large molecules. For example, the atoms could be colored on element type or atomic charge to highlight the chemistry of the molecule or different parts of the system could be colored to illustrate packing in a complex. Second, color can be used on

**FIGURE 15** Ray traced figure of van der Waals representation of the peptide.

molecular surfaces or other representations to highlight the properties of the molecule, such as electrostatic potential or hydrophobicity.

It is vitally important to be able to produce hardcopy plots as a record of molecular modeling work and for publication. The hardware options for producing presentation material was discussed in a previous section. Although color printing devices are becoming increasingly available, the majority of output is still produced as black and white. Traditionally, programs such as PLUTO or ORTEP were used to produce diagrams such as that shown in Fig. 16. More recently it has become increasingly straightforward to capture other types of representations from a working session of the graphics screen. These can then be converted into some usable format which can then be taken into some other program for tidying up or annotation. These programs are increasingly becoming available as part of the standard release of software on modern workstations or are available from public distribution centers on the network. The packages fall into two distinct categories. The first is one that can manipulate and modify the image in terms of postscript-like graphics primitives which allows the output of high quality, crisp line quality diagrams such as that of Fig. 16 which is suitable for publication. The second category is more for presentation purposes where images from the screen are directly captured and annotated.

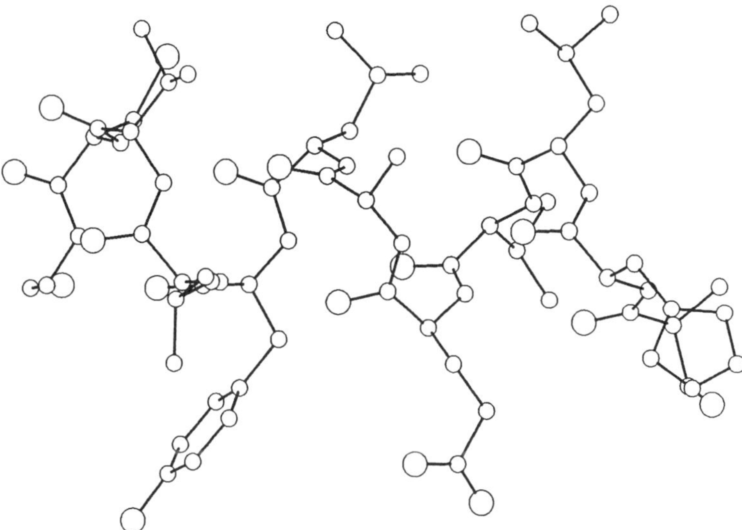

**FIGURE 16** PLUTO-like hardcopy representation of the peptide.

The latter suffer from the lack of resolution due to the limitations of the raster on which the systems work, but have the advantage of allowing color to be used and the incorporation of images, as used in this chapter in Fig. 4.

## VI. FUTURE PERSPECTIVES

This chapter focused on the generic features of the hardware and software requirements for molecular graphics and modeling systems. Clearly, rapid development will continue in both of these areas. Workstations will continue to get more powerful and relatively cheaper. Although the computing power at the present time may be limiting for some of the more ambitious calculations and database searches, it is sufficient to perform most of the routine molecular modeling tasks on a small workstation. There is a major constraint, however, in the power of the graphics and the way in which the chemist interacts with the data and its representation. It is hoped that developments in the hardware performance and in the techniques available will continue in this area.

Whatever the technologies that do develop, it is probable that for some time to come the computational techniques will fall well short of being able to predict conformation and properties for all but the simplest system. So, as in many skilled professions, the most important component in a successful molecular modeling study will be the quality and experience of the scientist involved. The equipment and techniques described in this chapter merely form the tools with which such modeling tasks can be attempted.

## REFERENCES

Allen, F. H., Bellard, S. H., Brice, M. D., Cartwright, B. A., Doubleway, A., Higgs, H., Hummelink, T., Hummelink-Peters, B. G., Kennard, O., Motherwell, W. D. S., Rodgers, J. A., and Watson, D. G. (1979). The Cambridge Crystallographic Data Centre: Computer based search, retrieval, analysis and display of information. *Acta Crystallogr.* **B35**, 2331.

Bernstein, F. C., Koetzel, T. F., Williams, G. J. B., Meyer, E. F., Brice, M. D., Rodgers, J. R., Kennard, O., Shimanouchi, K., and Tasumi, M. J. (1977). A computer based archival file for macromolecular structures. *J. Mol. Biol.* **112**, 535.

Humblet, C., and Marshall, G. R. (1981). Three dimensional modeling as an aid to drug design. *Drug. Dev. Res.* **1**, 409.

Jones, T. A. (1978), A graphics model building and refinement system for macromolecules. *J. Appl. Crystallogr.* **11**, 268.

Jones, T. A. (1982). FRODO: A graphics fitting program for macromolecules. *In* "Computational Crystallography" (D. Sayre, ed.), p. 303. Oxford Univ. Press.

Langridge, R., Ferrin, T. E., Kuntz, I. D., and Connolly, M. L. (1981). Real time color graphics in studies of molecular interactions. *Science* **211**, 661.

Levinthal, C. (1966). Molecular model building by computer. *Sci. Am.* **214**, 42.

# 3

# Molecular Modeling of Small Molecules

**TAMARA GUND**
New Jersey Institute of Technology
Newark College of Engineering
Biomedical Engineering Program
Department of Chemistry
University Heights
Newark, New Jersey

## I. INTRODUCTION

Molecular modeling, like art, means different things to different people. To the theoretician, it is often considered to be insufficiently rigorous to be taken seriously. To the experimentalist, it may be too theoretical and too removed from experiment to be considered pertinent. To the administrator, it may be considered an expensive toy which cuts into the experimentalist's laboratory time and mostly is used to embellish studies after completion to increase their chances of being accepted for publication.

In this chapter, as in other chapters in this book, the attempt is made to demonstrate that molecular modeling, in particular, and simulation, in general, represent complementary alternatives to theory and experiment for solving scientific problems. In attempts to predict properties of new systems, experiment is expensive and often difficult, and needs direction to be productive; and theory is usually insufficient to get the prediction exactly right. In our experience, models—mathematical/statistical, molecular, and mechanistic—can harness theory and can indicate the correct experiments to run.

## A. History

The origins of molecular modeling have been reviewed a number of times.[1] Briefly, Hueckel, Mullikan, and others postulated that molecules could be built up of nuclei surrounded by swarms of electrons in delocalized "molecular orbitals"; further elaboration by Pauling, Roberts, Streitwieser, Pople, Dewar, and other giants led to the current generation of molecular orbital programs for calculating molecular structures and properties. Independently, building on theories of vibrating molecules to explain infrared spectral absorbancies, Westheimer, Hendrickson, Wiberg, Boyd, and many others developed the current "classical mechanical" programs for computing structures, conformations, interaction energies, molecular dynamics, and other properties. With the advent of modern scientific computing and commercial scientific software companies in the late 1970s, progress has been swift. The interested molecular modeler now faces a daunting set of hardware and software choices.

## B. Why Do Modeling?

Reasons for performing a modeling study are numerous. The best reasons are to understand an experimental system or results; to optimize information to be derived from a planned experiment; or to apply theoretical or structural information to a scientific problem. Molecules are incomprehensibly small, but scientists have used various mechanical models for decades to illustrate and understand such useful concepts as molecular chirality, ring conformation, and steric strain. Computer models can provide the same insights as hand-held models, while allowing manipulations, e.g., superposition, energy calculation, and vibrations, that are difficult or impossible with mechanical models. Computer modeling programs are also useful for performing systematic and exhaustive studies that most scientists would be incapable or impatient of performing by hand. However, it is almost always true that becoming too enamored of the model may cause a scientist to compromise his or her objectivity. Models are to be tested and used, but should never be literally believed.

## C. Who Should Do Modeling?

There is now a recognized profession of computational chemist or molecular modeler; we estimate the number of practitioners as between 5000 and 10,000. These practitioners exist in universities and research institutes as well as in the pharmaceutical, chemical, and computer hardware and software industries (among others). These professionals justify and acquire for their companies graphics workstations, supercomputers and compute servers, and various modeling software systems; they often also write their own methods and applications software.

As the software has become less costly and easier to use, many practicing chemists and other scientists have begun to consider doing molecular modeling studies themselves, to understand a system of interest or estimate a property of interest. With molecular display and geometry optimization software on desktop Macintosh and MS-DOS computers, the number of amateur modelers already exceeds the number of professionals by at least an order of magnitude. This growth has occurred because of the obvious importance of models to all scientists and the recognition that computerized models are easier to manipulate and compute than mechanical models. This trend is expected to continue and to broaden to other scientific disciplines.

Chemists, both computational and experimental, have been the primary focus of molecular modeling software offerings to date. Other disciplines such as crystallographers, spectroscopists, and biochemists have also utilized such software. Biologists also have often seen the value of understanding chemical structure and may profitably use molecular visualization and simulation. Information scientists also, while not traditional users of molecular modeling, are recognizing that if an answer to a scientist's question does not exist in the literature, a computed answer is often better than no answer.

## D. Getting Started

Modeling studies may be rigorous or trivial and they may be heavily numeric or primarily graphical. They may require hours or months of supercomputer time or they may require seconds or minutes on a personal computer, depending on the question to be answered, the complexity of the model, and the accuracy required. It is usually a good idea to begin with a relatively simple model system and increase complexity over time so that there is a context for the results that are being obtained.

### 1. Hardware

At this time, the predominant systems for molecular modeling calculations are UNIX workstations, particularly three-dimensional (3-D) graphics workstations such as those available from Silicon Graphics Inc. But the entire range of computer hardware—from desktop Macintosh and MS-DOS personal computers to compute servers and supercomputers such as Crays—has been utilized for molecular modeling and computational chemistry applications. On low end machines, studies of molecular structure, conformation generation with force-field energies, database search, structure superposition, and simple properties calculations are possible. Larger systems are generally required for high level ab initio quantum mechanical calculations, for molecular dynamics calculations for significant numbers of picoseconds, for Monte Carlo simulations, for computing solvent effects with explicit solvent molecules, and for exhaustive conformation search and superposition.

### 2. Software Packages

A variety of commercial packages are available, ranging from \$50–\$5000 for PC-based systems to \$100,000 or more for supercomputer-based systems. Unfortunatley, at present there is no one system that meets all the needs of the molecular modeler nor is there likely to be; with the current level of activity in this area, new methods are being developed worldwide, in university, institute, and commercial laboratories. The industry is desperately in need of an open, high-level programming environment that would allow diverse applications to work together, with a consistent "look and feel." Lists of computer programs are available in this area,[2–4] but any such list is soon out of date.

## II. MOLECULAR MODELING FUNCTIONS

There are many possible methods and applications of molecular modeling. This section attempts to give the reader a feel for the types of studies that can be done, and why they would be done. Other chapters in this volume explore many of these topics in more detail.

### A. Structure Generation or Retrieval

Molecular structures may be generated by a variety of procedures.[1] If a crystal structure exists, it can be found in the Cambridge Crystallographic Data File[5] and turned into molecular coordinates by standard methods. If a 2-D chemist's structure diagram exists, with chirality information, this can be converted into 3-D molecular coordinates by a number of methods including a program called CONCORD.[6] Alternatively, standard fragments may be stitched together to create a 3-D molecule; most commercial modeling systems include fragment libraries for such molecule construction. A powerful method of creating molecules is to modify a known structure; most current modeling systems allow such "molecule editing."

A method that has become popular is to generate novel structures quickly according to some algorithm, and then filter or order the resulting structures by expected properties, e.g., program LUDI.[7] Such "*de novo* design" of molecules can offer provocative suggestions to the synthetic chemist, but in practice suffers from the fact that many of the suggestions tend to be difficult or impossible to synthesize.

Once a structure is created, a number of techniques may be used to generate different conformations of the structure and to rate them. The most straightforward method, systematic conformation search, rotates around each rotatable bond by some increment; this method is quite time-consuming for large numbers of rotatable bonds, cannot account for ring flexibility directly, and can miss conformations if the rotation increments are too large. Commonly observed "isopotential energy maps" are derived from such a systematic search. Other techniques include Monte Carlo search, stochastic search, and distance geometry methods.

## B. Structure Visualization

One of the most popular uses of molecular modeling systems is to visualize molecular structures and interactions in order to calibrate the chemist's intuition. The best chemists develop a "feel" for how structural variation will change properties of interest; while this may be rationalized in terms of steric, electronic, and other effects, it remains at some level intuitive. There is evidence that only about 50% of the human brain is involved in the logical, deductive, linear reasoning that we normally associate with the scientific method. If the more intuitive, visual, interactive, pattern-perceiving facilities of the brain can also be put to work on the problem, the solution is likely to be arrived at sooner. Just as we can, in our mind, walk through our apartment's layout and estimate whether a new sofa will fit in one or another corner, so can molecular graphics help us calibrate our perception of a receptor-binding site or structure–activity pattern so that we can design active compounds.

All molecular displays are two dimensional. The third dimension is suggested by perspective, by fading distant parts, by forward objects occluding distant ones, by dynamic motion of the molecule, or by one of the stereoscopic viewing methods. The use of a joystick, a mouse, or other interactive devices allows the user to interact with the structure and thereby use hand-to-eye coordination to further analyze spatial characteristics of the molecule.

Different methods of representing structure and properties can enhance this learning experience:[1,8] (1) Stick figures[9] (Fig. 1) and "licorice" figures (Fig. 2) represent the bonding arrangements and closely resemble the widely used hand-held Dreiding models. Stick models are popular modes of molecular display and are incorporated into most modeling programs. They enable an easy viewing on both vector and raster graphics machines and facile manipulation of the molecular skeleton. (2) Ball and stick models (Fig. 3) combine the stick features of joining atoms by lines, but the atoms are represented by spheres. This type of representation allows easy viewing of the skeleton with the advantage of a clearer 3-D perspective since atoms farther away are obscured by atoms closer to the front of the image. (3) Space-filling representations (Fig. 4) join atoms by spheres whose radius depends on the atom type.[10–12] The radii of the atoms generally match the van der Waals radii of the atoms they represent so that bonded atoms intersect each other. The images obtained look like the Corey–Pauling–Koltun (CPK) models. Such representations with shading can show three dimen-

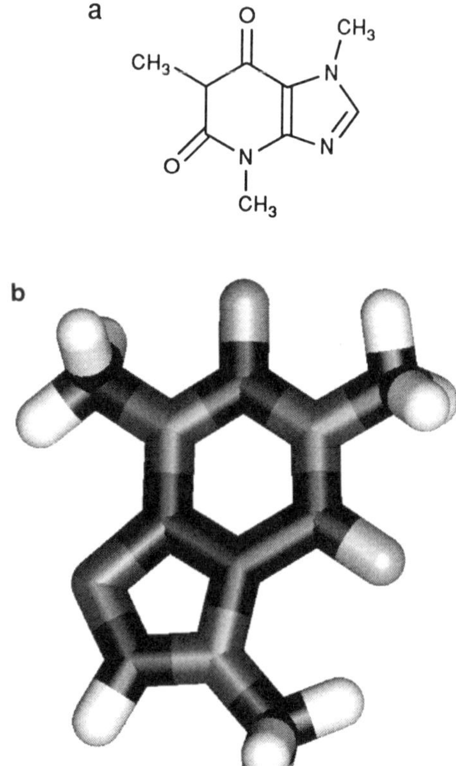


**FIGURE 1** Stick figure (a) or tubular (b) representation of caffeine.

sionality as well as the steric environment of the molecule. (4) Surface displays (Fig. 5) can represent the molecule as a stick or a ball and stick or some other combination of lines and solid atoms surrounded by dots,[13] lines,[14] or grids.[15] Such models can elucidate the steric environment, volumetric properties, or electronic properties of the molecule. Visual enhancements can be added by color coding atom types or charge distribution. Monochromatic enhancements can also be made where surfaces and charge distributions are represented by different surface textures.

## C. Conformation Generation

Most molecules can exist in different conformations. Conformations of the same molecule may be different in vacuum, in the liquid state, in various solutions, and in the crystal. In general, a population of conformations may exist in solution, and properties of interest (such as a biological activity)

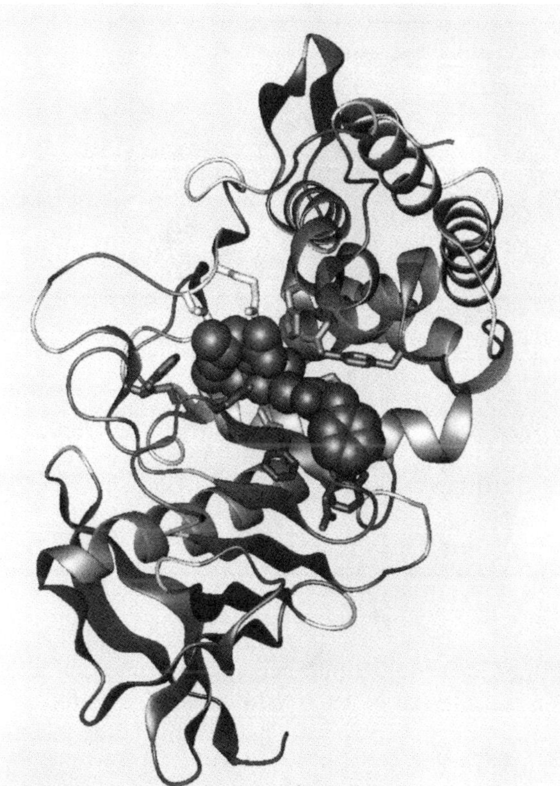

**FIGURE 2** Licorice figure representation of the protein thermolysin (Brookhaven file name 5 tmn.pdb) bound to the inhibitor, a phosphoramidate peptide analog shown in a space-filling mode. The nearby protein side chains are shown as "tubular" bonds, and the protein is shown as a ribbon structure where the $\alpha$ helices and $\beta$ sheets are wider ribbons.

may be associated with one or a small number of them. This property-specific conformation may not be the prevailing conformation in the solid state, in solution, or in vacuum. It thus becomes necessary to calculate possible conformations and deduce which conformation(s) is important for the property of interest.

Conformations may be defined in various ways. By holding bond angles and bond lengths constant and twisting about rotatable bonds, *torsion space* may be explored.[16] A problem occurs when rings are present; by temporarily breaking one ring bond, rotating other bonds, and reconnecting the broken bond, ring conformations can be explored. Torsion bonds may be systematically explored to find all possible conformations, but for large changes in torsion angle, conformations may be missed. Furthermore, when the number

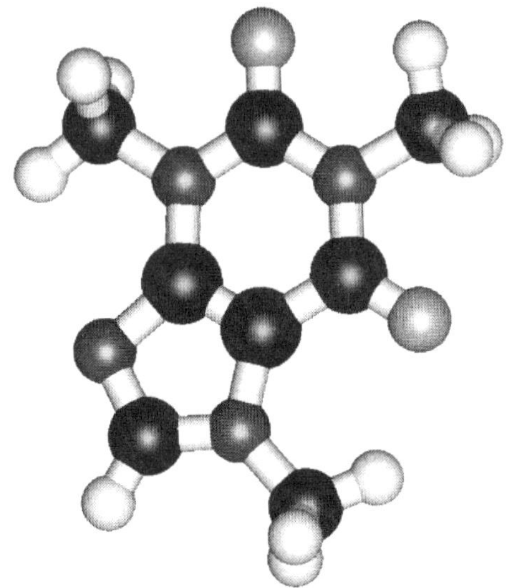

**FIGURE 3** Ball and stick representation of caffeine.

of rotating bonds is large, the number of conformations to be explored is quite large. As an alternative to *systematic* torsion exploration, there are various *stochastic* or random approaches. Monte Carlo techniques produce random conformations that can be analyzed statistically or energetically.[17]

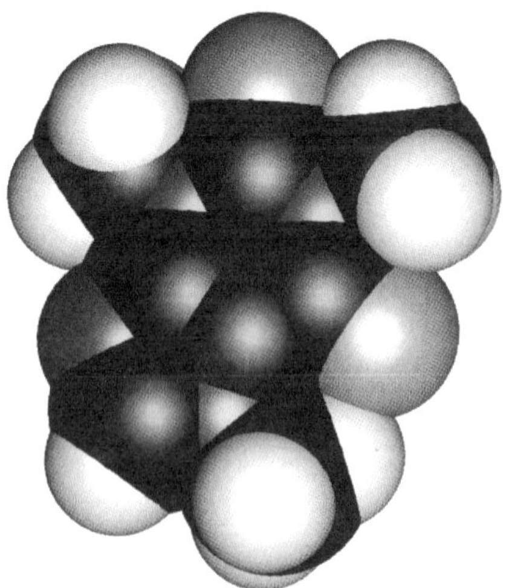

**FIGURE 4** Space fill representation of caffeine.

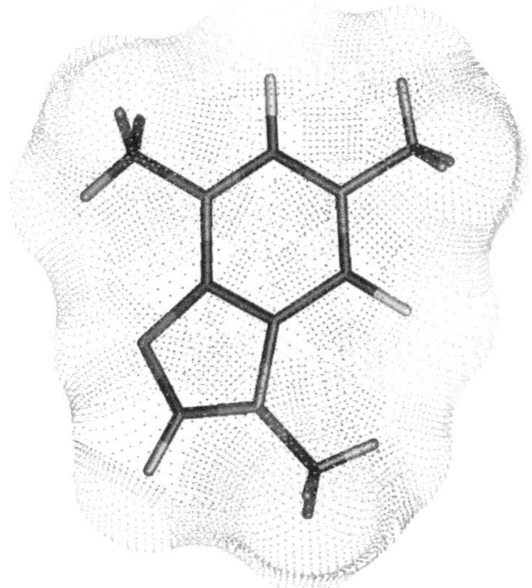

**FIGURE 5** Solvent accessible surface representation of caffeine. Color coding according to electrostatic energy would produce the electrostatic surface.

By moving atoms in *cartesian space*, geometries can be optimized and energies calculated, but only the local minimum energy conformation is found. With molecular dynamics calculations, atoms are moved according to the forces on the atoms, following Newton's laws of motion.[18] These mimic the natural movements of molecules, and over time (especially at high simulated temperatures) can explore all of conformational space, but inefficiently, since much time is spent exploring local conformations. A further advantage of dynamics calculations is that they can provide an estimate of molecular entropy and therefore free energy.

There is a third description of molecules, using *distance space*, or distance geometry; this provides an alternative methodology for exploring conformations.[19]

Simple conformational calculations give the energy *in vacuum*, but there are ways to account for solvent or crystalline environments. Solvent effects may be calculated as a bulk property or with explicit solvent molecules placed by Monte Carlo or molecular dynamics techniques. Conformation in solvent is more relevant to biological action.

Dynamics methods may be combined with Monte Carlo methods to more efficiently scan conformational space. Another way of making conformational search more efficient is by including experimental data such as NMR coupling constants and NOE close contacts determined by NMR spectroscopy; this is called "constrained conformation generation."[20]

## D. Deriving Bioactive Conformations

Given a molecule with biological activity, it is necessary to determine the conformation(s) of the molecule associated with its activity. Of the many possible conformations for a given molecule, only one or a small number of these actually bind to a receptor or enzyme. If one assumes that two or more molecules bind in a common mode, then the search for a common conformation may be made. The search is to find common conformations that are feasible and could bind. One approach involves the calculation of energetics since only low energy conformations will exist. The second approach involves the generation of all possible conformations in a systematic way so that all possible binding patterns can be explored. Many programs now possess automated conformation search where different criteria for searching can be stipulated. Criteria that one would be interested in include energy, allowed bonding and nonbonding distances, torsional angular increments, and other factors. Much work has been done in conformation searching. For example, Smith and Veber[21] have performed a systematic search of conformations of cyclic hexapeptides, constrained by NMR coupling constant and NOE data, and applied it to somatostatin analogs. Gund *et al.* have performed systematic conformation searching coupled with distance and energy parameters to determine bioactive conformations of a series of semirigid nicotinic agonists.[22] Paine and Scheraga[23] have used Monte Carlo methods to derive bioactive conformations of Met-enkephalin. Hagler[24] and Struthers *et al.*[25] have used molecular dynamics approaches to determine the bioactive conformations of antagonists of gonadotropin-releasing hormone (GnRH). This work showed that a linear analog could assume the same proposed bioactive conformation as a constrained, cyclic GnRH antagonist at a reasonable cost in energy, using a technique called template forcing to force the two molecules to superimpose. In another example, a combination of NMR and molecular dynamics studies on a cyclic enkephalin analog led to an active conformation model which is guiding the search for novel rigid peptide opiates.[26]

## E. Molecule Superposition and Alignment

Computing properties of molecules often involves comparisons across a homologous series. Many such techniques require superposition or alignment of the molecules so that their differences become obvious and interpretable. When molecules possess a large, rigid common substructure, their alignment is relatively easy. In the general case, the molecules are sufficiently different in structure or in conformation that their alignment is unobvious and perhaps not unique. There is an expanding literature on how to align molecules in such cases, and the field has recently been reviewed.[27] One method of alignment involves manually defining common binding groups.

This involves knowledge of important binding sites or binding parameters such as hydrophobic points or hydrogen bonding points. Alternatively, electrostatic calculations can be performed to define positive, neutral, and negative regions of a molecule and sites can then be superimposed on this basis. At least three sites must be defined and the best fit calculated. If one has a relatively rigid active ligand, then this could serve as a template for further fitting. Before fitting, low energy conformations are generated and relative distances between fitting groups are defined. In some cases, template forcing is employed to arrive at the best fit.

Many groups have been interested in developing methods to superimpose two defined 3-D structures when one does not know the alignments.[28–37] Brint and Willett[28] showed that clique detection methods based on interatomic distances rapidly find the best way to match two 3-D structures. Brint and Willett[28] were able to extend the method to find the maximum 3-D substructure common to all members of a set of molecules. In graph theory a clique may be defined as a subgraph in which every node is connected to every other node. For 3-D structure comparisons the nodes of the graph are the points for superposition labeled according to type. An edge between two nodes is the distance between them. The clique-detection algorithms find cliques in an input graph that match cliques in a reference graph. Thus they find corresponding points in the two 3-D structures. These corresponding points are of the same type in the two structures and all corresponding interpoint distances are identical.

A more automated method of fitting involves using the program DISCO (DIStance COmparisons) developed by Martin *et al.*[27] The program has the ability to propose superposition rules for structurally diverse compounds. When a compound has many low energy conformations and several points of superposition, DISCO can be useful in finding the best conformations and points of superposition. The method was successfully applied to some $D_2$ dopamine agonists and showed that some new 2-aminothiazoles fit the traditional points of superposition. The method involves generating and optimizing all low energy conformations by any suitable method. For each conformation, a program ALADDIN is used to calculate the location of points to be considered as part of the superposition. These points include atoms in the molecule and projections from the molecule to hydrogen-bond donors and acceptors or charged groups in the binding site. These positions and the relative energy of each conformation are input to the program DISCO. It uses a clique detection method to find superpositions that contain at least one conformation of each molecule and user-specified numbers of point types and chirality.

Other methods of superposition are available. Crippen[29] superimposed molecules of different structures using 3-D quantitative structure–activity relationship (QSAR) methods based on distance geometry. His algorithm is a brute force method considering all pairing distances between points of similar nature. He found that some compounds have multiple possible binding modes (different pairings and conformations). Other methods involve the active analog approach of Marshall (1) which looks for interpoint distances common to all active molecules by systematic rotation of bonds. A problem with this method is that the number of conformations and superpositions is dependent on the user's specification of similarity criteria. Another approach is the ensemble distance geometry method of Sheridan *et al.*[38] This method generates ensembles that contain one superimposed conformation of each molecule. The disadvantage of the method is that ensembles may contain high energy conformations and when the structures are minimized the superposed conformations may no longer match.

### F. Deriving the Pharmacophoric Pattern

A pharmacophore may be defined as the essential geometric arrangement of atoms or functional groups necessary to produce a given biological response. A given set of biologically active molecules which produce activity by the same mechanism are assumed to have the same essential pharmacophore. Pharmacophores have been described as *topologic* (graph-theoretic or connectivity-based structural fragment) and *topographic* (geometric, usually 3-D) patterns.[39] Topographic patterns are associatd with molecular conformation and should be more useful for predicting drug activity, even for chemically dissimilar compounds which possess the same type of bioactivity. Much research in pharmaceutical chemistry is concerned with deriving pharmacophoric patterns in molecules and designing new analogs based on the derived pharmacophores. Topographic pharmacophoric patterns may be defined by specifying atom types and interatomic distances.[39] Properties of pharmacophoric patterns that have been defined include chirality (stereochemistry), orientation (which face of a 3-D pattern is presented to the receptor), steric (which atoms are accessible to the receptor), and electrostatic (this could include indefiniteness where a lipophilic group in a drug can engage more than one hydrophobic region on the receptor[39–41]). The techniques of excluded volume,[42] shape analysis,[43] and distance geometry,[44,45] among others, have been used to derive pharmacophoric patterns. Thus distance geometry has been used to substantiate the pharmacophore for a series of nicotinic agonists.[38] The utilization of a pattern searching program has then been used to search for other possible nicotinic agonists of varying structural features from the Cambridge Data Bank.[46]

The technique of developing a pharmacophore is based on first deriving preferred conformations via a conformational searching scheme, then defining the common groups in terms of specific atom types, general atom types (e.g., hydrogen accepting), atoms or surfaces with a certain charge property, functional groups, or some other property in common. The molecules are then aligned and superimposed at the specified points in a defined way. The pharmacophore is then derived by joining the sites in common and calculating distances by averaging sites of superposition in a least-squares fit calculation. A pharmacophore needs at least three points of connection but it can have more. The complete pharmacophore can be represented geometrically as a triangle if three points are used or some other geometric figure if more points are available. The figure further describes the properties of the vertices as to the type of interaction with the receptor.

Automatic generation of pharmacophoric patterns is the goal of several modeling packages. This is a difficult procedure, however, because many bioactive molecules are flexible and their minimum energy conformations need not correspond to the receptor-bound conformation.[47] Automated pharmacophore mapping is now available in programs from Biocad, Biosym, Chemical Design Ltd., and Tripos.

## G. Receptor Mapping

The structures of most receptors are not known; however, 3-D structures of most ligands that interact with the corresponding receptors may be known. The lock and key concept holds that a drug fits into a receptor cavity to produce a complex which then produces the required biological response; however, we must be aware that both the ligand and the receptor during binding can alter their conformations to accommodate each other. Nevertheless, knowledge of a binding cavity is essential for understanding receptor–ligand interactions. When the structure of the receptor is not known, a "receptor map" may be constructed using the principle of receptor–ligand complementarity.

In the so-called active analog approach, bioactive conformations are derived for the active and inactive analogs and the volumes of each set are superimposed. The active analog volume is then subtracted from the combined inactive volume and what remains is the receptor map with a cavity where the active analogs fit. The inactive analogs are assumed to be inactive because of unfavorable steric interactions with the receptor. A number of algorithms are available to compute unions of molecular volumes. Pseudoelectron density functions calibrated to reproduce van der Waals radii[15] have been mapped to a 3-D grid to compute the union, intersection, and subtraction of volumes. Analytical volume representation by Connolly[48,49] may be an alternative which would allow optimization of volume overlap.

Receptor maps have been developed for many different types of systems. Sufrin *et al.*[50] developed a volume map for amino acid analogs of methionine which inhibited the enzyme methionine : adenosyl transferase; Hibert *et al.*[51] used receptor mapping to increase the selectivity of a lead compound for 5-$HT_{1A}$ receptor over the $\alpha_1$-adrenoreceptor; Ortwine *et al.*[52] used the mapping approach to design selective adenosine receptor agonists; and Gund[148] used this approach to explain the inactivity of a sterically confined nicotinic agonist. Marshall has been a pioneer in the excluded volume technique,[53] and others such as Beckett and Casy,[54] Kier,[55] Humber *et al.*,[56] Olsen *et al.*,[57] and Crippen[58] have used this technique successfully as an aid in designing more specific drugs or for explaining activities or inactivities of other drugs.

## H. Estimating Biological Activities

Discussions so far have focused on drugs being classified as active or inactive. In reality, drugs vary in activity from being very active or potent to being inactive. QSAR is a technique that quantifies the relationship between structure and biological data and is useful for optimizing the groups that modulate the potency of a molecule.[59–61] The field was started in the early sixties by Hansch *et al.*,[62] Fujita *et al.*,[63] Franke,[64] and others. The first attempts to extract correlations from three-dimensional data were based on congeneric series and regression analysis. QSAR started as a simple comparison of properties for two or more molecules using single numbers[65–67] and has ended up as a complex multivariable treatment of properties versus structure based on statistical analysis and relying on the extraordinary power of modern computers. QSAR has been useful for rationalizing compound activity and for rational design of new compounds.[61,67,68]

A typical QSAR contains a data table in which each row contains the properties of the compound and each column a different physical or biological molecular property. Typical properties include hydrophobicity, lipophilicity, electronic nature, and steric factors such as size of group on a series of related molecules. These descriptors are included in a multiple regression analysis to obtain a best-fitting equation. Choices as to the descriptors used, the mathematical form of the relationship between the potency and the descriptors, and the statistical method used to analyze the relationship are part of the "art" of QSAR practice.

A QSAR equation can provide information on the mechanism of action of the compounds or the atomic environment of the biological binding site. QSAR can indicate that substituents at one position interact with a hydrophobic center whereas substituents at another position do not.

QSAR can provide an estimate of the highest potency expected of a molecule in a series or can provide information on whether all parts of the molecule are in close contact with the binding site.[69] Relationships between potency and hydrophobicity, although not linear, can be fit either with a parabola or a differential equation.[70,71]

When two or more properties are correlated, colinearity can occur. Hansch *et al.*[72] suggested that if one employed cluster analysis on all possible substitutents and included one member from each of the clusters of similar substitutents, the designed ligands would show more independent variation in physical properties.[73]

Certain properties of molecules are intrinsically correlated. An example is the partition coefficients of compounds between water and various organic solvents such as octanol. Wold *et al.*[74] suggested that one analyze QSARs with principal components or partial least squares (PLS).[75] Both methods extract a set of independent vectors that are linear combinations of the individual properties. Each physical property contributes to each eigenvector.

Many different QSAR programs are available. For example, from Pomona College, the CLOGP program uses chemical information tools to calculate the octanol–water log P from an input 2-D structure of the molecule. CLOGP recognizes the substructures in the molecule and looks up the log P increments and corrections due to these substructures.[76,77] It also reports the measured value of the log P if this is in the database. The group also has a system for creating and searching a database of QSAR equations and the raw data behind them by various criteria.[78] Traditional QSAR studies are performed using 2-D structural data of molecules, but with the emergence of 3-D databases, relationships are obtained for 3-D data as well. Of the various 3-D QSAR approaches, CoMFA, developed by Cramer *et al.*,[79] is among the most popular. CoMFA combines 3-D conformational features and energies with biological potency and identifies the quantitative influence on potency of specific chemical features at particular regions in space. Input is by way of probe atoms which calculate the steric and electrostatic interactions energies at hundreds of points on a lattice surrounding the molecules. Additions by Martin *et al.* also include hydrogen-bonding interactions. PLS identifies the relationship between these energies and biological potency. CoMFA equations forecast the affinity of proposed molecules. The coefficients of the interaction energies in a CoMFA can be displayed as contours to aid new compound design.

## I. Molecular Interactions: Docking

Modeling the interaction of a drug with its receptor is a complex problem. Many forces are involved in the intermolecular association: hydrophobic, dispersion or van der Waals, hydrogen bonding, and electrostatic (ion pairing). The major driving force for binding appears to be hydrophobic interactions, but the specificity of the binding appears to be controlled by hydrogen bonding and electrostatic interactions.[80–82] Modeling the intermolecular interactions in a ligand–protein complex is difficult because there are so many degrees of freedom and insufficient knowledge of the effect of solvent on the binding association. The interaction of the drug and receptor site is stereospecific in that the receptor recognizes certain groups on the ligand, particularly intermolecular distances and/or molecular shape. This type of interaction is determined by the structure or fit of the drug molecule to the receptor site and induces a common biological response.

The process of "docking" a ligand to a binding site tries to mimic the natural course of interaction of the ligand and its receptor via a lowest energy pathway. Usually the receptor is kept rigid while the conformation of the drug molecule is allowed to change. In favorable cases,[83] the two molecules may be brought together by first assigning planes that contain the sites of interaction and then moving the planes while calculating binding energies of interaction. The molecules are physically moved closer to one another and the preferred docked conformation is minimized. The complex attains its own set of properties that are a composite of the individual molecules. When solvent properties are neglected in the docking process, ionic interactions and hydrogen-bonding energies are overestimated. These effects can be calculated by using free energy perturbation methods. Mertz and Kollman[84] estimated the free energy of thermolysin binding to a new inhibitor. Other workers[85,86] warn that it is difficult to verify when a simulation has converged and that calculation of a binding process is inaccurate. It seems, however, that ligand–macromolecule relative binding free energy can be calculated to within $\pm 1.5$–2 kcal/mol (about 10- to 30-fold in binding affinity) by this method.[85,86]

Determining the optimum fit and interaction energy of a ligand in a known receptor site remains difficult. There are simple methods for docking rigid ligands with rigid receptors[87,88] and flexible ligands with rigid receptors,[89] but general methods of docking conformationally flexible ligands and receptors are problematic. Flexible systems are best studied interactively, using chemical intuition as a guiding force.

### J. Calculation of Molecular Properties

Molecular properties are important indicators of utility of various molecules for use as detergents, semiconductors, building composites, agricultural agents, and pharmaceuticals. Many (but not all) properties can be measured once pure material is in hand. But when properties can be predicted, it can provide useful guidance as to what new chemicals to obtain or synthesize for testing. Consequently, much effort has been expended over the years in estimating or computing various molecular properties.

Properties may be categorized as macroscopic or microscopic. Microscopic properties are the same for the bulk material and for the smallest divisible component, usually a molecule, e.g., molecular weight. Macroscopic or bulk properties, however, only exist for the bulk material, e.g., melting point. Properties are normally categorized as physical (electronic, thermodynamic, physical state), chemical (reactivity, solubility, dynamics, explosivity), or biological [enzyme inhibition, receptor (ant)agonism, toxicity, metabolism]. Bulk properties of pure substances are largely their physical properties and may be computed from the corresponding molecular structures in many cases. Microscopic properties may normally be computed from the structures and conformations of the individual molecules or be estimated from properties of related structures.

Methods of estimating or computing properties include:[90]

*Interpolating properties:* a correlation is found between the desired property and another property or characteristic of related molecules; in this case the desired property may be computed from within the range of application of the correlation, and the interpolated property should be accurately estimated.

*Extrapolating properties:* in this case the correlation does not extend to include the molecule of interest, but by extending the correlation, it is possible to estimate the desired property. Since the validity of the correlation in the extrapolated region is usually not known, the accuracy of the extrapolated property is difficult to estimate.

*Computing properties:* in many cases it is possible to compute a property, directly or indirectly, with varying levels of accuracy. Such computed properties can be quite comparable to experimental accuracy and indeed may substitute for the experiment in cases where the experiment would be difficult or impossible to perform.

Table I summarizes some properties that may be computed and the computational technique used.

Such properties as heats of formation; enthalpies; activation energies; dipole moments; reaction paths; molecular, localized, and frontier orbitals; electronegativities; polarization; delocalization; atomic and bond population; molecular electrostatic properties; internal energies; electronic charge distribution; hydrophobic character; and other properties such as log P (octanol/water partition coefficient) and MR (molar refractivity)[119,120] can be derived by combinations of calculational methods including molecular mechanics, semiempirical and ab initio quantum mechanics, and other empirical methods.[121]

## K. Energy Calculations

One of the fundamental properties of molecules is their energy content and energy levels. Table II lists some types of energies and how they may be computed.

The three major theoretical computational methods of calculating properties of molecules include empirical (molecular mechanics), semiempirical, and ab initio (QM) methods. Molecular mechanics and semiempirical methods rely on embedded empirical parameters, whereas ab initio quantum mechanical methods are potentially capable of reproducing an experiment

**TABLE I** Some Molecular Properties and Methods of Computation

| Property | Method of computation | Reference |
|---|---|---|
| Boiling point (BP): monomethyl alkanes | Hosoya's index | 91 |
| BP: alkanes | Topological index | 92 |
| BP: olefins | Topological index | 93 |
| Critical T, P, V | | |
| Molar volume | | |
| Viscosity | | |
| Solubility | | |
| Heat capacity (Cp) | | |
| Thermal conductivity (k) | | |
| Density | | |
| Liquid-thermal conductivity | | |
| Heat of fusion | Rao from (T, K) or ($T_{b.p.}$, K) or ($T_{cr}$, K) | 94 |
| | Yalkowsky from $S_{fus}$ and chemical structure | 95 |
| Heat of sublimation | Chickos from $H_{vap}$ and $H_{fus}$ | 96 |
| Entropy of fusion | Martin, Yalkowski heat of fusion and melting point | 95, 97 |
| Solubility parameter (delta) | Hildebrand | 98 |
| Refractive index $n_d$ | Rathjen and Straub | 90, 99 |
| Molecular refraction for liquids ($R_M$) | Lorentz–Lorentz equation | 100–103 |
| Magnetic susceptibility | Bondi, Motoc, and Balaban | 104, 105 |
| Radius of gyration (R) | Thompson, Thompson, and Braun | 106 |
| Compressibility of liquids and bulk modulus ($K_{iso}$, $K_{adiab}$, $M_B$) | From spec. V and app. P for $K_i$ from velocity of sound ($U_s$) and liquid D for $K_{adiab}$ from Kiso | 107 |
| | NS P1, p2, V1, V2 for $M_B$ | 108–110 |
| Dipole moment (u) | Tykodi, Gasteiger, and Guillen | 111, 112 |
| | Molecular mechanics (Dosen-Micovic) | 113 |
| Partial atomic charges | EHT, IEHT, CNDO/2, PPP, STO-3G, charge 2, MNDO, HMO + Del Re, Gastieger, Kollman, and Karplus | 114 |
| Ionization potential ($I_p$) | | 115 |
| Electrostatic potential (P) | | 115 |
| Electric field (F) | | 116 |
| Heat of formation ($H_f$) | Pedley | 117, 118 |

without such parameters. The first two methods, which rely on availability of reliable experimental data, are best applied in situations where interpolations between experimental quantities are needed. The programs used are parameterized based on experimental facts.

### 1. Molecular Mechanics

Molecular mechanics methods are less complicated, fast, and are able to handle very large systems including enzymes.[122,123] Molecular mechanics

**TABLE II** Some Types of Molecular and Intermolecular Energies

| Energy type | Observation/computation method |
|---|---|
| Electronic energy | Ultraviolet spectrum, quantum mechanics (QM) |
| Protonation energy | p*K*, electron capture, QM |
| Conformational energy | NMR, electron diffraction, molecular mechanics (MM), QM |
| Solvation energy | NMR, thermodynamics, Monte Carlo MM, molecular dynamics, QM |
| Energy of sublimation | QM, MM |
| Crystallization energy | Thermodynamic measurements, MM |
| Enzyme binding energy | $K_i$, enzyme kinetics, MM, QM, free energy perturbation |

may give extremely accurate energies if the proper parameters are available. A disadvantage of molecular mechanics is that parameters are derived for ground state systems and are consequently unable to adequately represent geometries involved in bond making and bond breaking processes. Molecular mechanics methods are widely used to give accurate structures and energies for molecules. The method employs fundamental principles of vibrational spectroscopy as well as the idea that bonds have natural lengths and angles and molecules will adopt geometries that can best reach these natural values. Cases where natural values cannot be completely achieved result in strain, and molecular mechanics methods have van der Waals potential functions that can measure the amount of strain energy. The basic ideas go back to 1930, but serious use began in 1946. The calculations use the Born–Oppenheimer approximation, which describes the energy of the molecule in terms of the nuclear positions. This is the so-called potential energy surface. Molecular mechanics calculations then employ an empirically derived set of equations for the Born–Oppenheimer surface whose mathematical form is derived from classical mechanics.[124] This set of potential functions is called the force field and it contains adjustable parameters that are optimized to obtain the best fit of calculated and experimental properties of molecules, such as geometries, conformational energies, heats of formation, or other properties. The assumption is made that corresponding parameters and force constants are transferable from one molecule to another.

The general molecular mechanics force field is based on the Westheimer method and includes functions for bond stretching, angle bending, torsion, and van der Waals interactions, thus

$$V = V_{\text{stretch}} + V_{\text{bend}} + V_{\text{torsion}} + V_{\text{VDW}},$$

where each component may be represented by its own potential function such as:

$$\text{bond stretch} = V_r = \frac{1}{2} k_r (r - r_o)^2$$

$$\text{angle bend} = V_0 = \frac{1}{2} k_0 (0 - 0_o)^2$$

$$\text{torsional energy} = V_{tor} = k_w (1 - \cos 3\,w)$$

$$\text{nonbonded or VDW interactions} = V_{VDW} = E[(r_o/r)^{12} - 2(r_o/r)^6]$$

The sum extends over all bonds, bond angles, torsion angles, and nonbonded interactions between all atoms not bound to each other or to a common atom. More elaborate force fields may include Urey–Bradley terms (1,3-nonbonded interactions), cross interaction terms, electrostatic terms, and so on. The sum of all these terms gives the steric energy of the molecule. The individual functions may vary and contain more elaborate terms. In some molecular mechanics programs the energy equation can also have terms for electrostatic and hydrogen bonding contributions.

Most modeling programs have some form of molecular mechanics built in. The most popular programs for small molecules are based on Allingers MM2 and $MM_3$3 force fields.

Molecular mechanics energy minimization involves successive iterative computations where an initial conformation is submitted to full geometry optimization. All parameters defining the geometry of the system are modified by small increments until the overall structural energy reaches a local minimum. The local minimum, however, may not be the global minimum. Searching methods are then employed to search other conformations. These methods may involve systematic search,[125,126] which increments all rotatable bonds in turn to explore the complete conformation space of the molecule; distance geometry[127,128]; and other random sampling approaches.[129–131]

### 2. Semiempirical Calculations

These types of calculations are basically quantum mechanic but some experimental values are used as well. Unlike the molecular mechanics methods, semiempirical methods use mathematical formulations of the wave functions which describe hydrogen-like orbitals. The types of wave functions used include the Slater-type orbitals (STO) and Gaussian-type orbitals (GTO). Additional variations are added to the abbreviations. Semiempirical methods treat the linear combination of orbitals by iterative computations which establish a self-consistent field (SCF) and minimize the energy of the system.[132]

Semiempirical calculations differ in the approximations that are made concerning repulsions between electrons in different orbitals. The approximations are adjusted by parameterizing values to correspond to either ab initio data or available experimental data. The earliest methods used were the extended Huckel theory[133,133a] and CNDO[134] (complete neglect of differential overlap). These methods are not reliable for calculating geometry but are useful for calculating shapes of molecular orbitals and charges. These methods also do not give reliable energies. Improved methods include MINDO-3,[137] MNDO,[138] and AMI.[139,140] MOPAC contains the MINDO/3 and MNDO programs and has been parameterized for lithium, beryllium, boron, fluorine, aluminum, silicon, phosphorus, sulfur, chlorine, zinc, germanium, bromine, iodine, tin, mercury, and lead. MOPAC allows various geometric operations such as geometry optimization, constrained and unconstrained, with and without symmetry, transition state localization by use of a "reaction coordinate" gradient minimization,[142,143] and vibrational frequency calculations. MOPAC also allows quantities such as atomic charges, dipole moments, ionization potentials, and bond order to be calculated. MNDO has been applied for predicting polarizabilities, hyperpolarizabilities, ESCA, nuclear quadrupole resonance, and numerous other properties. MNDO, however, has some limitations which include production of a spurious repulsion that is apparent at distances just outside chemical bonding. MNDO thus does not reproduce hydrogen bonding or heats of formation of some systems very well.

### 3. Quantum Mechanics: Ab Initio

The solution of the Schrodinger equation with approximations is the basis of semiempirical calculations, whereas the solution of the Schrodinger equation without approximations is the basis of the highest level quantum mechanical calculation: the so-called "ab initio" method. There are two different approaches to ab initio methods. One is the "calibrated" approach favored by Pople and co-workers and the other is the "converged" approach favored by chemical physicists. In the first approach the full exact equations are used without approximations. The basis set is fixed in a semiempirical way by calibrating calculations on a variety of molecules. The error in a new application of the method is estimated based on the average error obtained compared with experimental data on the calibrating molecules. In the second approach, a sequence of calculations with improving basis sets is done on one molecule until convergence is reached. The error is estimated from the sensitivity of the result to further refinements in the basis set. The calibrated method is the more practical for routine computational chemistry use. The convergence technique is used for small molecules and for test cases.[144]

Ab initio methods are most useful for cases where there is no experimental data to draw from, but suffer from the disadvantages that much computer power is needed and therefore the method is not routinely useful for systems with more than 50 heavy atoms. Every method described has advantages and disadvantages, and a useful strategy is to use several methods and compare the results for consistency.

In general, ab initio calculations are iterative procedures based on SCF methods. Electron–electron repulsion is specifically taken into account. Normally, calculations are approached by the Hartree–Fock closed shell approximation which treats a single electron at a time interacting with an aggregate of all the other electrons. Self-consistency is achieved by a procedure in which a set of orbitals is assumed and the electron–electron repulsion is calculated. This energy is then used to calculate a new set of orbitals and these in turn are used to calculate a new repulsive energy. The process is continued until convergence occurs and self-consistency is achieved.[145]

The sophistication of the ab initio calculation is dependent on the number of basis set orbitals used. In general, the more orbitals used the more accurate the calculation, but also the slower the calculation. The individual ab initio calculations are further identified by abbreviations for the basis set of orbitals that are used, e.g., STO-3G, 4-31-G, 6-31G, and so forth.[145] Standard ab initio programs give good results for ground state molecular geometry, energy, and charge distribution for reasonable sized molecules. Results can also be obtained for description of reaction processes and for calculation of transition states and intermediates. The most popular ab initio programs include the Gaussian[146] and Hondo series.[147]

## III. EXAMPLES OF SMALL MOLECULE MODELING WORK

The following sections summarize some modeling work done on various small molecule systems.

### A. Nicotinic Ligands

The nicotinic receptor has been widely studied. Although its three-dimensional structure is not known, the receptor has been isolated and sequenced. The natural transmitter for the receptor is acetylcholine which is flexible and not a good model for deriving the pharmacophore of binding. There are, however, examples of more rigid ligands that act as agonists for the acetylcholine receptor that are better suited for deriving the pharmacophore of binding. A series, which has been extensively studied via modeling coupled with experiment, is represented by isoarecolone and its congeners. The modeling on these molecules illustrates several techniques which were described earlier; conformation generation, superposition and alignment, derivation of a pharmacophore for binding, use of electrostatics (isopotential maps) to rationalize activity and pharmacophore generation, and receptor mapping and its use in rationalizing activity of ligands.[148–152]

The molecules to be discussed are shown in Fig. 6. The most potent of

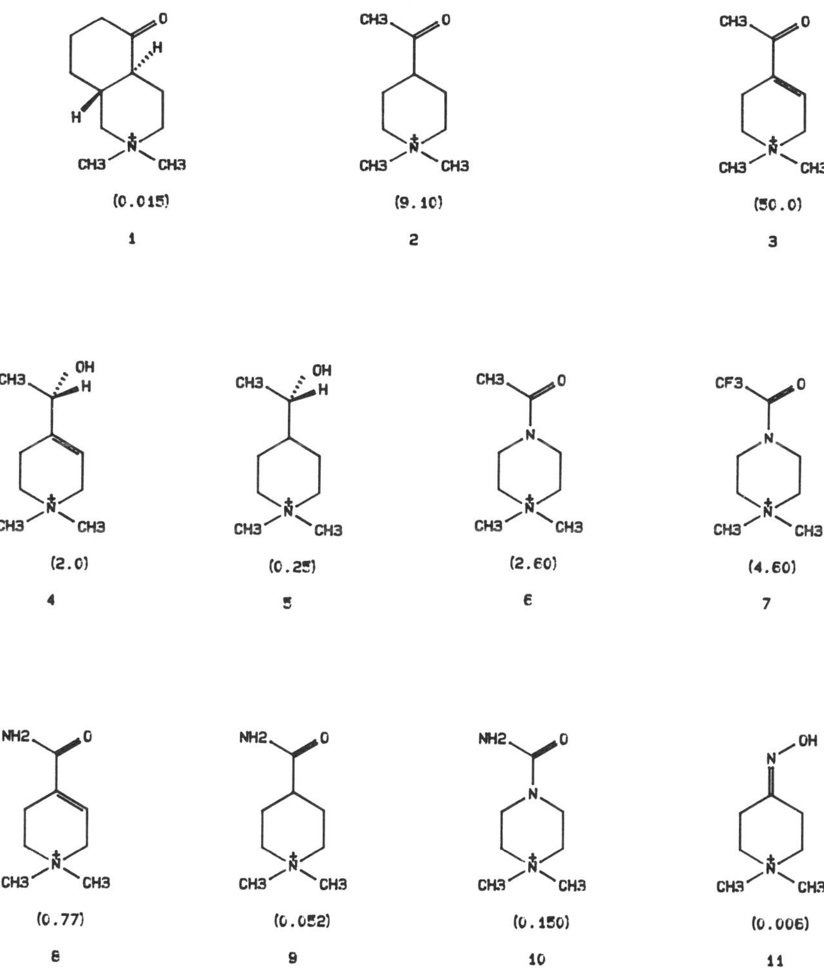


**FIGURE 6** Nicotinic agonists with potency activities relative to carbamylcholine.

these was isoarecolone (50 times more active than carbamylcholine) and one of the least active was (+)octahydro-2-methyl-*trans*-(1*H*)-isoquinolone methiodide (0.015 as active as carbamylcholine). Conformational searching was performed using MM2, varying the dihedral angle of the acetyl side chain. The important groups for binding were established to be the tetramethylammonium cationic group for a coulombic interaction, the carbonyl oxygen for a hydrogen bonding interaction, and the methyl group of isoarecolone for a hydrophobic interaction with the receptor. The optimal distance between the nitrogen and carbonyl oxygen at the van der Waals distance was determined to be 6.0 Å. Low energy conformations were computed and isoarecolone was selected as the template on which all other ligands were superimposed. The choice was made to superimpose three points: N, O of carbonyl, and C of carbonyl. Figure 7 shows these superpositions and clearly indicates that the more potent agonists, those with carbonyl and double bonds rather than hydroxyl groups, superpose better with isoarecolone, especially at the acetyl methyl group. The isoquinolone derivative, which has all the components of isoarecolone, but is more sterically constrained, was also superimposed on isoarecolone and its saturated derivative. Superposition did not indicate the reason for lower activity. These rationalizations were sought in electrostatic and receptor mapping studies. The pharmacophore was derived and consists mainly of the ammonium group and the carbonyl oxygen and carbon. The acetyl region was brought in as a secondary binding site that controls activity. The data leading to these conclusions came from activity versus structure studies where variation of the acetyl group produced more or less active agonists. For example, if the acetyl group is replaced with a trifluoroacetyl group, as in the piperazine derivative, activity increases; if it is replaced by an amino group, activity decreases; or if the acetyl group is tied in the ring, activity decreases. Partial charges were calculated using MOPAC MNDO, and electrostatic potentials were calculated using ARCHEM or CHEMX. Variation of the acetyl group indicated a change in electrostatic properties of the molecules. Namely, the trifluoroacetyl group is electron withdrawing and the electrostatic potential shows an electron-rich surface on the acetyl group which presumably increases potency. With an amino group this cap becomes positive, thus reducing potency; with a methyl group it is more neutral, thus showing intermediate potency. The fact that isoquinolone was much less potent was explained by looking at the receptor maps for these compounds. The excluded volume technique of Marshall was used. First, the volumes of active and inactive agonists were calculated and then the difference between the inactive and active volumes was taken. Figure 8 shows the derived receptor map. The active analogs fit inside the map, whereas the inactive analogs do not completely fit. The map was used to fit cytisine and to calculate the amount of steric hindrance produced by this moderately weak agonist[148,152] (Fig. 9).

These types of studies were thus used effectively to show that binding of nicotinic agonists to the nicotinic receptor depends on the ammonium group, the hydrophobic groups, and the hydrogen bonding groups of a ligand. Molecular modeling has offered another view at potency versus structure and was effective in discovering that another important binding site may be present for these agonists. The factors that seem important in controlling activity are (1) ground state conformation, (2) closeness of fit between the ground state (or local minimum energy) conformation and the template, (3) high positive potential around the cationic head, (4) high electron density around the acetyl group, and (5) steric environment around the acetyl region.

**FIGURE 7** Superposition of nicotinic agonists 2–7 at N, C (carbonyl or alcohol), and O (carbonyl or alcohol). (Stereoscopic representation.)

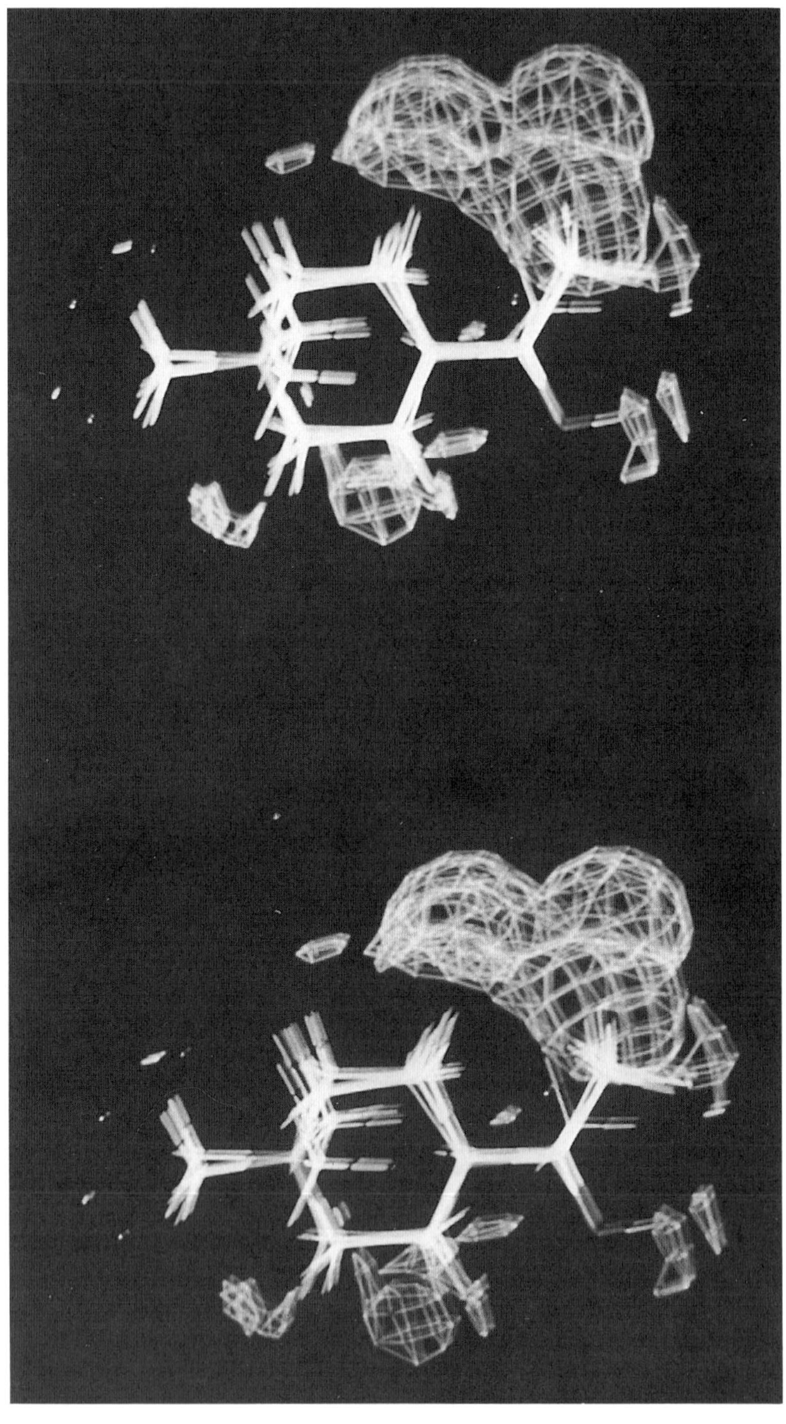

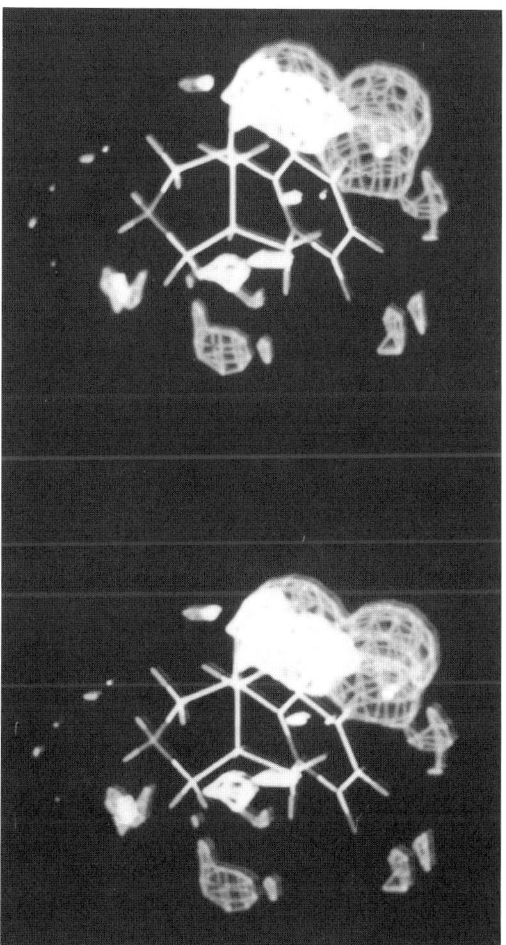

**FIGURE 9** Receptor map for nicotinic agonists with cytisine fitting inside the cavity. The white portion of the map indicates repulsion caused by cytisine. (Stereoscopic representation.)

## B. Sigma Ligands

Modeling was successful in determining a possible pharmacophore model of binding for several sigma selective ligands shown in Fig. 10. Of these, haloperidol is most active, but PRE084 is most selective and, in fact, is the most selective sigma ligand known.[153–158] Previous studies established

---

**FIGURE 8** Receptor map for nicotinic agonists with active ligands (2–8) fitting inside the cavity. (Stereoscopic representation.)

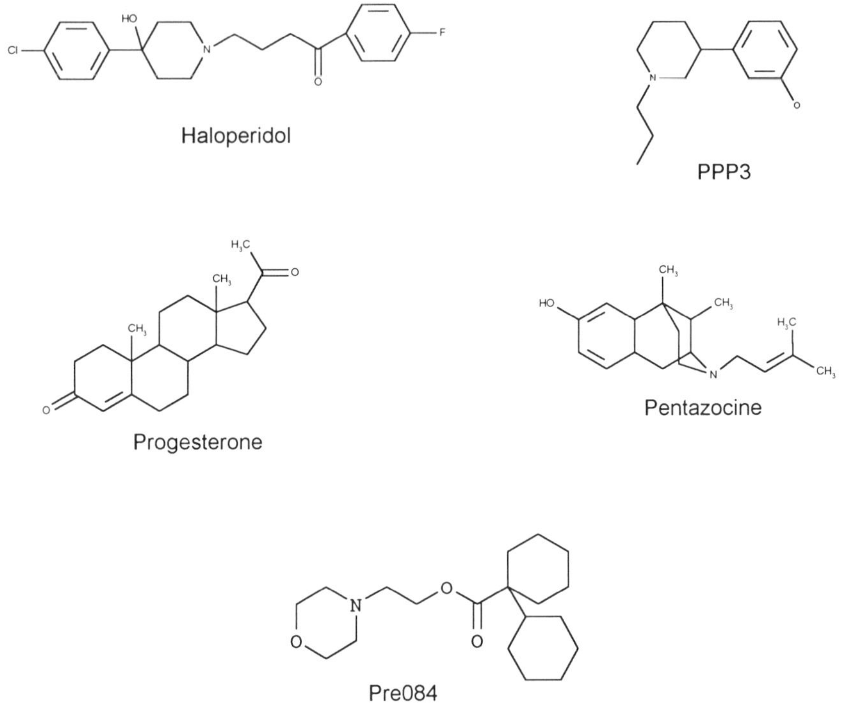

Haloperidol

PPP3

Progesterone

Pentazocine

Pre084

**FIGURE 10** Sigma selective ligands.

that for sigma binding, the ligand must posses a nitrogen as an electron-rich site, the corresponding lone pair, and a hydrophobic site represented usually by a phenyl group. These moieties are present in the classic sigma ligand haloperidol. However, not all sigma ligands possess these groups. For example, progesterone does not have a nitrogen or a phenyl group. Through molecular modeling studies, a pharmacophore has been derived for binding of molecules which do and do not possess nitrogen or phenyl rings. (+)PPP and (+)pentazocine are sterically constrained with conformational freedom present in the rotation of the side chain and in the boat/chair interconversions of the cyclohexane rings. Progesterone is very rigid, whereas haloperidol and PRE084 have conformational freedom around the single bonded side chains.The pharmacophore of binding was derived from the superimposed structures at the sites of binding (Fig. 11). To derive the sites of binding, electrostatic calculations were performed. Maxima and minima of electrostatic potential were calculated, and the three sites with corresponding potentials (from −5 to −40 kcal/mol) were then superimposed. The first site of superposition included the center of the phenyl rings with ring juncture AB or progesterone, for the hydrophobic region; the second site of superposition included the nitrogens with C-8 of progesterone [EP(kcal/mol) = −40 (N), −5(C-8)]; and the third site of superposition

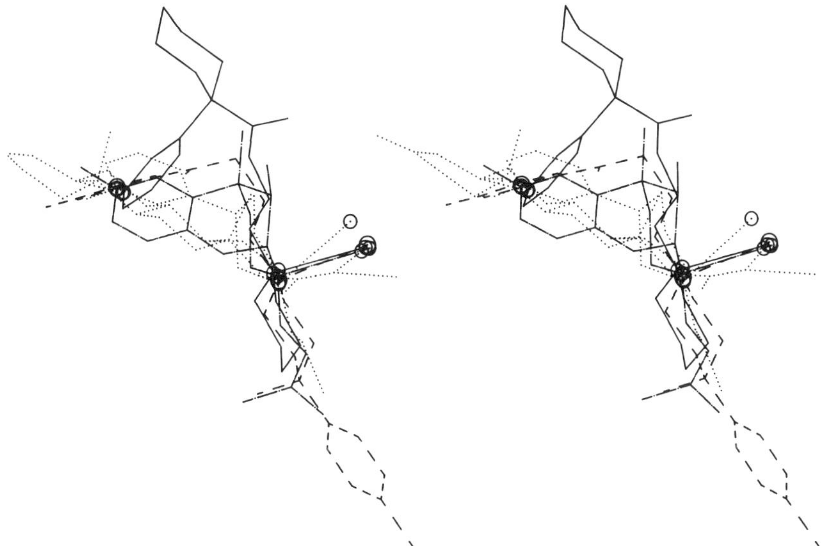

**FIGURE 11** Superposition of five sigma ligands from Fig. 10. Circles show common points of fitting: N, C, and lone pair of oxygen or nitrogen. . . ., progesterone; —, Pre084; . . . . , PPP3; - - -, haloperidol; - · · ·, pentazocine.

included the lone pairs of the nitrogen-containing compounds with a lone pair of progesterone oxygen on ring D. Joining the three points produced the pharmacophore (Fig. 12) represented by a triangle whose distances are: 5.3–5.77 Å for the N . . . C distance (sites 1 . . . 2), 7.3–7.9 Å for C . . . lone pair (sites 2 . . . 3), and 2.7–3.0 Å for N or C . . . lone pair (sites 1 . . . 3). To achieve the superposition of all five ligands with the fluorophenyl group of haloperidol constituing the hydrophobic site of binding, gauche conformations that were 2 kcal higher in energy were invoked.

The high selectivity of PRE084 was attributed to the amine moiety and to the cyclohexyl group. The cyclohexyl group can represent another binding site or can cause steric hindrance to the interaction at the other receptors.

**FIGURE 12** Triangle representation of the pharmacophore for binding of sigma ligands. . . ., progesterone; —, Pre084; . . . . , PPP3; - - -, haloperidol; - · · ·, pentazocine.

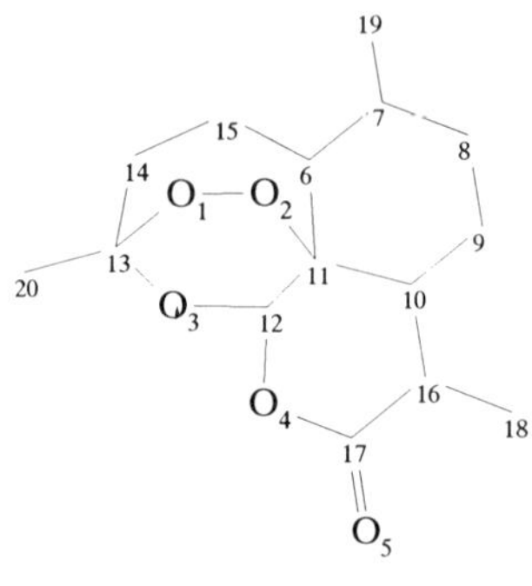

QINGHAOSU(QHS)

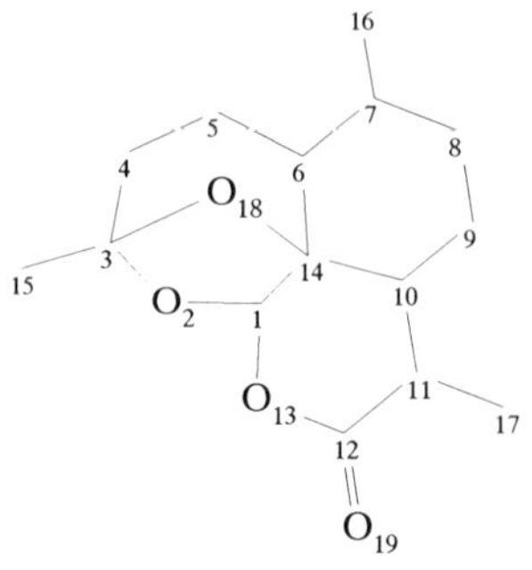

DESOXYARTEMISININ(DQHS)

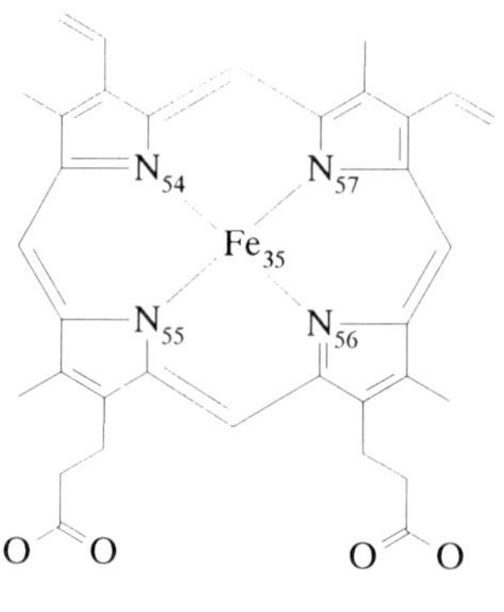

HEMIN

**FIGURE 13** Antimalarial agents.

## C. Antimalarial Agents

Docking studies were performed on the antimalarial agent artemisin (QHS, 1) and its inactive analog desoxyartemisinin (DQHS, 2) with hemin (3) to determine the most favorable mode of binding (Fig. 13).[83] It is supposed

**FIGURE 14** Space fill representation of the lowest energy-docked configuration of hemin [Fe(III)] and QHS. Peroxide bridge oxygens ($O_1$, $O_2$) and oxygen ($O_5$) interact preferentially with the hemin iron. Oxygens are black and iron is represented by large white surrounded by small black nitrogens. (Stereoscopic representation.)

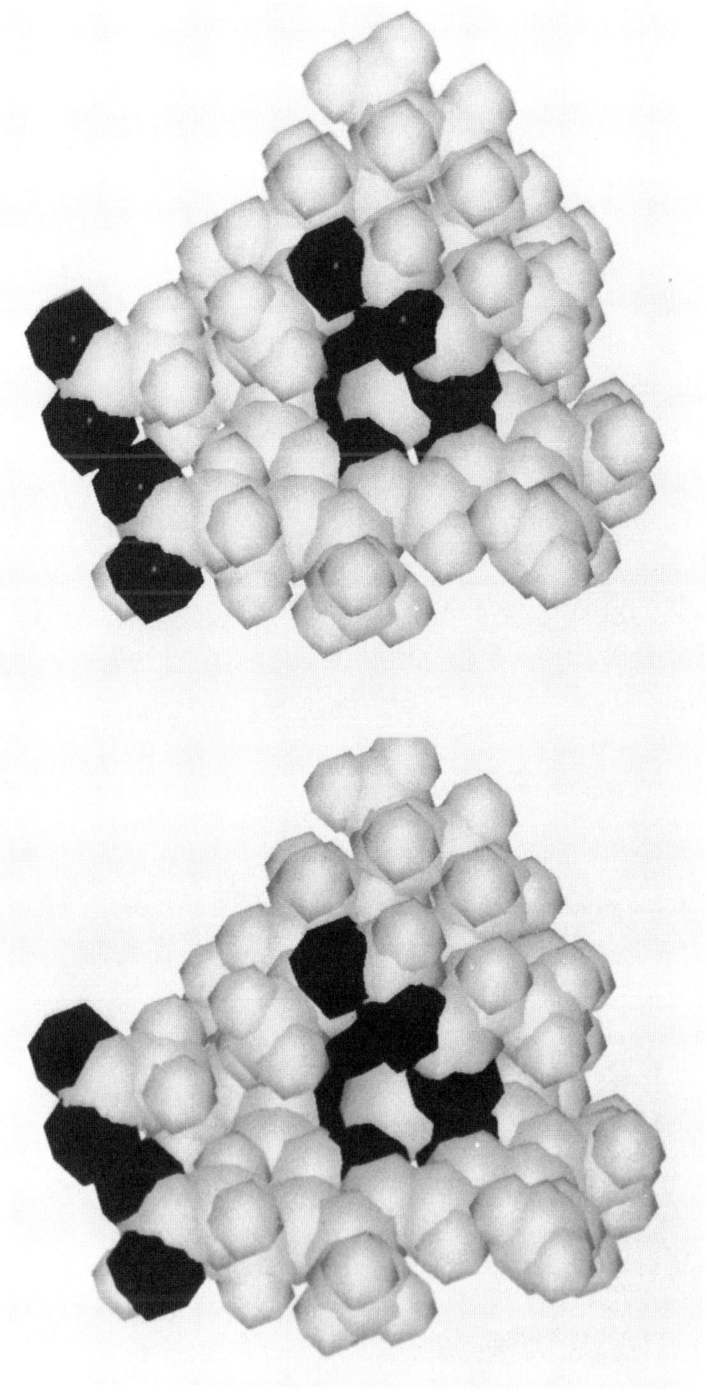

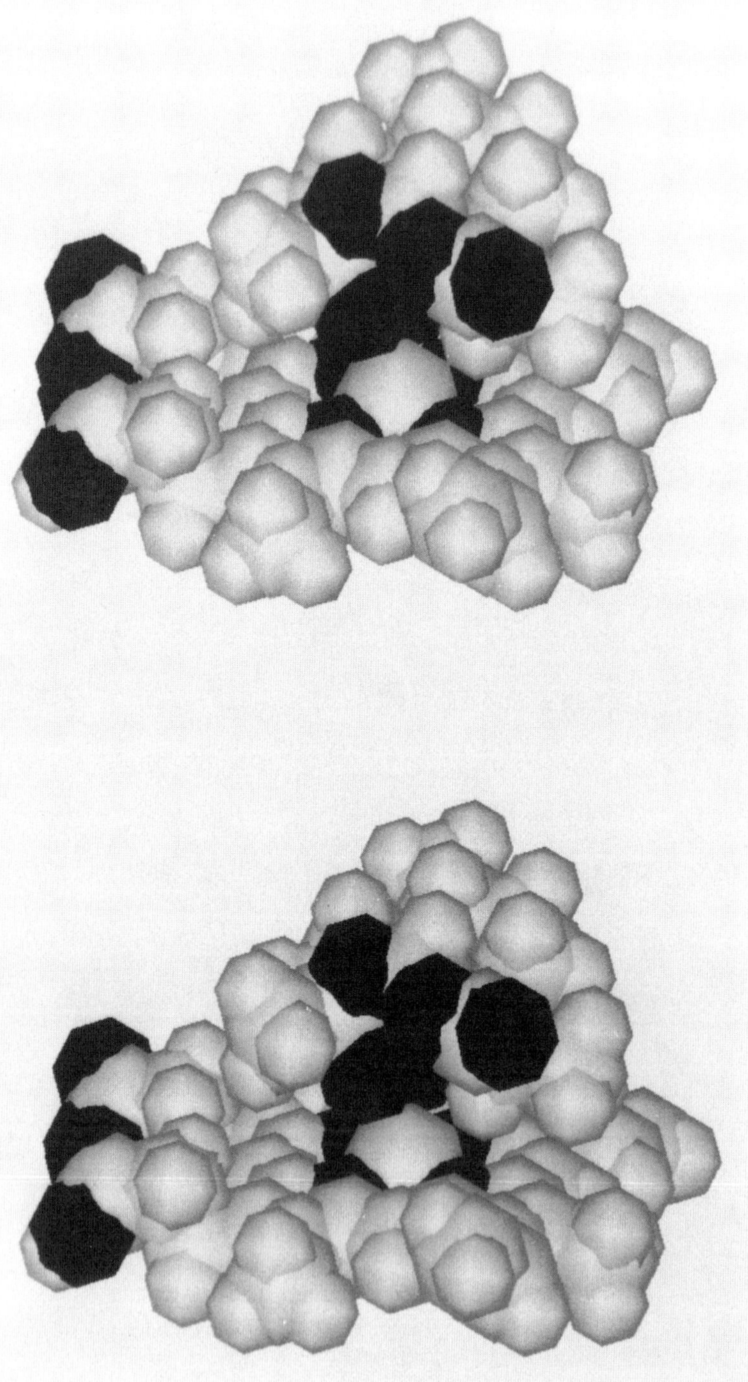

that the antimalarial activity of QHS results by binding to hemin as a first course of action. QHS has five oxygens and it is suspected that the activity is due to binding with the peroxide bridge. The inactive analog DQHS does not have a peroxy bridge but only four oxygens. After extensive binding studies the most favorable for QHS was found to involve the interaction of the peroxide bridge with the hemin iron. The desoxy derivative appeared to bind in a quite different orientation involving the nonperoxide oxygens. Figure 14 shows the most favorable QHS–hemin interaction with iron in the Fe(III) oxidation state. Three oxygen atoms from QHS interact preferentially with the hemin iron, the peroxide oxygens $O_1O_2$, and the nonperoxide oxygen O. Figure 15 shows the most favorable docked configuration of QHS and hemin with iron in the Fe(II) oxidation state. Here only the peroxide bridge oxygens $O_1O_2$ are involved in the interaction.

## ACKNOWLEDGMENTS

We thank Donald Gregory of Molecular Simulations Inc., Burlington, Massachusetts, for supplying Figs. 1–5 and Kanhiya Shukla of New Jersey Institute of Technology, Newark, New Jersey, for supplying the other figures in this chapter.

## REFERENCES

[1] Gund, T., and Gund, P. (1987). *In* "Molecular Structure and Energetics" (A. Greenberg and J. Liebman, eds.), Vol. 4, p. 319. VCH, Weinheim/New York.
[2] D. B. Boyd, (1990). *In* "Reviews in Computational Chemistry" (K. B. Lipkowitz and D. B. Boyd, eds.), Vol. 1, p. 355. VCH, New York.
[3] Cohen, N. C., Blaney, J. M., Humblet, C., Gund, P., and Barry, D. C. (1990). *J. Med. Chem.* **33,** 883.
[4] Sadek, M., and Munro, S. (1988). *J. Comp.-Aid. Mol. Design* **2,** 81.
[5] Kennard, O., Watson, D., Allen, F., Motherwell, W., Town, W., Rodgers, J. (1975). *Chem. Brit.* **11,** 213.
[6] Perlman, R. S. (1987). *Chem. Des. Autom. News* **2,** 1.
[7] Boehm, H. J. (1993). *In* "Trends in QSAR and Molecular Modeling 92" (C. G. Wermuth, ed.), p. 27. ESCOM, Leiden.
[8] Marshall, G. R., and Maylor, C. B. (1990). *In* "Comprehensive Medicinal Chemistry" (C. Hansch, P. Sammes, J. Taylor, and C. Ramsden, eds.), p. 431. Pergamon, Elmsford, NY.
[9] Levinthal, C. (1966). *Sci. Am.* **214,** 42.
[10] Feldman, R. J., Bing, D. H., Furie, B. C., and Furie, B. (1978). *Proc. Natl. Acad. Sci. U.S.A.* **75,** 5409.
[11] Smith, G. M., and Gund, P. (1978). *J. Chem. Info. Comput. Sci.* **18,** 207.
[12] Warme, P. K. (1977). *Comput. Biomed. Res.* **10,** 75.

**FIGURE 15** Space fill representation of the lowest energy-docked configuration of hemin [Fe(II)] and QHS. Peroxide bridge oxygens ($O_1$, $O_2$) interact preferentially with the hemin iron. (Stereoscopic representation.)

[13] Langridge, R., Ferrin, T. E., Kuntz, I. D., and Connolly, M. L. (1981). *Science* **211,** 661.
[14] Palmer, R. A., Tickle, J. H., and Tickle, I. J. (1983). *J. Mol. Graph.* **1,** 94.
[15] Marshall, G. R., and Barry, C. D. (1979) "Proceedings American Crystallographers Association," Honolulu.
[16] DeClercq (1984). *Tetrahedron* **40,** 3717–3729.
[17] Hagler, A. T., Osguthorpes, D. J., and Robson, B. (1980). *Science* **208,** 559.
[18] Hagler, A. T., Osguthorpes, D. J., Dauber-Osguthorpe, P., and Hempel, J. C. (1985). *Science* **227,** 1309.
[19] Havel, T. F., Kuntz, I. D., and Crippen, G. M. (1983). *Bull. Math. Biol.* **45,** 665.
[20] "Applications of NMR Spectroscopy to Problems in Stereochemistry and Conformational Analysis," Vol. 6. VCH, Weinheim/New York, 1986.
[21] Smith, G. M., and Veber, D. F. (1986). *Biochem. Biophys. Res. Commun.* **134,** 907.
[22] Spivak, C. E., Waters, J. A., Hermsmeier, M., Yadav, J. S., Liang, R. F., and Gund, T. M. (1988). *J. Med. Chem.* **31,** 545.
[23] Paine, G. H., and Sheraga, H. A. (1985). *Biopolymers* **24,** 1391.
[24] Struthers, R. S., Hagler, A. T., and Rivier, J. (1984). *In* "Conformationally Directed Drug Design: Peptides and Nucleic Acids as Templates or Targets" (J. A. Vida and M. Gordon, eds.), p. 239. Amer. Chem. Soc., Washington, D.C.
[25] Struthers, R. S., Solmajer, T. J., Campbell, K. B., Tanaka, G., Rivier, J., and Hagler, A. T. (1986). *In* "Computer Graphics and Molecular Modeling" (R. Fletterick and M. Zoller, eds.), p. 109. Cold Spring Harbor, NY.
[26] Mammi, N. J., Hassan, M., and Goodman, M. (1985). *J. Am. Chem. Soc.* **107,** 4008.
[27] Martin, Y. C., Bures, M. G., Donaher, E. A., DeLazzer, J., Lico, I., and Pavlik, P. (1993). *J. Comput.-Aid. Mol. Design* **7,** 83.
[28] Brint, A. T., and Willet, P. (1987). *J. Chem. Info. Comput. Sci.* **27,** 152.
[29] Crippen, G. J., *J. Med. Chem.*, 1979, **22,** 988; 1980, **23,** 599; 1981, **24,** 198.
[30] Crandell, C. W., and Smith, D. H. (1983). *J. Chem. Info. Comput. Sci.* **23,** 186.
[31] Dauzwiger, D. J., and Dean, P. M. (1985). *J. Theor. Biol.* **116,** 215.
[32] Namasivayam, S., and Dean, P. M. (1986). *J. Mol. Graph.* **4,** 46.
[33] Kato, Y., Itai, A., and Itaka, Y. (1987) *Tetrahedron* **43,** 5229.
[34] Chau, P. L., and Dean, P. M. (1987). *J. Mol. Graph.* **5,** 88, 97.
[35] Dean, P. M., and Chau, P. L. (1982). *J. Mol. Graph.* **5,** 152.
[36] Dean, P. M., Callow, P., and Chau, P. L. (1988). *J. Mol. Graph.* **6,** 28, 38.
[37] Hermann, R. B., and Herron, D. K. (1991). *J. Comput.-Aid. Mol. Design* **5,** 511.
[38] Sheridan, R. P., Nilakantan, R., Dixon, J. S., and Venkataraghavan, R. (1986). *J. Med. Chem.* **29,** 899.
[39] Gund, P. (1977). *In* "Progress in Molecular and Subcellular Biology" (F. E. Hahn, ed.), Vol. 5, p. 117. Springer-Verlag, New York.
[40] Humblet, C., and Marshall, G. R. (1981). *Drug Dev. Res.* **1,** 409.
[41] Marshall, G. R., and Simkin, N. A. (eds.) (1978). *In* "Medicinal Chemistry VI," p. 225. Proc's 6th Int. Symp. Med. Chem., Brighton, UK.
[42] Marshall, G. R., Barry, C. D., Bosshard, H. E., Dammkoehler, R. A., Dunn, A., Olson, E. C., and Christoffersen, R. E. (eds.) (1979). "Computer Aided Drug Design," Vol. 112, p. 205. ACS Symposium Series.
[43] Hopfinger, A. J. (1980). *J. Am. Chem. Soc.* **102,** 7196.
[44] Ghose, A. K., and Crippen, G. M. (1984). *J. Med. Chem.* **27,** 901.
[45] Ghose, A. K., and Crippen, G. M. (1985). *J. Med. Chem.* **28,** 333.
[46] Sheridan, R. P., and Venkataraghavan, R. (1987). *J. Comput.-Aid. Mol. Design* **1,** 243.
[47] Martin, Y. C., Bures, M. G., Danaher, E. A., DeLazzer, J., Lico, I., and Pavlik, P. A. (1993). *J. Comput.-Aid. Mol. Design* **7,** 83.
[48] Connolly, M. L. (1983). *Science* **221,** 709.
[49] Connolly, M. L. (1983) *J. Appl. Crystallogr.* **16,** 548.

[50] Sufrin, J. R., Dunn, D. A., and Marshall, G. R. (1981) *Mol. Pharmacol.* **19,** 307.
[51] Hibert, M. F., Gittos, M. W., Middlemiss, D. N., Mir, A. K., and Fozard, J. R. (1988). *J. Med. Chem.* **31,** 1087.
[52] Ortwine, D. F., Bridges, A. J., Humblet, C., and Trivedi, B. K. (1988). Abstracts of Papers, American Chemical Society, No. 92.
[53] Humblet, C., and Marshall, G. R. (1981). *Drug. Dev. Res.* **1,** 409.
[54] Beckett, A. H., and Casy, A. F. (1954). *J. Pharm. Pharmacol.* **6,** 986.
[55] Kier, L. B. (1968). *In* "Fundamental Concepts in Drug-Receptor Interactions" (J. F. Danielli, J. F. Moran, and D. J. Triggle, eds.), p. 15. Academic Press, New York.
[56] Humber, L. G., Bruderlein, F. T., Philipp, A. H., Gotz, M., and Voith, K. (1979). *J. Med. Chem.* **22,** 761.
[57] Olsen, G. L., Cheung, H.-C., Morgan, K. D., Blount, J. F., Todaro, F., Berger, L., Davidson, A. B., and Boff, E. (1981). *J. Med. Chem.* **24,** 1026.
[58] Crippen, G. M. (1982). *Mol. Pharmacol.* **22,** 11.
[59] Martin, Y. C., Kim, K.-H., and Bures, M. G. (1992). *In* "Medicinal Chemistry for the 21st Century" (C. G. Wermuth, N. Koga, H. Konig, and B. W. Metcalf, eds.), p. 295. Blackwell Sci. Oxford.
[60] Martin, Y. C. (1978). "Quantitative Drug Design." Dekker, New York.
[61] Martin, Y. C. (1981). *J. Med. Chem.* **24,** 229.
[62] Hansch, C., Maloney, P. P., Fujita, T., and Muir, R. M. (1962). *Nature* (*London*) **194,** 178.
[63] Fujita, T., Iwasa, I., and Hansch, C. (1964). *J. Am. Chem. Soc.* **86,** 5175.
[64] Franke, R. (1984). "Theoretical Drug Design Methods," Vol. 7. Elsevier, New York.
[65] Balaban, A. T., Chiriac, A., Motoc, I., and Simon, Z. (1980). *Lect. Notes Chem.* **15,** 1.
[66] Hopfinger, A. J. (1980). *J. Am. Chem. Soc.* **102,** 7196.
[67] Hopfinger, A. J. (1985). *J. Med. Chem.* **28,** 1133.
[68] Fujita, T. (1990). *In* "Comprehensive Medicinal Chemistry" (C. Hansch, P. G. Sammes, and J. B. Taylor, eds.), p. 497. Pergamon, Oxford.
[69] Kim, K. H., Martin, Y. C., Otis, E., and Mao, J. (1989). *J. Med. Chem.* **32,** 84.
[70] Martin, Y. C., and Hackbarth, J. J. (1976). *J. Med. Chem.* **19,** 1033.
[71] Martin, Y. C., and Hackbarth, J. J. (1977). *In* "Chemometrics: Theory and Application" (B. Kowalski, ed.), p. 153. American Chemical Society, Washington, D.C.
[72] Hansch, C., Unger, S. H., and Forsythe, A. B. (1973). *J. Med. Chem.* **16,** 1217.
[73] Martin, Y. C., and Panas, H. N. (1979). *J. Med. Chem.* **22,** 784.
[74] Wold, S., Albano, C., Dunn, W. J., Edlund, U., Esbensen, K., Geladi, K., Hellberg, S., Johansson, E., Lindberg, W., and Sjostrom, M. (1984). *In* "Chemometrics: Mathematicas and Statistics in Chemistry" (B. B. Kowalski, ed.), NATO ASI Series C, Vol. 138, p. 4 D. Reidel, Dordrecht, Holland.
[75] Berntsson, P., and Wold, S. (1986). "Quantitative Structure Activity Relationships," p. 45.
[76] Leo, A. (1990). *In* "Comprehensive Medicinal Chemistry" (C. Hansch, P. G. Sames, and J. B. Taylor, eds.), p. 295. Pergamon, Oxford
[77] Daylight Chemical Information Systems, Inc., Claremont St., Irvine, CA 92714.
[78] Hansch, C. (1989). *In* "QSAR: Quantitative Structure Activity Relationships in Drug Design" (J. L. Fauchere, ed), p. 23. A. R. Liss, New York.
[79] Cramer, R. D., Patterson, D. E., and Bunce, J. D. (1988). *J. Amer. Chem. Soc.* **110,** 5959.
[80] Fersht, A. R. (1984). *TIBS* **9,** 145.
[81] Fersht, A. R., Shi, J., Knill-Jones, J., Lowe, D. M., Wilkinson, A. J., Blow, D. M., Brick, P., Carter, P., Waye, M. M. Y., Winter, G. (1985). *Nature* **314,** 235.
[82] Street, I. P., Armstrong, C. R., and Withers, S. G. (1986). *Biochemistry* **25,** 6021.
[83] Gund, T. M., Shukla, K., and Meshnick, S. (1994). *J. Mol. Graph.*, in press.
[84] Merz, K., and Kollman, P. A. (1989). *J. Am. Chem. Soc.* **111,** 5649.
[85] van Gunsteren, W. F. (1989). *In* "Computer Simulation of Biomolecular Systems: Theoretical

and Experimental Applications" (W. F. van Gunsteren and P. K. Weiner, eds.), p. 27. ESCOM Science Publishers B. V., Leiden.
[86] Pearlman, D. A., and Kollman, P. A. (1989). *In* "Computer Simulation of Biomolecular Systems: Theoretical and Experimental Applications" (W. A. van Gunsteren and P. K. Weiner, eds.), p. 101. ESCOM Science Publishers B. V., Leiden.
[87] Langridge, R., and Klein, T. E. (1990). *In* "Comprehensive Medicinal Chemistry" (C. Hansch, P. Sammes, J. Taylor, and C. Ramsden, eds.), p. 413 Pergamon, Elmsford, NY.
[88] Blaney, J. M., Jorgensen, E. C., Connolly, M. L., Ferrin, T. E., Langridge, R., Oatley, S. J., Burridge, J. M., and Blake, C. F. (1982) *J. Med. Chem.* **25**, 785.
[89] DesJarlais, R. L., Sheridan, R. P., Dixon, J. S., Kuntz, I. D., and Venkataraghavan, R. (1986). *J. Med. Chem.* **29**, 2149.
[90] Horvath, A. L. (1992). "Molecular Design," Vol. 75. Elsevier, Amsterdam/New York.
[91] Gutman, I., and Polansky, O. E. (1986). "Mathematical Concepts in Organic Chemistry," p. 212. Springer-Verlag, Berlin.
[92] Filip, P. A., Balaban, T.-S., and Balaban, A. T. (1987). *J. Math. Chem.* **1**, 61.
[93] Hansen, P. J., and Jurs, P. C. (1987). *Anal. Chem.* **59**, 2322.
[94] Rao, M. B. (1975). *Indian J. Technol.* **13**, 571.
[95] Yalkowski, S. H. (1979). *I and EC Fundam.* **18**, 108.
[96] Chickos, J. S., Annunziata, R., Ladon, L. H., Hyman, A. S., and Liebman, J. F. (1986). *J. Organ. Chem.* **51**, 4311.
[97] Martin, E., Yalkowski, S. H., and Wells, J. E. (1979). *J. Pharm. Sci.* **68**, 565.
[98] Hildebrand, J. H. (1919). *J. Am. Chem. Soc.* **41**, 1067.
[99] Rathjen, W., and Straub, J. (1977). *In* "Proc. 7th Symp. on Thermophysical Properties" (A. Cezairliyan, ed.), p. 839. Am. Soc. Mech. Eng.
[100] Balaban, A. T. (1980). MATCH **8**, 159 (CA 93:238486).
[101] Bhatnagar, R. P., Singh, P., and Gupta, S. P. (1980). *Indian J. Chem.* **19**, 780.
[101a] Motoc, I., Balaban, A. T., Mekenyan, Ov., and Bonchev, D. (1982) MATCH **13**, 369.
[102] Motoc, I., and Balaban, Rev., Roum., Chim., 1981, **26**, 593 (CA 95:61216)
[103] Exner, O. (1985). "Structure and Physical Properties of Organic Compounds," p. 280. SNTL, Prague.
[104] Mulay, L. N., (1972). *In* "Techniques of Chemistry" (A. Weissberger and B. W. Rossiter, eds.), Vol. 1, p. 431. Wiley-Interscience, New York.
[105] Beran, J. A., and Kevan, L. (1969). *J. Phys. Chem.* **73**, 3860–3866.
[106] Thompson, W. H., and Braun, W. G. (1968). Proc. Div., Ref. Am. Pet Inst. **48**, 477 (CA 69:110001h).
[107] Mansker, L. D., Criser, A. C., Jangkamolkulchai, A., and Luks, K. D. (1987). *Chem. Eng. Commun.* **57**, 87.
[108] Cohen, M. L. (1985). *Phys. Rev.* **32**, 7988.
[109] Cohen, M. L. (1986). *Science* **234, 549.**
[110] Cohen, M. L. (1989). *Nature* **338**, 291.
[111] Tykodi, R. J. (1969). *J. Chem. Educ.* **66**, 1007.
[112] Gasteiger, J., and Guillen, M. D. (1983). *J. Chem. Res.* 304.
[113] Dosen-Misovic, L., Jeremic, D., and Allinger, N. L. (1983). *J. Am. Chem. Soc.* **105**, 1716.
[114] Horvath, A. L. (1992). "Molecular Design," Vol. 75, p. 469. Elsevier, Amsterdam/New York.
[115] Richards, W. G., and Hodgkin, E. E. (1988). *Chem. Brit.* **24**, 1141.
[116] Pedley, J. B., Naylor, R. D., and Kirby, S. P. (1986). "Thermochemical Data of Organic Compounds," 2nd Ed. Chapman and Hall, London.
[117] Vitale, D. E. (1986). *J. Chem. Educ.* **63**, 304.
[118] Yaws, C. L., and Chiang, P.-Y. (1988). **95**, 81.
[119] Hansch, C., and Leo, A. (1979). "Substitution Constants for Correlation Analysis in Chemistry and Biology." Wiley, New York.
[120] Leo, A., Weininger, D., and Weininger, A., "CLOGP, CMR, Medicinal Chemistry Project,

Pomona College: Claremont, CA 91711; version 3.54, distributed by Daylight Chemical Information Systems, 1989.
[121] Clark, T. (1985). "A Handbook of Computational Chemistry." Wiley, New York.
[122] Buckert, U., and Allinger, N. L. (1982). "Molecular Mechanics." American Chemical Society, Washington, D.C.
[123] Osawa, E., and Musso, H. (1981). *In* "Topics in Stereochemistry" (N. L. Allinger, E. L. Eliel, and S. H. Wilen, eds.), Vol. 13, p. 117. Wiley, New York.
[124] Allinger, N. L. (1976). *Adv. Phys. Organ. Chem.* **13,** 1.
[125] Dammkoehler, R. A., Darasek, S. F., and Berkley Shands, E. F. (1989). *J. Comput.-Aid. Mol. Design* **3,** 3.
[126] Motoc, I., Dammkoehler, R. A., and Marshall, G. R. (1986). *In* "Mathematical and Computational Concepts in Chemistry" (N. Trinajstic, ed.), p. 222. Horwood, Chichester.
[127] Crippen, G. M. (1981). *In* "Distance Geometry and Conformational Calculations" (D. Bawden, ed.), Research Studies Press, Wiley, New York.
[128] Crippen, G. M., Havel, T. F. (1988). *In* "Distance Geometry and Molecular Conformation" (D. Bawden, ed.), Research Studies Press, Wiley, New York.
[129] Billeter, M., Havel, T. F., and Wuthrich, K. J. (1987). *J. Comp. Chem.* **8,** 132.
[130] Billeter, M., Havel, T. F., and Kuntz, I. D. (1987). *Biopolymers* **26,** 777.
[131] Crippen, G. M. (1987). *J. Phys. Chem.* **91,** 6341.
[132] Stewart, J. J. P., *In* "Reviews in Computational Chemistry" (K. B. Lipkowitz and D. Boyd, eds.), p. 46. VCH Weinheim/New York.
[133] Hoffman, R. (1963). *J. Chem. Phys.* **39,** 1397.
[133a] Hoffman, R. (1989). *Curr. Contents Phys. Chem. Earth Sci.* **29,** 20.
[134] Pople, J. A., and Beveridge, D. L. (1970). "Approximate Molecular Orbital Theory." McGraw Hill, New York.
[135] Pople, J. A., and Segal, G. A. (1965). *J. Chem. Phys.* **43,** S136.
[136] Pople, J. A., and Segal, G. A. (1966). *J. Chem. Phys.* **44,** 3289.
[137] Bingham, R. C., Dewar, M. J. S., and Lo, D. H. (1975). *J. Am. Chem. Soc.* **97,** 1285.
[138] Dewar, M. J. S., and Thiel, W. (1977). *J. Am. Chem. Soc.* **99,** 4899.
[139] Dewar, M. J. S., Zoebisch, E. G., Healy, E. F., and Stewart, J. J. P. (1985). *J. Am. Chem. Soc.* **107,** 3902.
[140] Stewart, J. J. P. (1988). *Comput. Chem.* **13,** 157.
[141] Thiel, W. (1982). *Quant. Chem. Prog. Exch.* Prog. 438.
[142] Komornicki, A., and McIver, J. W. (1971). *Chem. Phys. Lett.* **10,** 303.
[143] Komornicki, A., and McIver, J. W. (1971). *J. Am. Chem. Soc.* **94,** 2625.
[144] Feller, D., and Davidson, E. R. (1980). *In* "Reviews in Computational Chemistry" (K. B. Lipkowitz and D. Boyd, eds.). VCH, Weinheim/New York.
[145] Boys, F. S., and Bernardi, F. (1970). *Mol. Phys.* **19,** 553.
[146] Faegri, K., and Almolf, J. (1986). *J. Comp. Chem.* **7,** 396.
[147] Partridge, H. (1987). *J. Chem. Phys.* **87,** 6643.
[148] Spivak, C. E., Waters, J. A., Yadav, J. S., Shang, W. C., Hermsmeier, M., Liang, R. F., and Gund, T. M. (1991). *J. Mol. Graph.* **9,** 105.
[149] Waters, J. A., Spivak, C. E., Hermsmeier, M., Yadav, J. S., Liang, R. F., and Gund, T. M. (1988). *J. Med. Chem.* **31,** 545.
[150] Spivak, C. E., Hermsmeier, M., Yadav, . S., Liang, R. F., and Gund, T. M. (1989). *J. Med. Chem.* **32,** 305.
[151] Hermsmeier, M., and Gund, T. M. (1989). *J. Mol. Graph.* **7,** 150.
[152] Gund, T. M., and Spivak, C. E. (1991). *In* "Methods in Enzymology" (J. Langone, ed.), Vol. 203, p. 677. Academic Press, San Diego.
[153] Gund, T. M., Shukla, K., and Su, T. P. (1991). *J. Math. Chem.* **8,** 309.
[154] Su, T. P., Shukla, K., and Gund, T. (1990). "Steroid and Neuronal Activity" (D. J. Chadwick, ed.), pp. 107–116. Ciba Foundation Symposium 153, Wiley, Chichester.
[155] Gund, T. M., Shukla, K., and Su, T. P. (1990). *Chem. Des. Autom. News* **5,** 1–15.

[156] Su, T. P., Wu, X. Z., Cone, E. J., Shukla, K., Gund, T. M., Dodge, A. L., and Parish, D. W. (1991). *J. Pharmacol. Exp. Ther.* **259**, 543.
[157] Gund, T. M., Shukla, K., Su, T. P., and Parish, D. (1992). *In* "Multiple Sigma and PCP Receptor Ligands: Mechanisms for Neuromodulation and Neuroprotection" (J. Kamenka and E. F. Domino, eds), p. 53, NPP Books, Ann Arbor, MI.
[158] Su, T. P., Wu, X. Z., Cone, E. J., Shukla, K., Gund, T. M., Dodge, A. L., and Parish, D. (1992). "Multiple Sigma and PCP Receptor Ligands: Mechanisms for Neuromodulation and Neuroprotection" (J. M. Kamenka and E. F. Domino, eds.), p. 147. NPP Books, Ann Arbor, MI.

# 4

# Computer-Assisted New Lead Design

**AKIKO ITAI*, MIHO YAMADA MIZUTANI, YOSHIHIKO NISHIBATA, AND NOBUO TOMIOKA***
Laboratory of Medicinal Molecular Design
Faculty of Pharmaceutical Sciences
The University of Tokyo
Hongo, Bunkyo-ku,
Tokyo, Japan
* Present address:
Institute of Medicinal Molecular Design
4-1-11 Hongo, Bunkyo-ku
Tokyo 113, Japan

## I. INTRODUCTION

The choice of lead structures is very important for success in drug development. As we have had no rational means to discover new leads until recently, there are many diseases for which effective therapeutic agents are still not available, simply because we have no appropriate leads. Most drug development so far has been done based on lead compounds derived from primary lead compounds which were found fortuitously or by random screening of natural and synthetic compounds. These primary lead molecules, in addition to natural ligand molecules, have been evolved to secondary leads through a repetition of chemical synthesis and evaluation. Such trial and error processes are labor intensive and time-consuming, with no guarantee of success. Our goal has been to establish a principle as well as methods for designing lead compounds logically. These difficult processes of drug development have been increasingly aided by newly developed techniques,

*Guidebook on Molecular Modeling in Drug Design*

using computers to create new leads without relying on chance or trial and error.

These advances have been made possible by remarkable progress in related fields, such as biochemistry, molecular biology, biotechnology, protein crystallography, and computers. Studies of the biological or biochemical mechanisms of diseases have provided clues to the macromolecules to be targeted or the types of molecules needed for new drugs. Due to striking progress in various techniques for protein isolation and purification, many proteins and other biological macromolecules have been isolated and their structures and functions investigated. Furthermore, more than 1000 three-dimensional (3-D) structures of proteins and protein complexes with their specific ligands have been elucidated crystallographically. Those structures provide reliable and detailed images of the so-called drug–receptor interaction, which can be regarded conceptually as a "lock and key." Structural aspects of molecular recognition by many biological macromolecules have thus been made clear at the atomic level. Increasing numbers of protein structures are being elucidated at present, and the process is being accelerated by tremendous progress in various techniques for protein crystallography.

A drug molecule manifests its activity through binding to a receptor macromolecule, whether the interaction is selective or not. As the interaction arises from the 3-D structures of both molecules, any method based on 2-D structures cannot be sufficient to interpret structure–activity relations or to design new active molecules. In particular, knowledge of the 3-D structure of the receptor is crucial to drug design or ligand design. Computers and computer graphics are indispensable for handling 3-D structures. They are used to visualize or simulate stable interactions between drug and receptor, as well as to analyze stable structures and the physical properties of each molecule, or to predict them theoretically. Due to the progress in developing software for molecular graphics, molecular modeling, and theoretical calculations, most such computer studies can be performed in any research institute in the world, even by nonspecialists without detailed knowledge of the software. Computers and computer graphics have become essential tools for rational drug design. Nevertheless, new active molecules with quite different structures could not be designed by such computer methods, which just facilitate rational structural modifications of known ligands through docking studies.

At present, the vast majority of new drugs are derived from discoveries in large-scale screenings or from analog syntheses. If a small amount of the receptor protein is isolated or available, thousands of compounds can be tested by means of radioactive labeling or spectroscopic detection techniques. Thus, laboratory screening can be a powerful means to find new active molecules with unexpected structures from an in-house compound library or among commercially available compounds, although samples of a great many compounds and much labor are required. Although this is a very powerful approach when the receptor is in hand, the results depend on the number and variety of compounds available for testing. Therefore, it would be helpful for creating new lead structures to develop rational methods using computers, in addition to experimental methods.

## II. BASIC CONCEPTS

### A. Molecular Recognition by Receptor and Ligand Design

Receptors do not recognize their ligands in terms of molecular frameworks or atom positions or connectivities between them, but in terms of free energy difference between unassociated and associated states of the ligand with the receptor. Between a drug and target receptor, hydrogen bonds (H-bonds), electrostatic interactions, and hydrophobic interactions are supposed to play important roles, together with van der Waals interactions, in stabilizing associated structures enthalpically. It is the 3-D complementarity in molecular shape and submolecular properties between a drug and the drug-binding site of the receptor that is the most important for specific binding to the target receptor. This means that molecules which satisfy the requirements three dimensionally can be specific ligands to the receptor. As a result, ligand molecules with a variety of skeletal structures can bind to a target receptor. Many examples of natural internal ligands, agonists, antagonists, and inhibitors that can bind to a common receptor competitively despite dissimilarity of their structures are known. Although only a few skeletal structures of ligands for a receptor are known in most cases at present, this is merely because no other efficient means to find other potential active structures exist so far. Rational lead generation, not relying on chance or trial and error, could be realized by establishing new methods to find such potential active structures.

### B. Active Conformation

Knowledge of the active conformations of known drugs or ligands is very useful for the rational design of new ligands. Even if the 3-D receptor structure is unknown, a rational approach can be made on the basis of the structures of known ligands, if the active conformation of one active molecule is known or if there is one active molecule with a rigid structure. Since the 1980s, many computer chemists have constructed working hypotheses on the assumption that the active conformation of a drug is the same as a stable conformation found in the crystal or in solution or is simply the lowest energy conformation. However, it is now accepted that the active conformation is not necessarily the same as any of these stable conformations since it is the total stability of the system that is important. All computer studies handling 3-D molecular structures for drug design, such as docking study, molecular superposition, and database search for lead discovery, should be done in terms of the active conformation. Therefore, active conformations of flexible molecules should be searched over all possible conformational space without assumptions. Generally, this has not been done or has been done manually based on visual judgments on graphic displays due to the extreme difficulty of the problem. Since the degrees of freedom for ligand conformation are completely linked with other degrees of freedom, such as rotation and translation (in other words, binding modes or superposing modes), the number of possible states that should be considered becomes enormous. Even in a docking study, which is relatively simple and straightforward, it is difficult to obtain the most stable complexed structure by simple use of molecular mechanics calculation or even by molecular dynamics calculation, without preparing a vast number of starting structures. For all purposes in rational drug design, it is necessary to develop an effective search method that can cover all the possibilities within a practical CPU time, for obtaining correct or reliable results. How to handle conformations of flexible molecules is the most difficult problem in rational drug design, even at present, when the speed of computers has been greatly improved.

### C. "Functon"

Binding to the target receptor is the primary requirement for drugs from the viewpoint of finding new leads, whereas reaching the receptor site is a secondary consideration at this stage. Then, typically, a drug molecule has not only a core part essential for receptor binding but also some subsidiary parts related to physical properties, such as solubility and drug delivery. Not all the atoms in the core part need necessarily be in direct contact with the receptor, but the framework of the core structure plays an important role in maintaining functional groups at the required position and orientation as well as in allowing the molecule to have suitable shape and flexibility. For expressing such core structures generally, the existing terms "lead," "pharmacophore," or "pharmacophoric group" seem to be inadequate. The term "lead" clearly expresses actual compounds, although the definition is not precise. One definition of lead is that it is the active compound which was found first in a family of active compounds. Another definition is that a lead is an active compound with a sufficiently potent activity to be a starting point for optimization. According to such definitions, a lead seems to be a much more advanced stage than the core structures we conceive. The term "pharmacophore" or "pharmacophoric group" is defined as atoms or functional groups or their relative positions that are (supposed to be) essential for the activity. These terms do not express actual chemical entities, but express conceptual states.

Thus, we would like to propose a new term "functon" for the core structure. Functon is defined as the key structure to bind to the target receptor, consisting of a framework and functional groups. Many functons for a given target receptor should exist and all existing active structures should have their own functon. A functon is an actual chemical entity like a lead, different from a pharmacophore. A functon can be regarded as a potential lead because it might be derived into a true lead through appropriate structural modification.

Estimation of functons from known active compounds is done by superposing the compounds in an appropriate way when they are known to bind to the same receptor site or by chemical synthetic studies to modify the active structures. Molecular superposition is one of the most important means to interpret relations between 3-D structures and activities of known active compounds in the case where the receptor structure is unknown. By superposing molecules with dissimilar structures on the assumption of a common receptor, the active conformation and functional groups important for the activity can be derived for each compound, in addition to the functon. It goes without saying that correct superposition is essential for reliable results. Molecules should not be superposed in terms of atomic positions or chemical structures, as in the conventional methods, but in terms of physical and chemical properties involved in the intermolecular interaction with the receptor, for this purpose.

## D. Approaches to Discover New Functons

In order to exhibit biological activities, molecules should be able to reach the receptor site in the body and bind to the target receptor there. These two factors should be considered separately in drug design for the purpose of creating new leads in a rational way, although they have often not been separated in classical studies of structure–activity relationships. As different categories of molecular properties are involved in the two factors, they have different categories of structural requirements. The physical properties concerning the former factor (reaching the site) are rather macroscopic and measurable ones inherent in the molecule, i.e., oil–water partition coefficient, solubility, and dipole moment. No specific structural features are required for controlling this factor. However, specific structural or chemical features are required for controlling the latter factor: that is, there must be a significant complementarity in molecular shape and also in submolecular properties between a drug molecule and the drug-binding site of the receptor. Although consideration of drug delivery and toxicity is very important for drug development, design focusing on receptor binding seems to be most important for obtaining new lead compounds. If we could somehow find new functons which satisfy the latter requirement, we would be able to accomplish rational lead generation by modifying the structures so as to satisfy the former requirements as well as to improve the potency.

Computer-aided techniques can assist in both discovery and optimization of lead compounds. It goes without saying that reliable 3-D structural information either on the receptor or on known drug compounds is necessary not only for discovery of new functons, but also for all rational approaches.

The 3-D structure of the receptor is far more useful for rational drug design than drug structures. As the receptor structure determines all the requirements for shape and properties of ligands, relations between structures and activities can be interpreted with least speculation. *De novo* ligand design can be performed based on the receptor structure, as well as minor modification of known ligands through docking study. However, receptor-based methods cannot be applied if an X-ray, NMR, or high-quality model structure of the receptor is not available. At present, this represents a limitation to the approach. There are many interesting targets for drug development where the 3-D receptor structure is not known. In most cases, the receptors have not even been isolated.

When the receptor structure is not available, the only approach is to depend on a hypothesis based on known active structures. It is supposed that structures with some structural features common to a known active molecule may have the same activity, assuming that the 3-D features are important for receptor binding. As it is difficult to establish the 3-D features, comprising the active conformation, functional groups important for the activity, and their relative position and orientation, it is essential to construct a working hypothesis, i.e., a pharmacophore hypothesis. Superposing active and inactive molecules can assist in constructing more plausible hypotheses. A receptor cavity model built based on the superposed active molecules might be made use of as a substitute for the actual receptor structure. In searching for or designing new active molecules based on the pharmacophore hypotheses, some disappointment is unavoidable, but such studies are useful to test the validity of the hypothesis and to improve it. Many new active structures have been obtained based on such pharmacophore hypotheses.

For discovering possible functons by using computers, there are two conceptually different approaches. One is searching for them from a vast number of existing compounds in 3-D structural databases and the other is constructing them computationally.

## E. Approaches to the Cases with Known and Unknown Receptor Structure

Different approaches are taken for the cases with known and unknown receptor structure, for interpreting 3-D structure–activity relations, and for designing new active structures.

### 1. Approaches with Receptor Structure

Approaches used when the receptor structure is available are shown in Fig. 1. Simulation of stable docking structures between receptor and known drugs, natural ligands, or imaginary compounds, the so-called docking study, is one of the most important techniques in computer-aided drug design. Before the appearance of *de novo* ligand design techniques, the docking

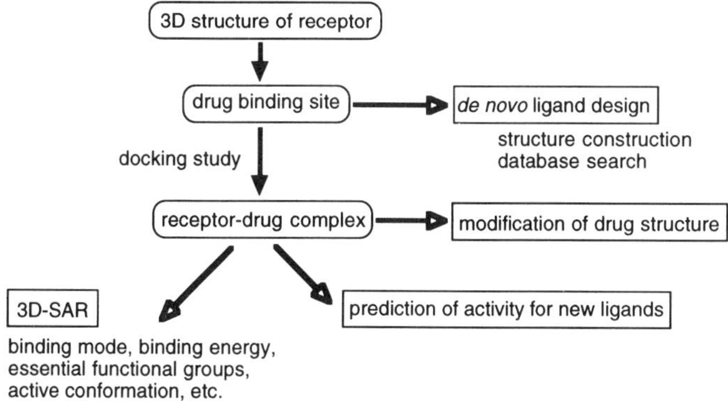


**FIGURE 1** Approaches with receptor structure.

study was the only way to make use of the receptor structure for the purpose of rational drug design. First, it is very useful for interpreting activities of known ligands. Based on the stable docking model, the active conformation and the role of each functional group in the ligand can be estimated. H-bond donor–acceptor pairs, electrostatically positive–negative pairs, and hydrophobic pairs between receptor and ligand can also be predicted. Activities of unknown compounds can also be predicted from the viewpoint of conformity to the receptor cavity and complementarity in properties involved in specific interactions. Furthermore, structures of known ligands can be properly modified based on the stable docking models. The receptor binding of docked ligands can be improved by substituting or adding atoms (or groups) in the ligand structure so as to form additional H-bonds, to get favorable electrostatic interactions or to fill vacant spaces inside the binding site.

In addition to such approaches based on the docking study, techniques of *de novo* ligand design using the receptor structure have greatly advanced in recent years. By *de novo* design techniques, we can find molecules that may bind to the target receptor without using any information on existing ligands. There are two approaches to discover new ligands with structures not analogous to the existing ones. One is searching 3-D databases and the other is constructing new structures. Both approaches are described in later sections.

### 2. Approaches without Receptor Structure

In the case where the 3-D receptor structure is unknown, we have no choice but to design new molecules based on 3-D structures of natural ligands or known drugs. For this purpose, it would be helpful if we can properly presume the 3-D requirements for the activity based on known active molecules. It is the active conformation, functional groups, or atoms essential for the activity and the shape and volume occupied by ligands that are meaningful as 3-D requirements. Although synthetic efforts using conformationally restricted analogs or functionally converted analogs are very important, they are not easy in most cases. Therefore, it seems to be indispensable to proceed by designing new structures, based on assumptions about the 3-D requirements.

Assumptions on chemical groups or atoms essential for the activity are called "pharmacophore hypotheses." Pharmacophore was initially a 2-D concept, but is now a 3-D one. Molecules with pharmacophore groups arranged at three-dimensionally appropriate positions are considered to have higher probability of exhibiting the activity. So, 3-D database searching in terms of geometrical relations between pharmacophore groups or structure construction to generate molecules which satisfy such relations has become an important means to discover new active structures. Of course, the most important point is how to presume or assume the 3-D requirements for the activity, and this determines the success or failure of the work.

Molecular superposition is a useful means to estimate the 3-D requirements rationally. Use of a single active molecule can provide these requirements, if it is conformationally rigid and a significant structure–activity study has been performed. However, any single ligand does not completely reflect the environment in the drug-binding site of the receptor. Competitive ligand molecules with structurally different features would reflect it much more faithfully by complementing each other, if they are appropriately superposed. Design and synthesis of new molecules based on such a receptor model or pharmacophore hypotheses might produce active compounds and would at least be useful to confirm the hypothesis.

## III. DOCKING PROBLEM AND DOCKING METHODS

Although receptor and ligand molecules exist in an equilibrium between associated and unassociated states, it is supposed that they adopt the energetically most stable docking structure in the associated state. So, the ultimate goal of the docking study is to obtain the global-energy-minimum docking model as well as its precise structure and stability. The docking model should be searched covering all possibilities. All degrees of freedom for rotation and translation of the ligand molecule (docking mode) should be taken into account together with those for ligand conformation. As these degrees of freedom are completely interlinked, the correct docking structure cannot be obtained unless all of them are correctly assumed. Despite the enormous number of models to be examined, the stability of each docking structure should be evaluated quantitatively.

In the docking of flexible ligands, there are degrees of freedom for ligand conformation in addition to six degrees of freedom for rotation and translation of one molecule. Because the number of conformations to be examined is so enormous, docking simulation is usually performed as follows: several likely docking models are constructed interactively by using computer graphic displays, and then the models are compared with each other after energy minimization. We have developed a program named GREEN in order to perform docking study as efficiently as possible (described later). However, by such a manual method, it is not easy to find the global-energy-minimum structure, even in the case where either the binding mode or ligand conformation is known or can be assumed properly based on experiments with analogous compounds. In order to obtain reliable docking results without any presumptions, an automatic search method for docking study is needed in place of interactive methods. Problems in such an automatic search method are (1) how to cover all possible binding modes and ligand conformations and (2) how to choose the energetically most stable one among them. We have succeeded in developing an automatic docking program, named ADAM, which is also described.

### A. Program GREEN: Three-Dimensional Grid Description of Binding Site Environment and Energy Calculation

The program GREEN was originally developed for interactive docking study on a graphic display (Tomioka *et al.*, 1987a,b; Tomioka and Itai, 1984, 1994). The characteristic feature of the program is the 3-D grid description of binding site environment. Various potential data which describe physical and chemical properties inside the ligand-binding site are tabulated on a 3-D grid. They are used for visualizing the binding site environment and for rapidly calculating intermolecular energy. Both the ADAM and LEGEND programs (described in later sections) also use the same algorithm for rapid estimation of intermolecular interaction energy. Therefore, we will first explain the algorithm of the GREEN program before describing the two *de novo* ligand design programs ADAM and LEGEND.

Calculation of the "3-D grid data" are as follows. First, the program adds hydrogen atoms to each protein atom and assigns atom types and atomic charges to all atoms, using the values of the AMBER force field (Wiener *et al.*, 1984, 1986). Inside the ligand binding region specified by a rectangular box, a 3-D grid with an interval of 0.3–0.5 Å is generated. On each grid point, the program calculates van der Waals potential ($G_{vdw}$), electrostatic potential ($G_{elc}$), and H-bonding flags ($G_{hbd}$) using the equation shown in Fig. 2. The van der Waals term is calculated between a probe atom placed on the grid point and all the protein atoms by using an empirical potential function. Energy values for several probe atoms such as carbon, hydrogen, nitrogen, and oxygen are stored separately. The electrostatic potential $G_{elc}$ is calculated using atomic charges on the protein atoms. The value of $G_{elc}$ is equivalent to the electrostatic interaction energy, assuming

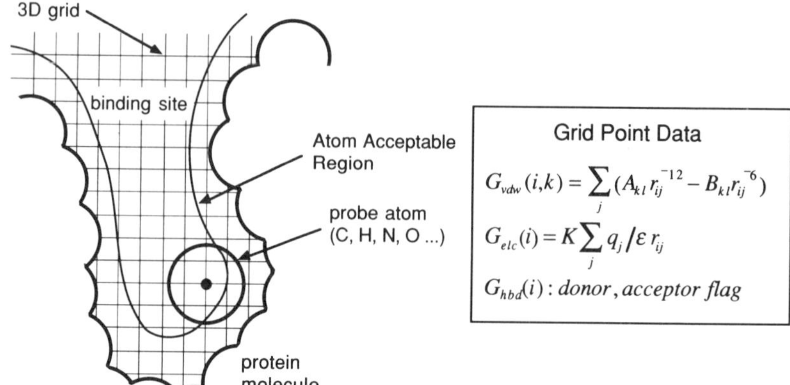


**FIGURE 2** Representation of the binding site environment with 3-D grid point data by the GREEN program.

that the probe atom bears a positive unit charge. As regarding H-bonds, the expected positions of the H-bond partners are calculated automatically for each H-bond group in the protein, based on the distances and directions of lone pairs and hydrogen atoms in the H-bond group. If a grid point is at a favorable location to form an H-bond to a functional group in the protein, the H-bond flag $G_{\text{hbd}}$ is turned on with the H-bond character (donor, acceptor, or ambivalent) of the functional group.

With the values of $G_{\text{vdw}}$, the "atom acceptable region" is defined as a region where $G_{\text{vdw}}$ is below a given threshold. The atom acceptable region describes a region that can accommodate the centers of ligand atoms, different from van der Waals volume and other conventional volume definitions. Graphical representation of the atom acceptable region facilitates fitting of a ligand to the receptor and examining the extent of conformity of the ligand.

Calculation of the interaction energy is greatly accelerated by the grid approximation compared to the conventional method. When a ligand molecule is placed in the grided region, the intermolecular interaction energy between protein and ligand ($E_{\text{inter}}$) is calculated using the equation shown in Fig. 3. The van der Waals energy is calculated simply by summing up the $G_{\text{vdw}}$ value on the grid point nearest to the ligand atom (or the value interpolated from the eight neighboring grid points), taking the values for the corresponding probe atom type $k_m$. Electrostatic interaction energy is calculated by summing the product of the atomic charge of ligand atom $q_m$ and the $G_{\text{elc}}$ value of the nearest grid point. As for the H-bonding energy, the program does not use empirical energy functions, but simply counts the number of H-bonds ($n_{\text{HB}}$) that are judged by the matching of ligand heteroatoms to the $G_{\text{hbd}}$ flag and gives constant energy $K_{\text{HB}}$ to each H-bond

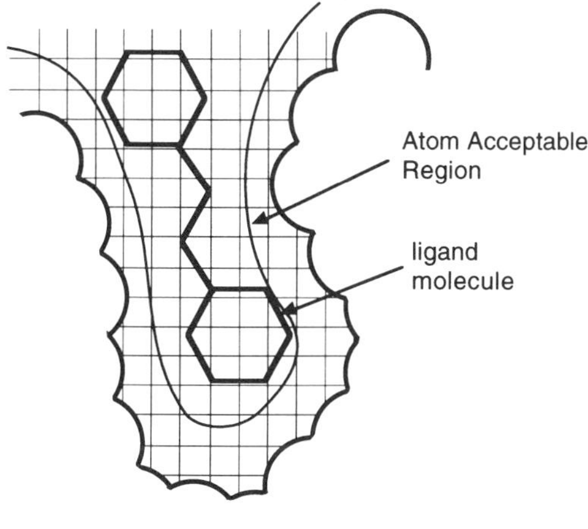


$$E_{inter} = \sum_{m} (G_{vdw}(m, k_m) + G_{elc}(m)\, q_m + K_{HB} n_{HB})$$

**FIGURE 3** Calculation of the interaction energy using the 3-D grid point data.

formed. Intramolecular energy of the ligand is obtained by conventional empirical energy calculation.

Program GREEN provides various functions for facilitating the docking study such as graphical representation of binding site environment, real-time monitoring of interaction energy during docking operation, model building of ligand structure, energy minimization, and Monte Carlo simulation. GREEN also provides functions to assist design of new structures by the LEGEND programs. It provides interface functions for preparing the grid point data used by LEGEND and for setting up input parameters of LEGEND calculation which is executed as a batch-style job. GREEN is also used for the analysis of the results of LEGEND and ADAM calculations.

## B. Automatic Docking Method

Methods for automated search for stable docking structures of protein–ligand or protein–protein have been investigated by many researchers (Ghose and Crippen, 1985; Goodsell and Olson, 1990; Jiang and Kim, 1991; Bacon and Moult, 1992; Kasinos *et al.*, 1992). Applications of various algorithms to this problem are still being attempted. Among them, the DOCK program developed by Kuntz and co-workers (1982) has been practically used. The characteristics of DOCK are description of the shape of the receptor cavities in terms of dozens of spheres which come into contact with the solvent-accessible surface of the cavity. The program finds specific orientations of a ligand to dock to the receptor site by using a fast vector-matching algorithm, using vectors made from the centers of the binding-site spheres and vectors between ligand atoms. The program covers possible docking modes by testing vector matching in various orientations and ranks them by a score that represents shape complementarity between two molecules. The method is very elegant and efficient, but it has the disadvantage that the algorithm is too complicated to consider the conformational degrees of freedom for the ligand.

As mentioned in the section describing the basic concept, the active conformation of each molecule does not necessarily coincide with the conformation found in the crystal or in solution, or the energetically most stable one, or any of the local minimum ones. So whatever structure is used as the input structure for docking, all possible conformations should be examined for the ability to form the most stable complex in the docking study. In the conformation-rigid docking of a conformationally flexible molecule, it is unlikely that the correct docking model would be obtained. From our unpublished calculation on the stability of active conformations based on several crystal structures of enzyme–inhibitor complexes, the energy differences between the active conformation and the most stable conformation were within the range of 3–7 kcal/mol, although the value varied depending on the force field and dielectric constant used. Therefore, all conformations within about 7 kcal/mol of the minimum must be considered in searching active conformations of flexible molecules, whether conformations are stored in advance or newly generated.

We have developed an efficient automatic docking method that can search for stable docking structures of the protein–ligand complex automatically, overcoming the conformational problem. The method was named ADAM (Yamada and Itai, 1993a,b; Yamada *et al.*, 1993; Mizutani Yamada *et al.*, 1994). The goal of the development of the method was to search for the most stable docking model among all possible ones for a single ligand, and also to identify molecules that may stably bind to the receptor from a vast number of molecules in 3-D structural databases.

### 1. Algorithm of ADAM

The characteristics of the algorithm of ADAM are: (1) stable docking models are searched while covering all possible binding modes and ligand conformations; (2) favorable H-bond schemes are utilized as a clue in the initial selection of likely docking modes and ligand conformations at the same time; (3) total energy (sum of intramolecular and intermolecular energies) consisting of van der Waals, electrostatic, and H-bond interactions is used as a criterion in all pruning processes; (4) the procedure includes three structure optimization steps to produce accurately stable docking structures, changing torsion angles continuously; and (5) several to dozens of promising docking models ranked by total energy are output from the program completely automatically.

An outline of the method is shown in the flow chart (Fig. 4).

In the first step of docking, possible H-bonding schemes are searched by covering all combination sets of correspondences between H-bonding dummy atoms in the protein and heteroatoms in the ligand instead of covering all possible binding modes by systematic rotation and translation of the ligand. Each combination set is examined to see whether the expected H-bonding scheme can be formed or not, for all possible ligand conformations. If $p$ H-bonds are formed from $m$ dummy atoms in the protein and $n$ heteroatoms in the ligand, the number of combination sets $N(p)$ is given by the following equation. Each combination set consists of $p$ pairs of H-bonding atoms in both molecules. The combination sets in which the H-bonding characters are not complementary in all pairs are excluded:

$$N(p) = {}_{m}\mathrm{P}_{p} \times {}_{n}\mathrm{C}_{p}.$$

As regarding the ligand conformation, all rotatable bonds are rotated systematically. In each combination set, possible conformations in the H-bonding part, which involves H-bonding functional groups, are selected together with possible H-bonding schemes. In order to judge the feasibility of forming H-bonds between the corresponding atoms in two molecules placed in different coordinate systems, distances among dummy atoms are compared with those among corresponding heteroatoms in each conformation for each combination set. If deviations between all corresponding distances and intramolecular energy of the ligand structure obtained by minimizing the root mean square deviations of the distances are within given criteria, the H-bonding scheme is regarded as possible in that conformation. Then the

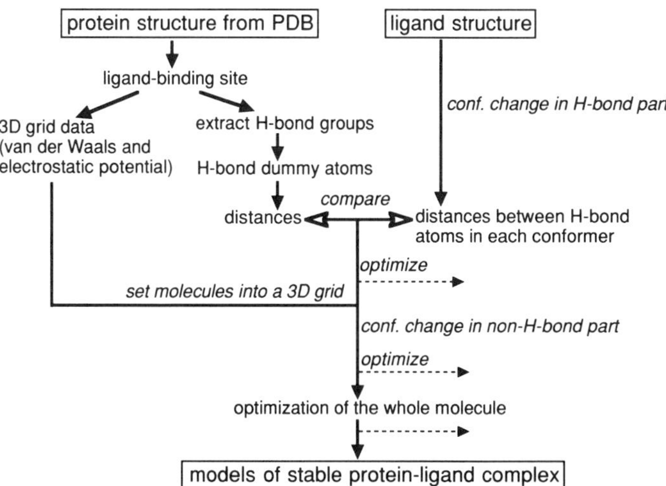


**FIGURE 4** Flow chart of the automatic docking method ADAM.

obtained ligand conformers with possible H-bonding schemes are put into the binding site, by a least-squares fitting of positions of heteroatoms with those of the corresponding dummy atoms. The 3-D grid-point data of the GREEN program (described earlier) are used to calculate or minimize interaction energy. The docked structures are pruned on the basis of total energy after energy minimization by the Simplex method, optimizing position and orientation as well as bond rotation in the H-bonding part of the ligand. Conformations in the remaining part are generated for each likely structure of the H-bonding part. After this step, low-energy models are subject to energy minimization by the Simplex method, taking into account the degrees of freedom for rotation, translation, and conformations of the whole ligand structure, and are further pruned. From several to several dozen stable docking structures, which we call "initial docking models," are output here. The number of output models varies depending on the energetic criteria for pruning. Then, the final docking models are obtained by energy minimization of the initial docking models, using the AMBER molecular mechanics program (Singh *et al.*, 1989) or other molecular mechanics programs. In this energy minimization, the positions of the protein atoms and the water molecules near the ligand-binding region are optimized together with those of the ligand atoms. The energy-minimized models are finally ranked by total energy.

***The prepruning algorithm*** By ADAM, the whole process of obtaining initial docking models is performed continuously and automatically for any molecule. However, in the case of a ligand molecule with many rotatable bonds or many H-bonding atoms, a straightforward search of all the possibilities is not a good policy since a vast number of possibilities needs to be considered. To finish the docking of such molecules in a short CPU time, we have developed a new method, namely the prepruning (PP) procedure. The method is based on the assumption that H-bonding schemes disallowed in a partial structure are also disallowed in the whole structure. The partial structure here need not be a conformationally rigid one, although a structure including more than two H-bonding groups is favorable. Choice of the partial structure does not affect the docking results of the whole molecule. Docking of the whole molecule proceeds automatically, dispensing with unnecessary searches, based on the information obtained by the preliminary step. The PP procedure greatly accelerates the docking process.

### 2. Applications of ADAM Docking

It is natural that the greater the number of H-bonding groups in the molecule, the more favorable the docking by our method becomes. But, the method can also handle systems with only one or two H-bonding groups, although the method is then less advantageous. The usefulness of the ADAM program was confirmed by applications to several enzyme–ligand systems whose complexed structures have been elucidated crystallographically. In all the examples, the crystal structures of the complexes were accurately reproduced by the most stable docking models output from the ADAM program. It was proved that our method offers excellent reliability, reproducibility, and accuracy of the docking results, as well as high speed and ease of use.

***Dihydrofolate reductase*** Here, we present examples of docking trimethoprim (TMP) and methotrexate (MTX) molecules to dihydrofolate reductase (DHFR). Atomic coordinates of the binary complex of *Escherichia coli* DHFR and MTX (Bolin *et al.*, 1982) were taken from the Protein Data Bank (Bernstein *et al.*, 1977). As the allowable region for the ligand to be docked, the substrate-binding site of the enzyme was prepared by removing the MTX molecule from the complex. A 3-D grid with an interval of 0.4 Å was generated inside the region, and various potential data were tabulated at each grid point. All the water molecules were removed except for two molecules strongly bound to the enzyme at the bottom of the binding site. The atomic charge on each protein atom was assigned by taking values for the corresponding atom of the amino acid provided in the AMBER program. An H-bonding group type number was also assigned to all H-bonding functional groups in the protein. There were 10 H-bonding groups exposed in the binding-site region, from which 13 dummy atoms were produced.

The 3-D structures of the isolated TMP and MTX molecules were taken from the Cambridge Structural Database (Allen *et al.*, 1979), and atomic charges were calculated by molecular orbital calculation for both molecules. H-bonding heteroatoms used for estimating possible H-bonding schemes and the rotated bonds are shown in Fig. 5.

As regarding the TMP molecule, H-bonding schemes involving three H-bonds were searched, between four heteroatoms in TMP and 13 dummy

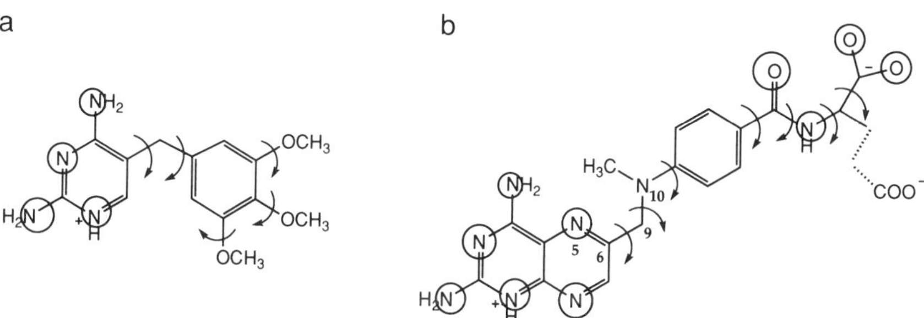


**FIGURE 5** Chemical structures of (a) trimethoprim and (b) methotrexate. Heteroatoms used for predicting possible H-bonding schemes are encircled. Rotated bonds are shown by arched arrows.

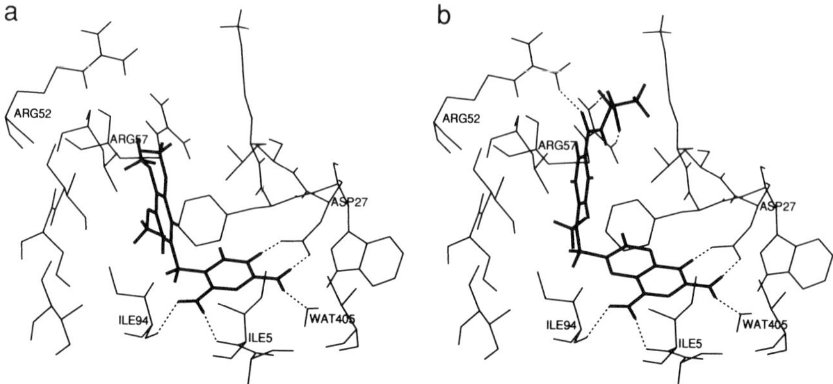


**FIGURE 6** Structures of the most stable docking models for trimethoprim (a) and methotrexate (b) in the binding site.

atoms in the protein. The number of combination sets searched was $N(3) = {}_{13}P_3 \times {}_4C_3 = 6864$. Since the H-bonding part of TMP (2,4-diaminopyrimidine moiety) does not include any rotatable bond, conformations in the five rotatable bonds were considered after searching possible H-bonding schemes of the H-bonding moiety. Each bond was rotated by 120° first, and later by 15° for searching likely conformations. The whole docking process proceeded automatically and gave nine initial docking models.

As regarding the MTX molecule, H-bonding schemes involving six and five H-bonds between 10 heteroatoms in MTX and 13 dummy atoms in the protein were searched, and conformations of seven rotatable bonds were considered for each of a large number of combination sets $N(5) + N(6) = {}_{13}P_6 \times {}_{10}C_6 + {}_{13}P_5 \times {}_{10}C_5 = 298{,}378{,}080$. In order to cover such an enormous number of possibilities in a short computational time, the PP procedure was adopted. As the partial structure for the procedure, the structure from the pteridine ring to the benzene ring was selected. The number of combination sets was reduced to 280 by excluding unused heteroatoms and impossible H-bonding schemes. Finally, the program output 11 initial docking models. Figure 6 shows the most stable docking model which was obtained by energy minimization of the initial docking models for MTX, together with that for TMP.

The most stable docking models for both molecules well reproduced the crystal structures. In both cases, the bound conformation differed very much from the input one, which was taken from the crystal structure of the isolated molecule (for MTX, see Fig. 7). The H-bonding scheme as well as the position and orientation of the TMP and MTX molecules were very similar to those in crystal structures in the literature, although the degree of similarity cannot be calculated in the case of TMP because the atomic

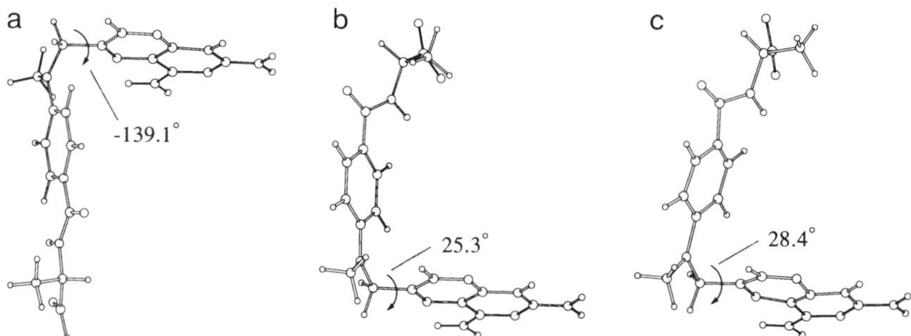


**FIGURE 7** Conformations of methotrexate in (a) input structure (the crystal structure of the isolated molecule), (b) the most stable docking model, and (c) the crystal structure in the complex with the enzyme. The pteridine rings are viewed from the same direction.

coordinates of the crystal structure are not available from PDB. As regarding the MTX molecule, the root mean square value of the most stable docking model to the crystal structure was 0.26 Å. It took 6 min for docking TMP and 10 min for MTX (using the PP procedure) on an IRIS 4-D workstation (CPU 150 MHz R4400). The required time varied depending on the energetic criteria for pruning.

## IV. THREE-DIMENSIONAL DATABASE SEARCH APPROACHES

A new lead does not necessarily mean a new compound. More than several million compounds are known to exist or to have existed on earth. Many may exhibit unknown biological activities. Generally, just a few biological activities or functions, for which assay systems happened to be available, have been examined for any given compound. It can therefore be said that almost all biological activities of almost all known compounds remain uninvestigated.

Random screening is one approach in finding new active molecules with unexpected structures from the known compounds. Most lead discoveries since the 1980s have been achieved through random screening in laboratories. However, laboratory screening requires a lot of labor and actual samples of compounds as well as the isolated receptor protein. On the other hand, computer screening is easily performed with a vast number of compounds, regardless of availability. Laboratory screening can be performed based on the actual receptor and radioactive ligands, even when there is no structural knowledge of the receptor, whereas computer screening requires significant information on receptor structure or the structures of known active compounds. But, in the case that such information is available, computer screening is useful because of its easiness and the variety of compounds that can be examined. Use of large 3-D databases of compounds is achieving significant success in finding molecules that may act as active ligands at receptor sites (Borman, 1992; Martin, 1992; Ho and Marshall 1993). The major advantage of finding new activities from in-house (company library) or commercially available compounds is that the activity can be assayed experimentally without synthetic effort.

Chemical databases have traditionally been made by depicting compounds as 2-D structures, similar to figures. Except for the Cambridge Structural Database (CSD), all commercial, academic, and proprietary chemical databases have been accumulated in 2-D format and used in 2-D, but many of these 2-D databases have been converted into a 3-D format and 3-D information is beginning to be used. How to deal with the many complexities inherent in 3-D chemical information handling remains a problem, but software for storing, searching, and manipulating 3-D structures has greatly advanced. Several programs are already available which convert 2-D database structures to 3-D automatically, and most of the 2-D structures in the databases can be converted to one of the stable local minima 3-D structures.

A variety of search programs are already able to identify molecules that are structurally related to known bioactive compounds, for the purpose of finding compounds that may act as ligands at receptor sites. Structural frameworks and other 3-D geometrical requirements such as distances and angles between atoms (or groups) are used as queries for searching a large database based on a pharmacophore hypothesis. The advantage of these queries is that they can be used for the cases where receptor structure is unknown, although the probability that their hits exhibit the activity is not high because of other unknown requirements, such as shape and size of the occupied volume to be allowed for each molecule, and the possibility that the hypothesis may not be correct. The geometrical queries can be more usefully used for the case where receptor structure is known. Lam and coworkers (1994) have succeeded in developing potent nonpeptide inhibitors of human immunodeficiency virus (HIV) protease, starting from a molecule identified by searching CSD in terms of distances among two hydrophobic groups and a hydroxyl group found in the crystal structure of the protein–peptidyl ligand complex. They chose the starting molecule among many hits based on the conformity and H-bonding possibility to the receptor cavity without assumptions.

Most 3-D search programs still focus on the properties of ligands. However, it is more rational to search molecules in terms of receptor fitting, if the receptor structure is available. Kuntz and co-workers applied their automatic docking program DOCK to search molecules in 3-D databases such as CSD and Available Chemicals Directory (ACD), making use of the properties of the receptors. 3-D databases are searched for molecules that are likely to conform to the 3-D shape of a receptor. The output of the program is a list of compounds ranked by a score based on the overall complementarity with the receptor. DOCK has been used successfully in the discovery of active molecules with novel structures in several enzyme and DNA systems (Kuntz, 1992; Shoichet *et al.*, 1993). The most widely known success was that an anti-depressant drug, haloperidol, was identified by DOCK as a ligand conforming to the active site of HIV protease (Des Jarlais *et al.*, 1990). Inhibitory activity of the compound on the enzyme was confirmed experimentally, and it was the first such inhibitor of nonpeptide structure to have been found. Furthermore, they reported that they could obtain promising new compounds with $K_i \sim 10\ \mu M$ by further similarity search procedures based on the molecules initially found by DOCK.

These successes were accomplished by conformation-rigid searching. If the program could deal with conformational flexibilities, many more compounds might be identified as promising ligands, although much more CPU time would be required for that purpose. As DOCK selects molecules using a simple score that shows conformity to the receptor cavity, the large number of molecules retrieved must be culled to a rather small number (e.g., 10 ~ 30) of candidates by manual inspection. The possibilities of H-bond formation, electrostatic interaction, and clash with the target receptor atoms must be examined for each molecule using computer graphic displays. In order to increase the efficiency of the selection process, attempts to improve the program have been made by calculating intermolecular interaction energy in addition to the DOCK score (Meng *et al.*, 1992). Another improvement has been made to deal with conformational flexibility in DOCK (Leach and Kuntz, 1992). However, it seems to be difficult to improve the method sufficiently for it to be able to handle the conformational freedoms of a vast number of molecules in a practical CPU time.

### Conformational Problem in Three-Dimensional Database Search

Most 3-D searches have been performed only on the conformations as registered in the databases. It is obvious that 3-D searching will become more useful for identifying new active molecules, if it takes into account other possible conformations. In a rigid search, only a part of the possible molecules are picked up, missing molecules that may interact with the receptor in conformations other than that in the database. In order to overcome this problem, an extremely large memory space or extremely long CPU time (or a very high-performance computer) is needed. At present, some 3-D searching systems take into account the conformational freedom to some extent by preparing several to several dozens of low-energy conformations, which are surveyed in advance, for a compound in their database. Although it is desirable to consider many low-energy conformations, it would require a vast CPU time and memory space. Even for intermediately flexible compounds with up to six rotatable bonds, the number of local minimum low-energy structures that should be considered becomes extremely large. The development of techniques to facilitate conformationally flexible searching is probably the most important problem in the field of 3-D database searching. The whole area of searching while taking account of conformational flexibility is still in its infancy.

The important point in conformationally flexible searching by means of receptor fitting is to select molecules while automatically searching the most stable docking structure for each molecule among the enormous number of compounds in the databases in a practical CPU time and with practical computer resources.

### Extension of ADAM to 3-D Database Search

We have succeeded in developing an excellent program, ADAM, for automatic docking which can handle conformational flexibilities of ligands, as shown in the previous section. The method, as it is, would greatly assist the design of new active structures rationally through docking study, but we have extended it to make it applicable for database searching by means of receptor fitting. The new database search system is named ADAM&EVE (M. Y. Mizutani and A. Itai, unpublished). For applying ADAM to a large number of compounds in databases one after another, we must consider how to minimize the time required for each compound and how to automate the setup for each compound. For the former purpose, the algorithm of the ADAM program was improved to reduce the time requirement, and various criteria and parameters were tuned up to speed up the docking. When a great many compounds must be searched, the required time is more important than accuracy. The required time was decreased to less than 30 sec for the TMP molecule (with 5 rotatable bonds) on an IRIS 4D workstation (CPU 150 MHz R4400), although the required time is greatly dependent on the number of rotatable bonds in addition to various conditions such as the numbers of dummy atoms and heteroatoms in compounds. As regarding the latter point, we decided to prepare a secondary database (ADAM-style database) to be searched by ADAM, and have developed another program (EVE-make) for this. The new program automatically adds some additional data required for ADAM search to each entry of the conventional 3-D databases which contain atomic coordinates and connectivity information between them. EVE-make produces atomic coordinates of hydrogen atoms and assigns a force-field atom type and atomic charge for each atom (used to calculate and minimize inter- and intramolecular energies), the H-bond group number to each heteroatom (used for estimating possible H-bond schemes), and the starting, end, and increment torsion angles for each rotatable bond to change conformations. Although any method can be accommodated in EVE-make to calculate atomic charges, the Gasteiger method is used to reduce CPU time at the cost of accuracy at present.

As a test to verify the usefulness of our new system ADAM&EVE, we have converted the ACD-3D (MDL Information Systems, Inc.), which contains about 110,000 compounds, into an ADAM-style database. In the application to *E. coli* dihydrofolate reductase, structures of many known inhibitors and their analogs were identified as possible ligands by ADAM search. Furthermore, the activity of a tricyclic compound, which was first predicted by the program, was confirmed experimentally. In the application to HIV protease, many promising molecules were identified as hits. Among 12 such compounds that we have obtained and tested so far, 5 compounds with various skeletal structures inhibited the enzyme more potently than haloperidol. The mean required time for one molecule with less than six rotatable bonds was about 4 sec on the IRIS workstation, although it varies depending on various conditions, such as the number of dummy atoms and heteroatoms and the size of the ligand binding site, in addition to the criteria for a hit. The program system is fast enough to be practically used for searching a large number of compounds. It will become an efficient tool for lead discovery in cases where the receptor structure is available.

## V. AUTOMATED STRUCTURE CONSTRUCTION METHODS

The database approaches seem to be very useful for finding new leads, making the best use of the vast number of accumulated compounds in hand. If compounds in hand or commercially available are picked up, their activities can be assayed without the need for chemical synthesis. Even if the compounds are not available, significant information on a synthetic route or on the physical properties of the compounds might be available from the literature. However, from the viewpoint of suggesting as wide a range of possible structures as possible, structure construction methods seem to be superior to database search methods. Structure construction methods can generate all possible structures which satisfy the requirements, if the software is sufficiently good. Further, structure construction methods are better able to output molecules with a highly favorable interaction with the receptor than database search methods. That is, a constructed structure can be regarded as a custom-made dress, whereas a searched structure resembles a ready-made dress, although both can be modified to fit better afterwards. *De novo* ligand design by structure construction methods has attracted much attention from medicinal chemists, as well as database search methods, and several methods for structure construction have been reported for cases where receptor structure is either known or unknown.

## A. Structure Construction Methods with Known Three-Dimensional Structure of the Receptor

Lewis and Dean (1989) have reported methods using spacer skeletons of appropriate size to span the binding sites in the enzyme active site. In the first method, they use spacer skeletons, which are assemblies of molecular subgraphs, to obtain molecular graphs and incorporate predicted ligand points at their vertices. Since then, Lewis (1990) has reported another method using a diamond lattice to determine favorable ways of spanning between distant regions of an active site. Later, Lewis and co-workers (1992) developed a method that treats all the atoms from the highest scoring molecules selected by DOCK as an irregular lattice which can be used to connect distant atoms and/or fragments. Chau and Dean (1992) have introduced the idea of combining small (three or four atoms) fragment structures with the chemical graph obtained by Lewis's method. Finally, Lewis (1992) developed a method to connect two isolated fragments with linear chains of atoms which can be composed of various atom types. The chains are generated by solving a series of trigonometric equations in torsion space. Lewis and co-workers have also examined various problems such as conformations and charge distributions in structure generation.

Moon and Howe (1991) have developed GROW, which uses a buildup procedure to determine the best peptidyl inhibitor or substrate for a given enzyme. A large predefined library of conformations of each amino acid is used in the construction process. Each conformation of each residue is tested according to a molecular mechanics force field. The program generates only peptide structures and does not generate general drug structures.

Nishibata and Itai (1991, 1993) have reported the program LEGEND, which constructs stable molecules that can fit well to an active site. The program grows molecules by adding atoms one by one, using random numbers and a force field. Random numbers are used to determine a root atom, an atom type, a bond type, and a torsion angle for the new atom. In addition to atom types corresponding to a single atom, the program uses atom types for fragment groups, such as aromatic ring, amide, and carbonyl, to grow molecules. Intra- and intermolecular stabilities are examined for each new atom using the force field, and if the energies for the atom are less stable than given criteria, the program regenerates that atom or reassigns a new atom. Finally, molecules are selected on the basis of intra- and intermolecular energies and some structural features. Details are described later.

The LUDI program, reported by Böhm (1992), is fragment based. The program works in three steps. First, it calculates interaction sites into which both hydrogen bonding and hydrophobic fragments can be favorably placed. The interaction sites are derived from distributions of nonbonded contacts generated by a search through the Cambridge Structural Database or are generated by the use of rules. The second step is the fitting of molecular fragments onto the interaction sites. Fragments can be placed at up to four different interaction sites. The program holds a library of about 1000 fragments for fitting. The final step is the connection of some or all of the fitted fragments to form a single molecule by using smaller bridging fragments such as —$CH_2$— and —COO—.

Rotstein and Murcko (1993) also reported a fragment-based method, GroupBuild. The program uses fragment groups (building blocks) in the predefined library to build up molecules. A user-selected enzyme seed atom(s) may be used to determine the area(s) in which structure generation begins. Alternatively, the program may begin with a predocked "inhibitor core" from which fragments are grown. For each new fragment generated by the program, several thousand candidates in a variety of locations and orientations are considered. Each of these candidates is scored based on a standard force field. The selected fragment and orientation are chosen from among the highest scoring cases.

Gillet and co-workers (1993) developed SPROUT, which uses artificial intelligence techniques to moderate the combinatorial explosion that is inherent in structure generation. Structure generation by this program is divided into two phases: primary structure generation to produce molecular graphs to fit the steric constraint, and secondary structure generation, which is the process of introducing appropriate functionality to the graphs to produce molecules that satisfy the secondary constraints, e.g., electrostatics and hydrophobicity.

Pearlman and Murcko (1993) developed CONCEPTS. The method places a group of atom-like particles in the site. The particles are free to move within the site to improve binding to the protein. Covalent connections among the particles are made in a stochastic and dynamically reversible manner. The changes in the topology are either accepted or rejected depending on their ability to improve the total energy of the enzyme–inhibitor complex.

Bohacek (1993) developed GROWMOL, which is very similar to LEGEND. It grows molecules by adding atoms one by one, using random numbers to determine root atom, atom type, bond type, and atom position (torsion angle) for each new atom. The major difference from LEGEND is the way that torsion angles are determined. In their method, the torsion angle is chosen by random number from several predefined low-energy ones, whereas in LEGEND, the torsion angle itself is determined by random number generally, except for cases where there is a limited number of possible torsion angles, such as 0° or 180°.

Karplus and co-workers have developed the MCSS method to find favorable sites for small functional groups in the active site of proteins (Miranker and Karplus 1991; Caflisch *et al.*, 1993). The energetically favorable sites and orientations of some functional groups in an active site are searched very rapidly by simultaneous simulation of multiple ligands based on a time-dependent Hartree approximation. For example, from an energy minimization of multiple small molecules (~1000) such as acetamide or acetic acid or methanol, a great many favorable positions of carbonyl or imino or hydroxyl groups are obtained at a time. The HOOK program, developed by Hubbard and co-workers, generates new molecules, making use of those stable positions of various functional groups. The program searches appropriate template groups from the template library in the program, which can covalently bond any number of functional groups positioned in the active site. Then it generates new molecules, assembling all the groups into a molecule.

### 1. LEGEND Program

Here, we describe the program LEGEND as a typical example of an automated structure construction method based on the 3-D structure of the receptor.

***Outline of the program*** LEGEND (Nishibata and Itai, 1991) generates many possible structures which can fit well to the ligand-binding site of the target receptor for *de novo* ligand design. It grows intra- and intermolecularly stable molecules by adding atoms one by one using random numbers and a force field. Random numbers are used to determine all unsettled quantities as well as to explore spatially in the binding region and structurally in the ligand structures. The force field is used to generate significantly stable structures, both for generation of each atom and for evaluation of the generated structure. The program makes use of tabulated potentials on a 3-D grid to calculate intermolecular interaction energy very rapidly and to judge heteroatom positions favorable for hydrophilic interaction with the receptor. The program uses special atom types to introduce moieties such as benzene, amide, and carbonyl groups, as fragment groups. If such an atom type is once assigned as a new atom by a random number through the usual atom generation process, the group is implicitly introduced into the growing structure with an orientation determined by the next atom position assigned by random number. If the aromatic ring is not completed within the maximum number of atoms to be generated for a molecule, the remaining ones are supplied in the last step of structure construction. An outline of the structure construction procedures in LEGEND is shown in Fig. 8.

***Basic algorithm of LEGEND*** The basic algorithm of LEGEND is as follows: As regarding the first atom in structure construction, it is the option of the user either to start from a starting atom which is automatically generated by the program or to start from an input structure (seed). In the former case, the program generates an H-bonding heteroatom so as to form an H-bond to one of the several anchor groups (H-bonding groups in the

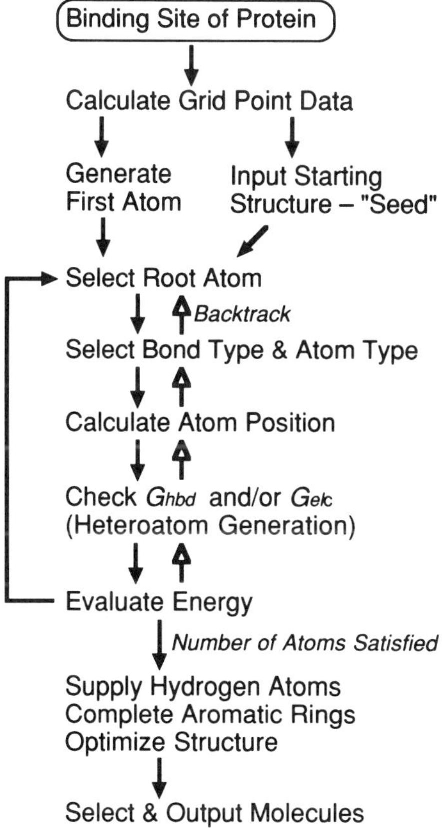


**FIGURE 8** Flow chart of the structure construction method LEGEND.

protein), which are assigned in advance. In the latter case, any fragment structure such as a partial structure of a known active compound can be input to be used as a starting structure. After the generation of the first atom or the placement of the seed structure, atoms are placed one by one, determining a root atom (to which the new atom is bonded) and atom type, bond type, and atomic position (torsion angle) of the new atom by means of random numbers, using standard bond lengths and angles for corresponding atom pairs. In addition to atom types consisting of a single atom of carbon, nitrogen, oxygen, and others, atom types consisting of fragment structures, such as aromatic ring, amide, and carbonyl groups, are used in the program. For every new atom, intra- and intermolecular van der Waals energies as well as conformational energies for torsion angle are evaluated using a force field. If the new atom is not energetically acceptable due to close contact with receptor atoms or previously generated atoms or an unacceptable torsion angle after a specified number of trials of regeneration, the program attempts to reassign the root atom. If the attempts fail after a given number of repeats, the program stops trying to add a new atom to the current generated structure and backtracks to the last step, i.e., it withdraws the last of the previously generated atoms and regenerates that atom. The tabulated van der Waals potentials on the 3D-grid are used to estimate intermolecular van der Waals energy rapidly. Thus, the program grows a stable molecule in various directions by means of random numbers, exploring within the space allowed for ligands. The latest version of LEGEND makes it possible to grow molecules to a specified direction by indicating several H-bond functional groups in the protein in the order of growth. This function also assists in generating ligand molecules with more H-bonds by forcing the molecule to grow near the wall of the receptor cavity.

When the molecule reaches the size specified by the user in advance, the program stops generating new atoms, completes fragmentary aromatic rings by adding missing carbon atoms, and supplies hydrogen atoms for all remaining valencies of nonhydrogen atoms. Therefore, the number of atoms in each completed molecule varies somewhat. Finally, the structure is optimized by the Simplex method, taking into account the intra- and intermolecular energies based on atomic charges calculated by the Del Re method. The quality of generated structures varies greatly depending on the energetic criteria, as well as criteria for each new atom. It is natural that the more severe the criteria, the greater the percentage of good structures and the greater the CPU time required. From the numerous structures generated from LEGEND, a small number of structures are selected on the basis of energetic or other structural criteria. The intermolecular interaction of each generated structure with the receptor cavity is well visualized by the grid representation of GREEN.

The program generates heteroatoms adequately so that they form H-bonds to the receptor. Heteroatoms are generated as follows: the program generates only carbon atoms ($sp^3$, $sp^2$, aromatic, carbonyl) by random number and they are converted into appropriate heteroatoms according to the electrostatic potential or the H-bond expectability of the position, which are tabulated in advance on the 3-D grid (Nishibata and Itai, 1993; Iwama *et al.*, in preparation). With this algorithm, the program generates ligands with heteroatoms that can form as many H-bonds as possible with the target receptor.

***Examples of LEGEND application*** We have applied LEGEND to several enzyme systems such as DHFR, trypsin, and HIV-1 protease. We show here the result with DHFR to exemplify the usefulness of the program. The atomic coordinates of *E. coli* DHFR with $NADP^+$ were taken from the crystal structure of the ternary complex (enzyme–coenzyme $NADP^+$–inhibitor folate) (Bystroff *et al.*, 1990) deposited in the Protein Data Bank. Hydrogen atoms were located at the geometrically appropriate positions, and atomic charge was assigned to each protein atom. In order to design inhibitor molecules that occupy the substrate-binding site competing with the substrate, the folate molecule bound in the crystal was removed. Two water molecules, which are highly stabilized at the bottom of the cavity by more than two hydrogen bonds to the enzyme, were left as a part of the protein structure, as well as the $NADP^+$ molecule. All other water molecules in the PDB data were removed.

The region allowed for newly generated molecules was indicated so as to fully include the substrate-binding site, and a grid with a 0.4-Å interval was generated inside the region. The van der Waals potentials ($G_{vdw}$) were calculated for carbon, nitrogen, oxygen, and hydrogen probe atoms. The electrostatic potential ($G_{elc}$) was calculated from the atomic charges of the protein atoms assuming the distance-dependent dielectric constant, $\varepsilon = 1r$. The side chain carboxylic acid of Asp-27 was assumed to be ionized to carboxylate anion. The environment of the ligand-binding site of *E. coli* DHFR is shown in Fig. 9.

Using this binding site environment, we first generated 200 structures, choosing 15 atoms as the number of atoms in a molecule and using a guanidino group as the starting seed structure. The guanidino group was located at the position of the group in the folate molecule in the crystal structure of the ternary complex. A rather small molecular size was selected in order to check whether the key structures of well-known inhibitors could be generated by the program. Among the 200 structures, many overlapped structures were found, which only differed in alkyl substituent groups. After combining such structures, a total of 20 independent skeletal structures were obtained. All of them fit well to the receptor cavity and form at least two H-bonds with the receptor in addition to those of the input seed structure.

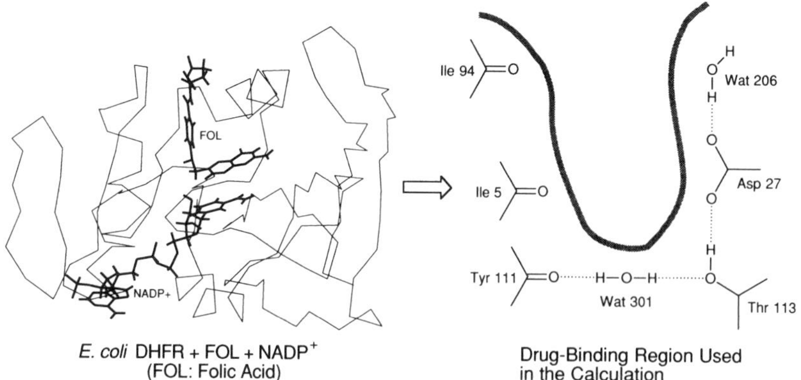


**FIGURE 9** Crystal structure (left) and binding-site environment (right) of *E. coli* dihydrofolate reductase (DHFR).

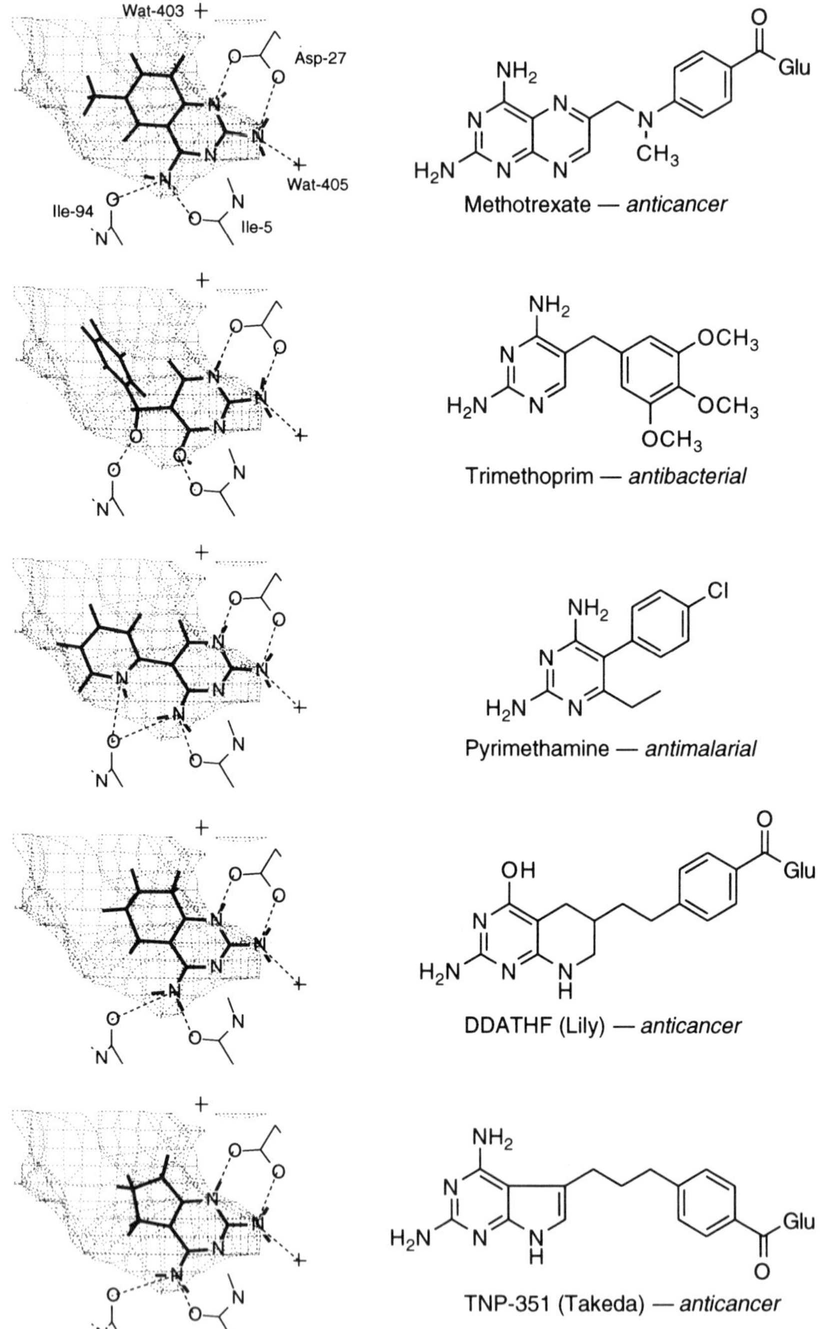


**FIGURE 10** Output structures from LEGEND program (left) corresponding to the core structures of well-known inhibitors of DHFR (right). The cage shows the allowed space for nonhydrogen atoms in ligands using the ligand-binding site representation of the GREEN program.

The core structures of five known inhibitors were reproduced in the generated structures, as shown in Fig. 10.

Output structures from an attempt to generate larger molecules by LEGEND are shown in Fig. 11. The number of atoms in a molecule to be generated by random number was set at 30 atoms in the input data. It is also a good approach to construct large ligands in a stepwise way using promising small structures output from LEGEND again as seed structures instead of constructing large structures from the beginning.

LEGEND generates a wide variety of structures using random numbers and provides stable structures using a force field and backtracking. In addition, the program maximizes H-bonds to reflect the receptor environment by taking a positive step to introduce heteroatoms at appropriate positions.

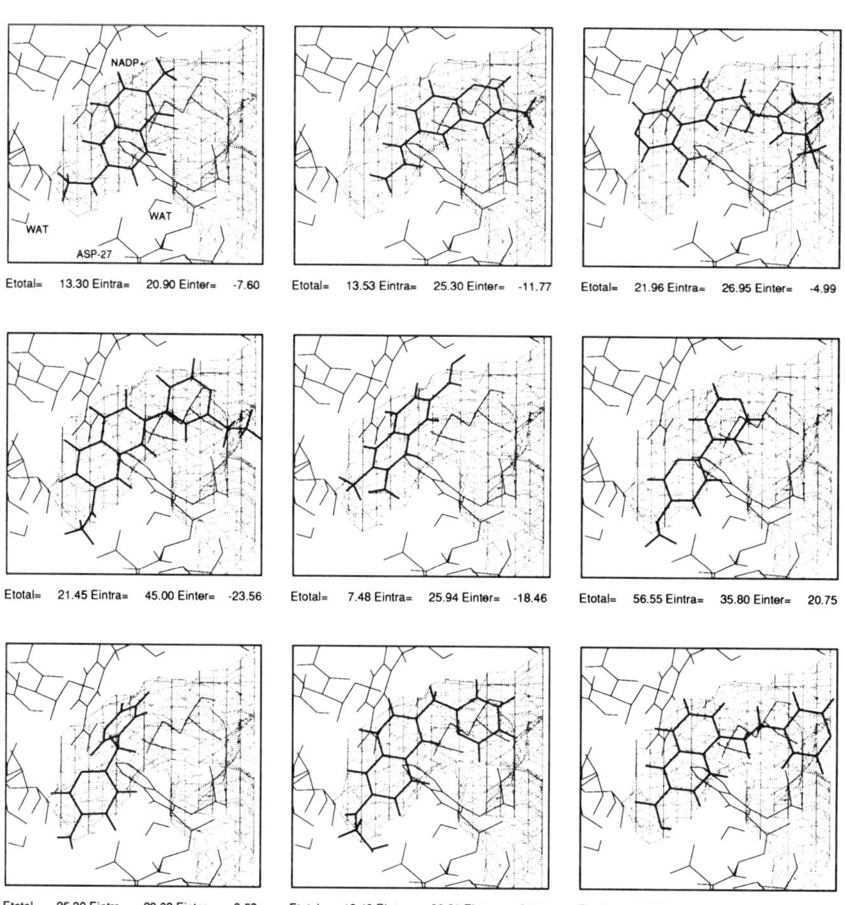


**FIGURE 11** Some of the output structures from LEGEND when the number of atoms in a molecule is chosen as 30.

At the same time, the program takes into account the hydrophobic environment of the binding site by generating only carbon atoms by random numbers and by converting some of them into heteroatoms only when a favorable hydrophilic interaction with the receptor is possible. The validity of our method was confirmed by the fact that the key structures of five known inhibitors were included in the output structures in the application to dihydrofolate reductase. The validity was further confirmed by synthesis of a new compound which was derived from one of the generated structures. Accordingly, the rest of the structures are also expected to be active if they are modified appropriately and synthesized.

One of the characteristics of our method is its flexibility in use. Structures can be constructed in various ways as required by the user. The range of allowed target site, molecular size, ratio of incorporation among various atom types, input of seed structure and its positioning, specification of growing direction, energetic criteria, number of molecules to be generated, and others greatly affect the variety of generated structures. As the program generates the structures considering only the receptor binding, it will be necessary to further modify the structures, taking into account physical properties such as solubility and membrane permeability, in addition to improving the receptor binding, in order to obtain actual lead compounds.

## 2. Advantages and Disadvantages of Various Methods

Structure construction methods reported so far can be classified into two categories. One is growing a molecule by adding atoms one by one, and the other is building a molecule using fragment groups prepared in the program. Most of the structure construction methods belong to the latter approach. The basic concept of fragment-based methods is placing fragment groups at the appropriate positions of the binding site one after another and linking them together. The major differences among the existing fragment-based methods are how to select suitable fragment groups, how to locate them, how to connect them, how to divide the binding site to accept each fragment, and so on. The differences greatly affect the variety of structures generated from each method. Although LEGEND belongs to the atom-by-atom approach fundamentally, it also uses atom types for special fragment groups, such as phenyl and carbonyl. The differences are that only a small number of atom types are assigned to fragment groups, and positions and orientations of the groups are determined by random number in LEGEND.

Structure construction and 3-D database search by receptor fitting are continuous in a sense because conformationally rigid groups are conventionally used as fragment groups in the fragment-based structure construction methods. So, the algorithms of fragment fitting to the receptor cavity in that case are the same as with those of conformationally rigid receptor fitting in a docking study or 3-D database search. The entire structure of a molecule is fitted to the entire space of the receptor cavity in the latter, whereas each fragment structure is fitted to a part of the allowed space or to a part of the possible positions of functional groups in the former.

One of the problems of fragment-based methods is that each fragment group need not be placed at the most stable position, rather, it is the stability of the whole molecule that is most important for possible ligands. If all the fragments that compose a molecule are placed at the optimum position independently, generation of a variety of structures that are sufficiently stable as whole molecules might be prevented.

In the atom-by-atom method, it is not necessary to select favorable fragment groups or to place them in the most stable position and orientation. As regarding aromatic groups, the use of fragment groups often found in drug structures may be more favorable than atom-by-atom methods. In the fragment-based methods, it seems rather easy to cover possible aromatic groups, including heteroatoms, and the number of such groups would not be enormous. But, as regards aliphatic groups, it seems that the atom-by-atom method is more favorable than the fragment-based method because conformational variations in aliphatic groups are much larger than those in aromatic groups. In addition to typical different conformations, there are many conformations with subtle differences around the typical standard conformations in aliphatic groups. For example, most cyclic alkane or alkene or alkanone rings can exist in several conformations, and the preferred conformation varies greatly depending on the fused ring and substituent groups. Even halfway conformations between typical stable conformations are found. Since the orientations of substituent groups and their relative positions are significantly varied by such frame structures, fragment-based methods that use only typical conformations or a small number of stable ones stored in the program to construct molecules might miss many promising structures. The use of many slightly different conformations for a fragment structure would greatly increase the computation time required, although the variety of output structures would greatly increase. Coverage of possible structures may not be compatible with speed of structure construction in any method.

One of the advantages of the simple fragment method is that each generated structure should be chemically acceptable because fragment structures familiar to medicinal chemists are used in the structure construction program. There are many cases where synthetic approaches to such fragment structures are established or such groups are synthetically favorable. It is natural that fragment structures which have geometrically valid bond lengths and angles and no excessive distortion are used for structure construction. Nevertheless, conformational stability is not always assured, even in fragment-based methods, because of new rotatable bonds between fragment groups and linking groups. In the case of the atom-by-atom method, the generated structures are not always chemically plausible or acceptable, unless many rules are used in the structure generation step. LEGEND uses only a simple rule that prohibits covalent bonding between two atoms with special atoms types.

A filtering step to eliminate geometrically implausible output structures is necessary in the atom-by-atom method, and this step drastically reduces the speed of structure generation. For example, LEGEND can generate structures in less than a half minute per molecule without filtering and structural optimization on the IRIS 4D workstation, but it requires several hours to generate 100 structures with rather severe energetic criteria. The required time per molecule is completely dependent on the number of atoms in a molecule, energetic criteria for each new atom, and energetic criteria for each generated molecule in LEGEND. Instead, structures with desired stability in terms of total energy or intra- or intermolecular energies can be obtained.

### B. Structure Construction in the Case of Unknown Receptor Structure

Most drug developments at present are performed without knowledge of receptor structure. There are many cases where new active structures or functions are needed for developing new drugs with different behaviors, and there are also many cases where working hypotheses must be confirmed by syntheses of different compounds bearing presumed pharmacophores. So, it is useful to suggest different structures based on the structures of known active compounds, either by 3-D database searching or by structure construction.

As regarding algorithms for structure construction based on pharmacophores or functional groups, the fundamental concepts are the same as in the receptor-based case. In some receptor-based methods, structures are constructed by locating several functional groups appropriately in the active site first and then connecting them using a template structure. A suitable template structure is searched from an internal template library or from an external database or built at the site. Functional groups in ligands favorable for binding to the receptor can be reliably selected and positioned in receptor-based construction, whereas they are selected and positioned based on speculation and assumption from an existing active molecule (or molecules) in pharmacophore-based construction. Therefore, most of the algorithms developed for receptor-based cases might be used for other cases, although some modification of the programs might be needed. The problem in constructing new structures when receptor structure is not available is that the validity of the generated structures must be evaluated in terms of some quantity other than intermolecular interaction energy. Although pharmacophore-based methods may generate only a part of the possible structures in receptor-based cases, new structures with functional groups or pharmacophores common to known active structures are easily accepted by synthetic chemists.

Bartlett and co-workers developed CAVEAT with the aim of designing rigid analogs for a flexible lead compound (Lauri and Bartlett, 1994). CAVEAT searches a database of cyclic compounds which can act as templates to connect any number of fragments already positioned. The fragments can be positioned in the active site based on receptor structure or as structural parts of an active molecule in a specified conformation. It uses vectors to identify suitable templates and generates new cyclic structures possessing the fragments as substituent groups. Other 3-D databases of compounds other than cyclic ones can also be searched by the program, although use of conventional 3-D databases including conformation-flexible molecules faces the problem of how to cover the degrees of freedom for possible conformations.

Tschinke and Cohen (1993) developed NEWLEAD, which generates candidate structures from pharmacophore hypotheses. In their method, pharmacophoric pieces are connected with spacers assembled from small chemical entities (atoms, chains, or ring moieties). Starting from selected functional groups obtained from the bioactive conformations of reference molecules, the program generates new structures that are chemically unrelated to the reference molecule, in addition to the expected structures related to the reference one.

### 1. LINKOR Program

We have developed the LINKOR program for constructing new structures based on given functional groups and a receptor cavity model (Inoue, Kanazawa, Takeda, Tomioka, and Itai, unpublished). The flow diagram of the LINKOR program is shown in Fig. 12. The functional groups and their relative positions are estimated or assumed from the structure–activity relationships of known active compounds. The program connects the functional groups by appropriate skeletal structures made of several atoms, inside a receptor cavity model which is represented on a 3-D grid, in place of the actual receptor structure. Atoms of the connecting structures are generated one by one in such a way that each new atom has valid bond lengths and angles with the root atom and related atoms. Allowed topologies of connecting structures are provided in the program as various lengths of atom sequences composed of various atom types. Possible atom positions for each connecting group are searched systematically, by checking whether each atom can exist inside the receptor cavity model or not. Ring structures can be generated when connecting groups include aromatic atom types or can be generated by connecting closely contacting nonbonded atoms covalently. In the ring closure, one or two more atoms are supplied if necessary. The created structures are optimized by molecular mechanics

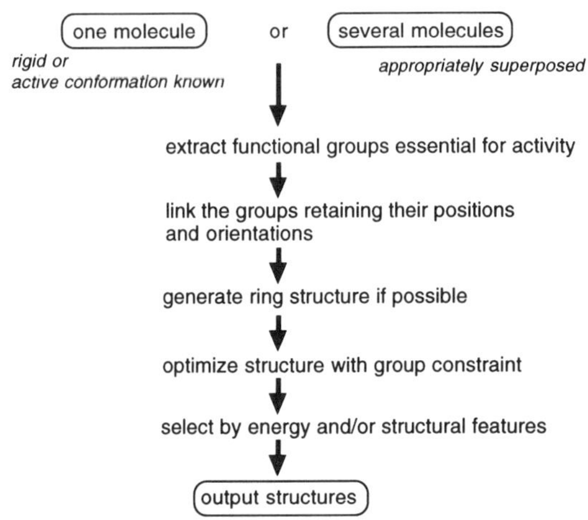


**FIGURE 12** Flow diagram of the LINKOR program.

calculation. The promising ones are selected based on criteria of intramolecular stability and the satisfaction of 3-D geometrical requirements.

Figure 13 shows an example of structure construction by LINKOR based on the functional groups in morphine molecule. In this case, a receptor cavity model was constructed from the crystal structure of morphine molecule alone, which is conformationally rigid. About 100 structures were constructed by connecting the phenyl ring (ring A) and cationic nitrogen atom in the morphine molecule. Among these structures, several molecules having the same skeletal structures as known analgesic compounds were found.

### 2. 3-D Receptor Cavity Model

The characteristic of structure construction by LINKOR is that the program is based on a 3-D receptor cavity model that can be an alternative to the 3-D structure of actual receptor. The receptor cavity model used for LINKOR is constructed based on a single active molecule or appropriately superposed plural active molecules using the program RECEPS (Kato *et al.*, 1987; Itai *et al.*, 1988). The model provides 3-D information on the physical and chemical properties of the receptor cavity, as well as the size and shape of the cavity, H-bond expectability, and electrostatic potential. The information can be presumed from a single active molecule if it is conformationally rigid or the active conformation can be appropriately assumed, together with the functional groups essential for the activity. However, in the case where there are two or more active molecules that bind to a common receptor, they probably reflect the environment of the receptor cavity more faithfully than any single molecule. Moreover, active conformations and

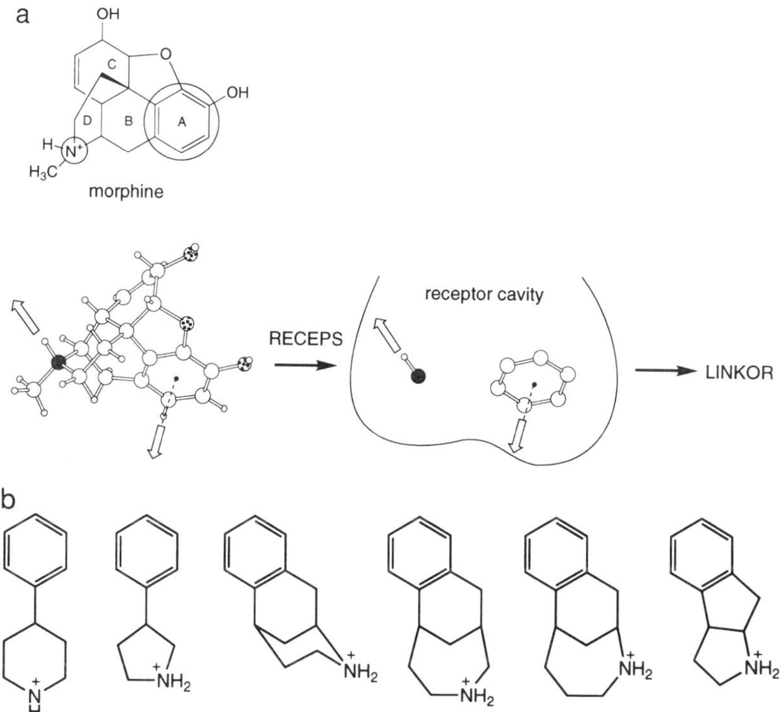


**FIGURE 13** (a) Scheme of structure construction of morphine-related molecules using the LINKOR program. (b) Structures constructed by LINKOR having the same skeletal structures as known analgesic compounds.

functional groups essential for the activity might be rationally estimated by molecular superposition if those molecules have different structural features.

Use of the receptor cavity model has the following advantages in constructing new active structures. First, the model can be used to restrict the size and shape occupied by the new structures. The activities of the new structures seem to be assured if their sizes and shapes are within those of the template structures. There are cases where different functional groups are supposed to play equivalent roles in receptor binding. Such equivalent functional groups can be used for structure construction in addition to those in the template active molecule. Without such a model, structure construction based on pharmacophore groups may afford a large number of new structures, many of which may not exhibit the activity due to inappropriate size or shape. Further, starting from the properties of a receptor cavity model, small functional groups necessary for specific binding can be placed interactively or automatically, without using the information from template molecules. However, the use of such models restricts the variety of the structures constructed by the program to within the shape and size of the template structures used, which are only a part of the possible active structures.

The RECEPS program was developed for the purpose of superposing molecules with dissimilar structures, which can bind to the same target receptor competitively, and constructing a 3-D receptor cavity model. RECEPS superposes molecules not in terms of atomic positions or frameworks as in the conventional methods, but in terms of physical and chemical properties, assuming a common receptor site. Molecules are superposed considering similarities in the properties involved in H-bonds, electrostatic potentials, molecular shapes, and so on. As a result of superposition, functional groups essential for the activity and the active conformation of each molecule can be estimated. As such information is essential for constructing the receptor cavity model, suitable assumptions must be made if they cannot be defined by superposition. The shape and size of allowed space, positions and characters of expected H-bond groups, electrostatic potentials, and so on are described on a 3-D grid. Although RECEPS was originally a program to superpose molecules interactively on graphic displays, an automatic superposing program, AUTOFIT, was developed later to superpose molecules computationally. The characteristic of the method is to superpose molecules in terms of matching H-bonding partner sites between two molecules. All superposing modes, which are produced from all correspondences of H-bond groups, are evaluated automatically, covering all possible conformations for

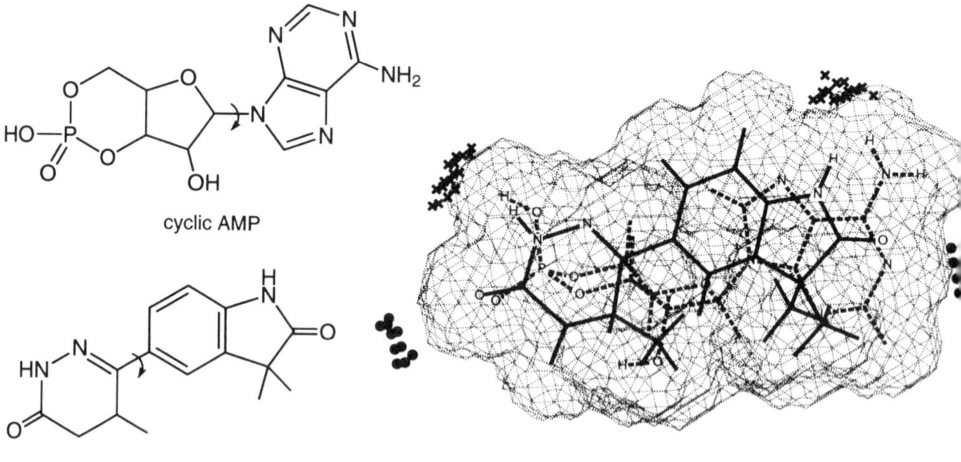


**FIGURE 14** Superposition of cyclic AMP (substrate) and an inhibitor of phosphodiesterase III (LY195115) by the RECEPS program. The cage (right) represents common volume occupied by both molecules, and markers represent H-bonding sites expected in the enzyme.

both molecules. The method ranks them by maximizing the matching of the H-bonding partner sites by a special least-squares calculation (Kato *et al.*, 1992). The AUTOFIT program is used to make a preliminary search for promising superposing models in RECEPS, and the preliminary results are refined and ranked considering indexes for various properties using 3-D grid point data. Figure 14 shows an example of molecular superposition and a receptor cavity model obtained with RECEPS, using the substrate cyclic AMP and an inhibitor of phosphodiesterase III.

## VI. SCOPE AND LIMITATIONS

### A. Points for Consideration in Structure Construction Methods

#### 1. Intramolecular Validity and Stability

The validity of the structure construction method should be evaluated from two viewpoints: the validity of each output structure itself and favorable interaction with the receptor protein. The definition of a valid structure is that it must be chemically plausible, geometrically proper, and conformationally not too unstable in the case of a flexible molecule. The chemical plausibility is realized by the introduction of experience-based rules and the proper geometries such as bond lengths and angles are realized by structure optimization by molecular mechanics in the atom-by-atom method. These are easily realized in the fragment-based method. The conformational stability is evaluated as the intramolecular part of the total energy of the protein–ligand complex. The problem is that the actual stability of compounds with different molecular size or quite dissimilar structures cannot be compared simply in terms of the energy values from force fields or molecular orbitals. Calculated heat of formation values is different between such molecules, even though they may stably exist after having been synthesized. How to handle the intramolecular energy still remains unsolved in the problem of ranking various molecules generated or identified by *de novo* ligand discovery.

#### 2. Stability of Active Conformation

As regarding the conformation, generated structures can be regarded as being in the active conformation, in the structure construction methods. Although the generated conformation need not be the same as that of the energetically global minimum or even that of any one of the local minima of a single molecule, it is not desirable that the active conformation should be highly unstable compared to the global minimum conformation. If the ligand structure is conformationally too unfavorable, the docking structure might become too unstable in total energy. It is therefore necessary to examine the relative stability of the generated conformation compared with all other possible conformations of the molecule. The situation is the same for compounds identified by searching databases. Conformations found by conformation flexible search should be reexamined to check whether they are not too unstable compared with the global minimum state. Thus, database search methods and structure construction methods have a common problem in that the energy difference from the global minimum structure, or the easiness of adopting the active conformation, should be considered. It is natural that the smaller the energy difference, the more favorable the receptor binding. Although the acceptable energy difference cannot be determined theoretically, we have tentatively adopted a maximum of 7 kcal/mol as the criterion.

### 3. Intermolecular Stability

Favorable intermolecular interaction with the target receptor is another requirement for the generated structure. As well as intramolecular stability, intermolecular interaction energy with the receptor is greatly dependent on the molecular size or the number of atoms in the molecule. The larger the molecule, the more stable the intermolecular energy, when there is no close contact between two molecules. Therefore, intermolecular interaction stabilities with a receptor cannot be simply compared between molecules with different molecular sizes, in a different sense from intramolecular stabilities. From a structural viewpoint, a significant complementarity in molecular shape and properties is required between the binding site and generated molecules. To form as many H-bonds and strong electrostatic interactions to the receptor as possible, complementarity of properties involved in such interactions is also required. LEGEND achieves the shape complementarity by monitoring the intermolecular potential energy at each atom generation step. Furthermore, the program takes a positive step to generate heteroatoms that form hydrophilic (H-bonding and/or electrostatic) interactions with the receptor. Only carbon atoms are generated by random number, then those that can form hydrophilic interactions with the receptor are converted into heteroatoms according to the receptor environment. As a result, the program achieves a complementarity in hydrophobic property by supplying only carbon atoms at the hydrophobic sites of the receptor cavity. LUDI achieves the complementarity in hydrophilic and hydrophobic properties by selecting and positioning fragments so as to satisfy such properties prepared as interaction sites in the binding site.

### 4. Coverage of Possible Structures

Validity of each generated structure is the minimum requirement for structure construction methods. Furthermore, possible structures should be covered as thoroughly as possible among the generated structures. Coverage of possible structures is an important advantage of the structure construction approach over the database-searching approach. There seem to be large differences in the variety of generated structures and coverage of possible structures among the structure construction methods reported so far. The efficiency as well as usefulness of structure construction methods depend on the variety and coverage, although it is greatly affected by many other factors, such as parameters, energetic criteria, and repertoire of fragments or template groups, in addition to the algorithm itself.

We have shown that random generation by LEGEND can efficiently sample various structures with various conformations and positions, although they may not cover all possible structures. It is still an open question how many molecules should be generated by LEGEND in order to cover possible structures adequately. Actually, it is difficult to cover all possible ones by random generation. But, it is also difficult using fragment-based methods since all possible fragment groups and all possible connections between them must be tried with all possible positions and orientations. If a limited variety of fragment groups are considered for constructing new molecules, not all possible structures would be covered. Furthermore, if the fit of fragment groups is tested in a partitioned space of the allowed region, all the possible ways to partition the space must be examined combinatorially with the possible fragment groups. So, it might be necessary to use several methods with different algorithms in order to generate as wide a range of structures as possible.

## 5. Selection of Structures

The ultimate criterion for selection from the theoretical viewpoint is binding affinity to the target receptor. Theoretically, the binding constant $K$ is predicted by the free energy change $\Delta G_{\text{bind}}$ via the equation $\Delta G_{\text{bind}} = -RT \ln K$. But, $\Delta G_{\text{bind}}$ cannot be calculated simply by conventional energy calculation from the structure of the receptor–ligand complex. In order to predict the binding constant correctly, the free energy difference between the associated state and the unassociated state must be estimated, taking into account the energies of conformation change from the unbound form to the bound form and of solvation and desolvation of the receptor and ligand, as shown in Fig. 15.

$\Delta G_{\text{bind}}$ is divided into parts, $\Delta G_{\text{conf}}$, $\Delta G_{\text{desolv}}$, and $\Delta G_{\text{assoc}}$, which are the free energy changes for conformation change (most stable in itself $\rightarrow$ active conformation), for desolvation, and for association. Each is divided into an enthalpic term and an entropic term:

$$\begin{aligned}\Delta G_{\text{bind}} &= \Delta G_{\text{conf}} + \Delta G_{\text{desolv}} + \Delta G_{\text{assoc}} \\ &= (\Delta H_{\text{conf}} - T\Delta S_{\text{conf}}) + (\Delta H_{\text{desolv}} - T\Delta S_{\text{desolv}}) \\ &\quad + (\Delta H_{\text{assoc}} - T\Delta S_{\text{assoc}}).\end{aligned}$$

Free energy cannot be correctly estimated, even if the enthalpic contribution can be calculated with satisfactory accuracy, because the entropic contribution to the free energy cannot be neglected in many drug–receptor interactions.

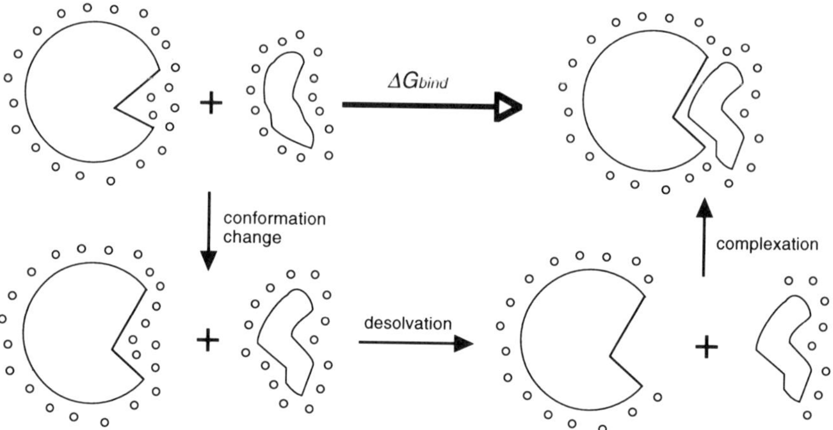


**FIGURE 15** Thermodynamic cycle of free energy of ligand binding.

Although the total energy is calculated for all structures in most structure construction methods, it is merely used to prioritize them. One order of binding constant corresponds to only 1.37 kcal/mol in free energy difference at room temperature. Even in very similar molecules in which all other factors can be neglected, the accuracy of force-field calculation is insufficient for predicting the relative binding constant quantitatively. Moreover, conventional energy calculation, whether with molecular mechanics or molecular orbitals, cannot be used to compare the stabilities of structures of different molecular size and composed of different atoms. The free energy perturbation method based on molecular dynamics calculation can calculate free energy difference between two molecules with very similar structures. But it cannot be used to obtain the absolute value of free energy or to compare the free energies of molecules with quite dissimilar structures.

Böhm (1994) reported a simple empirical function that takes into account different H-bond strengths, lipophilic contacts, and rotatable bonds in the ligand. It can be used for the evaluation of structures generated by *de novo* design methods. A fit of the adjustable parameters in the new scoring function to experimentally determined binding constants $K_i$, using a preliminary set of 45 protein–ligand complexes with known 3-D structures, yielded a scoring function with the following contributions to the free energy change of binding $\Delta G_{\text{bind}}$: +5.4 kJ/mol for overall loss of translational and rotational entropy; −4.7 kJ/mol for an ideal neutral H-bond; −8.3 kJ/mol for an ideal ionic interaction; −0.17 kJ/Å$^2$ for a lipophilic contact; and +1.4 kJ/mol for a rotatable bond.

Selection of promising structures from generated ones is most important for *de novo* lead generation by a structure construction program. However, selection should not be based on the generated structures as they are, but on the modified structures taking into account various factors in addition to receptor binding, such as solubility, chemical reactivity, physical stability, and synthetic ease. Candidates for chemical syntheses should be selected from the modified structures from both theoretical and synthetic viewpoints because it requires a lot of time and labor to synthesize any new compound.

## B. Handling of X-Ray Structures of Proteins

An X-ray structure of a protein is a good starting point for designing new structures. Although such structures are the most reliable structural information that can be used for drug design, there are many problems in using them. The following problems are common to all usages of the structures. First, in many proteins, a small number of water molecules bind to the protein especially strongly or are supposed to be specially stabilized by more than two H-bonds to the protein in the crystal structures. They are sometimes regarded as part of the protein structure. Whether such water molecules should be left in the active site or not has to be decided when the protein structure is used for *de novo* ligand design as well as docking study. In the lead discovery of HIV protease inhibitors using a 3-D database, a DuPont-Merck group found a very promising molecule by regarding such water as part of the ligand molecule rather than as part of the protein (Lam *et al.*, 1994). As the assumption greatly affects the results, structure construction must be performed with both assumptions in some cases, i.e., with and without the water molecule(s). In cases where the protein includes some cofactor molecules or some metal ions, it is also necessary to judge whether they should be left in the protein structure, considering their roles or functions.

Side chain groups of amino acids such as Asp, Glu, Lys, and Arg in the protein active site often play important roles in the interaction with ligands as charged groups. It is supposed that the contribution of such charged groups to the stabilization of the receptor–ligand complex is especially large because electrostatic interactions of H-bonds involving charged groups are much larger than those between neutral heteroatoms. But, confirmation of this is not easy due to the lack of information on hydrogen atoms from protein crystallography. So, it should be appropriately assumed whether each of such side chain groups exposed in the active site exists in ionized form or not. This assumption also affects the structure construction and filtering processes. Structure–activity studies on ligands or on point-mutated proteins might help us to make appropriate assumptions.

It would be energetically favorable for the generated molecule to have appropriate functional groups with counter charges at appropriate positions in the molecule, in cases where such charged groups are supposed to exist in the active site. LEGEND has a function optionally to generate appropriate charged groups at the appropriate positions automatically by detecting the existence of charged groups in the protein; for example, a carboxylate anion group for Lys and Arg side chains and a protonated amino or guanidino or amidino group for Glu and Asp side chains.

The most serious problem is the flexibility of protein structures. A number of pairs of protein structures with and without a bound ligand have been determined. They gave some indication of the conformation changes of proteins which can occur during the binding of a ligand. Conformation changes on ligand binding are negligible in some cases, but not negligible in most cases. Naturally, when the *de novo* ligand design is attempted, it is better to use the protein structure solved with typical ligands, if available. In several allosteric enzymes, structural differences have also been elucidated between structures with and without effectors. In the case of targeting such enzymes, structural changes on effector binding should be considered very carefully. In addition, there are some examples where structures of several complexes with different ligands have been elucidated for a protein. Apparently, the size and shape of the binding site of the protein varies according to each ligand structure to some extent. In our experience, it is sometimes difficult to dock even a natural substrate molecule stably in the site occupied by an inhibitor in the crystal structure. This means that it may be difficult to generate other types of ligand structure based on protein structure complexed with a particular ligand. How to deal with induced fit of protein structures is one of the most difficult problems still to overcome, and it is not merely the problem of *de novo* ligand design. In our strategies for *de novo* ligand design as well as for docking simulation, we take into account the flexibility of protein structure in the following way. First, the program generates possible ligand structures assuming the rigid receptor structure, but allowing a rather larger space for ligands than simple consideration of van der Waals radii for protein atoms would indicate. Then, the generated structures are energy minimized together with the receptor structure considering protein flexibility, and they are selected by comparing their total energies. It is necessary to establish a rational method to overcome this problem computationally.

## C. Future Perspectives

Lead discovery without relying on chance has long been the final goal of computer-assisted drug design. We now have various methods, newly developed or being developed, which may allow us to realize this goal. The increasing volume of information on receptor structures will accelerate the innovation and change the nature of drug development. Basic research to solve various difficult problems that still remain, such as handling protein structures or predicting binding constants for putative ligands, will become more and more important.

## REFERENCES


Allen, F. H., Bellard, S., Brice, M. D., Cartwright, B. A., Doubleday, A., Higgs, H., Hummelink, T., Hummelink-Peters, B. G., Kennard, O., Motherwell, W. D. S., Rodgers, J. R., and Watson, D. G. (1979). The Cambridge Crystallographic Data Centre: Computer-based search, retrieval, analysis and display of information. *Acta Cryst.* **B35**, 2331–2339.

Bacon, D. J., and Moult, J. (1992). Docking by least-squares fitting of molecular surface patterns. *J. Mol. Biol.* **225**, 849–858.

Bernstein, F. C., Koetzle, T. F., Williams, G. J. B., Meyer, E. F., Jr., Brice, M. D., Rodgers, J. R., Kennard, O., Shimanouchi, T., and Tasumi, M. (1977). The Protein Data Bank: A computer-based archival file for macromolecule structures. *J. Mol. Biol.* **112**, 535–542.

Bohacek, R. S. (1992). GROW: A new method for generating large numbers of novel guest molecules that are sterically and chemically compatible to a host molecule binding site: Application to an enzyme/inhibitor system. *In* "Abstracts of XIIth International Symposium on Medicinal Chemistry, Basel," p. 90.

Böhm, H.-J. (1992). The computer program LUDI: A new method for the *de novo* design of enzyme inhibitors. *J. Comput.-Aided Mol. Design* **6**, 61–78.

Böhm, H.-J. (1994). The development of a simple empirical scoring function to estimate the binding constant for a protein-ligand complex of known three-dimensional structure. *J. Comput.-Aided Mol. Design* **8**, 243–256.

Bolin, J. T., Filman, D. J., Matthews, D. A., Hamlin, R. C., and Kraut, J. (1982). Crystal structure of *Escherichia coli* and *Lactobacillus casei* dihydrofolate reductase refined at 1.7 Å resolution. I. General features and binding of methotrexate. *J. Biol. Chem.* **257**, 13650–13662.

Borman, S. (1992). New 3-D search and *de novo* design techniques aid drug development. *C&EN August 10*, 18–31.

Burkert, U., and Alliner, N. L. (1982). "Molecular Mechanics," ACS Monograph 177. American Chemical Society.

Bystroff, C., Oatley, S. J., and Kraut, J. (1990). Crystal structures of *Escherichia coli* dihydrofolate reductase: The $NADP^+$ holoenzyme and the folate-$NADP^+$ ternary complex. Substrate binding and a model for the transition state. *Biochemistry* **29**, 3263–3277.

Caflisch, A., Miranker, A., and Karplus, M. (1993). Multiple copy simultaneous search and construction of ligands in binding sites: Application to inhibitors of HIV-1 aspartic proteinase. *J. Med. Chem.* **36**, 2142–2167.

Chau, P.-L., and Dean, P. M. (1992). Automated site-directed drug design: The generation of a basic set of fragments to be used for automated structure assembly. *J. Comput.-Aided Mol. Design* **6**, 385–396.

DesJarlais, R. L., Seibel, G. L., Kuntz, I. D., Furth, P. S., Alvarez, J. C., Ortiz de Montellano, P. R., DeCamp, D. L., Babe, L. M., and Craik, C. S. (1990). Structure-based design of nonpeptide inhibitors specific for the human immunodeficiency virus 1 protease. *Proc. Natl. Acad. Sci. U.S.A.* **87**, 6644–6648.

Ghose, A. K., and Crippen, G. M. (1985). Geometrically feasible binding modes of a flexible ligand molecule at the receptor site. *J. Comp. Chem.* **6**, 350–359.

Gillet, V., Johnson, A. P., Mata, P., Sike, S., and Williams, P. (1993). SPROUT: A program for structure generation. *J. Comput.-Aided Mol. Design* **7**, 127–153.

Goodsell, D. S., and Olson, A. J. (1990). Automated docking of substrates to proteins by simulated annealing. *PROTEINS: Struct. Funct. Genet.* **8**, 195–202.

Ho, C. M. W., and Marshall, G. R. (1993). FOUNDATION: A program to retrieve all possible structures containing a user-defined minimum number of matching query elements from three-dimensional databases. *J. Comput.-Aided Mol. Design* **7**, 3–22.

Itai, A., Kato, Y., Tomioka, N., Iitaka, Y., Endo, Y., Hasegawa, M., Shudo, K., Fujiki, H., and Sakai, S. (1988). A receptor model for tumor promoters: Rational superposition of teleocidins and phorbol esters. *Proc. Natl. Acad. Sci. U.S.A.* **85**, 3688–3692.

Jiang, F., and Kim, S.-H. (1991). "Soft docking": Matching of molecular surface cubes. *J. Mol. Biol.* **219**, 79–102.

Kasinos, N., Lilley, G. A., Subbarao, N., and Haneef, I. (1992). A robust and efficient automated docking algorithm for molecular recognition. *Protein Eng.* **5**, 69–75.

Kato, Y., Itai, A., and Iitaka, Y. (1987). A novel method for superposing molecules and receptor mapping. *Tetrahedron* **43**, 5229–5236.

Kato, Y., Inoue, A., Yamada, M., Tomioka, N., and Itai, A. (1992). Automatic superposition of drug molecules based on their common receptor site. *J. Comput.-Aided Mol. Design* **6**, 475–486.

Kuntz, I. D. (1992). Structure-based strategies for drug design and discovery. *Science* **257**, 1078–1082.

Kuntz, I. D., Blaney, J. M., Oatley, S. J., Langridge, R., and Ferrin, T. E. (1982). A geometric approach to macromolecule-ligand interactions. *J. Mol. Biol.* **161**, 269–288.

Lam, P. Y. S., Ladhav, P. K., Eyermann, C. J., Hodge, C. N., Ru, Y., Bacheler, L. T., Meek, T. L., Otto, M. J., Rayner, M. M., Wong, Y. N., Chang, C.-H., Weber, P. C., Jackson, D. A., Sharpe, T. R., and Erickson-Viitanen, S. (1994). Rational design of potent, bioavailable, nonpeptide cyclic ureas as HIV protease inhibitors. *Science* **263**, 380–384.

Lauri, G., and Bartlett, P. A. (1994). CAVEAT: A program to facilitate the design of organic molecules. *J. Comput.-Aided Mol. Design* **8**, 51–66.

Leach, A. R., and Kuntz, I. D. (1992). Conformational analysis of flexible ligands in macromolecular receptor sites. *J. Comp. Chem.* **13**, 730–748.

Lewis, R. A. (1990). Automated site-directed drug design: Approaches to the formation of 3D molecular graphs. *J. Comput.-Aided Mol. Design* **4**, 205–210.

Lewis, R. A. (1992). Automated site-directed drug design: A method for the generation of general three-dimensional molecular graphs. *J. Mol. Graph.* **10**, 131–143.

Lewis, R. A., and Dean, P. M. (1989). Automated site-directed drug design: The concept of spacer skeletons for primary structure generation. *Proc. Roy. Soc.* **B236**, 125–140.

Lewis, R. A., Roe, D. C., Huang, C., Ferrin, T. E., Langridge, R., and Kuntz, I. D. (1992). Automated site-directed drug design using molecular lattices. *J. Mol. Graph.* **10**, 66–78.

Martin, Y. C. (1992). 3D database searching in drug design. *J. Med. Chem.* **35**, 2145–2154.

Meng, E. C., Shoichet, B. K., and Kuntz, I. D. (1992). Automated docking with grid-based energy evaluation. *J. Comp. Chem.* **13**, 505–524.

Miranker, A., and Karplus, M. (1991). Functionality maps of binding sites: A multiple copy simultaneous search method. *PROTEINS: Struct. Funct. Genet.* **11**, 29–34.

Mizutani Yamada, M., Tomioka, N., and Itai, A. (1994). Rational automatic search method for stable docking models of protein and ligand. *J. Mol. Biol.* **243**, 310–326.

Moon, J. B., and Howe, W. J. (1991). Computer design of bioactive molecules: A method for receptor-based *de novo* ligand design. *PROTEINS: Struct. Funct. Genet.* **11**, 314–328.

Nishibata, Y., and Itai, A. (1991). Automatic creation of drug candidate structures based on receptor structure: Starting point for artificial lead generation. *Tetrahedron* **47**, 8985–8990.

Nishibata, Y., and Itai, A. (1993). Confirmation of usefulness of a structure construction program based on three-dimensional receptor structure for rational lead generation. *J. Med. Chem.* **36**, 2921–2928.

Pearlman, D. A., and Murcko, M. A. (1993). CONCEPTS: New dynamic algorithm for *de novo* drug suggestion. *J. Comp. Chem.* **14**, 1184–1193.

Rotstein, S. H., and Murcko, M. A. (1993). GroupBuild: A fragment-based method for *de novo* drug design. *J. Med. Chem.* **36**, 1700–1710.

Shoichet, B. K., Stroud, R. M., Santi, D. V., Kuntz, I. D., and Perry, K. M. (1993). Structure-based discovery of inhibitors of thymidylate synthase. *Science* **259**, 1445–1450.

Singh, U. C., Weiner, P. K., Caldwell, J., and Kollman, P. (1989). AMBER 3.0 Revision A, University of California, San Francisco.

Tomioka, N., and Itai, A. (1984). GREEN: A program for rational docking study. *In* "Abstracts of Papers, 6th Symposium on Medicinal Chemistry," p. 105. Pharmaceutical Society of Japan, Tokyo.

Tomioka, N., and Itai, A. (1994). GREEN: A program package for docking studies in rational drug design. *J. Comput.-Aided Mol. Design* **8**, 347–366.

Tomioka, N., Itai, A., and Iitaka, Y. (1987a). Real-time estimation and visualization of protein-ligand interaction on 3D graphics display. *In* "Proceedings of the Symposium on Three-Dimensional Structures and Drug Design" (Y. Iitaka and A. Itai, eds.), pp. 186–194. University of Tokyo Press, Tokyo.

Tomioka, N., Itai, A., and Iitaka, Y. (1987b). A method for fast energy estimation and visualization of protein-ligand interaction. *J. Comput.-Aided Mol. Design* **1**, 197–210.

Tschinke, V., and Cohen, N. C. (1993). The NEWLEAD program: A new method for the design of candidate structures from pharmacophoric hypotheses. *J. Med. Chem.* **36**, 3863–3870.

Weiner, S. J., Kollman, P. A., Case, D. A., Singh, U. C., Ghio, C., Alagona, G., Profeta, S., and Weiner, P. (1984). A new force field for molecular mechanical simulation of nucleic acids and proteins. *J. Am. Chem. Soc.* **106**, 765–784.

Weiner, S. J., Kollman, P. A., Nguyen, D. T., and Case, D. A. (1986). An all atom force field for simulations of proteins and nucleic acids. *J. Comp. Chem.* **7**, 230–252.

Yamada, M., and Itai, A. (1993a). Development of and efficient automated docking method. *Chem. Pharm. Bull.* **41**, 1200–1202.

Yamada, M., and Itai, A. (1993b). Application and evaluation of the automated docking method. *Chem. Pharm. Bull.* **41**, 1203–1205.

Yamada, M., Tomioka, N., and Itai, A. (1993). A novel automatic docking method for flexible molecules. *In* "Trends in QSAR and Molecular Modelling 92" (C. G. Wermuth, ed.), pp. 404–406. ESCOM, Leiden.

# 5

# Experimental Techniques and Data Banks

**JOHN P. PRIESTLE**
Pharmaceuticals Division
Ciba-Geigy Limited
CH-4002 Basel, Switzerland

**C. GREGORY PARIS**
Research Division
Ciba-Geigy Pharmaceuticals
Summit, New Jersey

## I. EXPERIMENTAL TECHNIQUES

### A. X-Ray Crystallography

The bulk of our knowledge about three-dimensional (3-D) structures of biological macromolecules comes from X-ray crystallographic investigations. Macromolecular X-ray crystallography is a later extension of classical X-ray crystallography. This means that the theory and practice of X-ray diffraction had already been well worked out, although macromolecular samples have special problems seldom encountered in so-called "small molecule" crystallography, chief among them being the labile nature of the samples and the large number and relatively poor quality of the diffraction data. While it is far beyond the scope of this book to develop in detail the theory of X-ray diffraction or its rigorous mathematical derivation, a brief outline is necessary for the nonspecialist who uses the coordinates determined from a crystallographic investigation to be able to judge the quality of the results and to understand the limitations of the method. For a more detailed description of X-ray crystallography in general, the book "Crystal Structure Analysis, a Primer" (Glusker and Trueblood, 1985) is a good introduction, while

*Guidebook on Molecular Modeling in Drug Design*

"Fundamentals of Crystallography" (Giacovazzo *et al.*, 1992) is more detailed. The standard text for macromolecular crystallography is still "Protein Crystallography" (Blundell and Johnson, 1976), although it is somewhat out of date. "Crystallography Made Crystal Clear: A Guide for Users of Macromolecular Models" by Gale Rhodes (1993) is a newer book aimed less at crystallographers than at investigators who use crystallographically derived structures.

The heart of any X-ray crystallographic experiment is always a crystal. A crystal is a 3-D ordered array of atoms, ions, molecules, or groups of these. The basic unit of a crystal is a parallelepiped called the unit cell, which, when repeated in three dimensions, builds up the crystal. The three nonparallel edges of the unit cell parallelepiped are called the unit cell axes (a, b, and c). The intersection of these three axes defines the origin of the unit cell. The angles between these axes are the unit cell angles (called $\alpha$, $\beta$, and $\gamma$). The convention is that $\alpha$ is the angle between the b and c axes, $\beta$ is between a and c, and $\gamma$ is between a and b. The unit cell itself may be subdivided by symmetry. The presence of crystallographic symmetry means that only a part of the unit cell (called the asymmetric unit) must be investigated. Once the location of all atoms in the asymmetric unit has been determined, application of symmetry generates the full unit cell.

### 1. Symmetry

Symmetry elements are composed of rotation axes, mirror planes, inversion centers, or combinations of these with a translation. In addition to symmetry elements, there are different crystal lattices (primitive, face centered, and body centered). There are just 240 ways to combine the various possible symmetry operators and lattice types in three dimensions, giving rise to the 240 possible 3-D space groups. In a crystal the molecules themselves restrict the possible symmetry elements. If the molecules have chiral centers (as do proteins and nucleic acids), then symmetry operators that invert the hand of the molecule, such as mirror planes and inversion centers, cannot exist. This restricts the number of possible space groups for chiral molecules to 65.

The user of a coordinate set should be able to check if there are potential packing effects. In large proteins this is usually manifested as an external side chain interacting with a symmetry-related neighbor. Examples also exist where entire flexible loops appear more ordered due to interactions with neighboring molecules. In small, flexible molecules, this "crystal artifact" can be more serious since it may be that only one of many possible conformers will pack together properly ("freezes out") in the crystal.

### 2. Diffraction

The most useful property of crystals, from the structural scientist's point of view, is their ability to diffract X-rays. In the crystal, electrons and protons, being charged, are caused to oscillate by the fluctuating magnetic field of the impinging X-rays. The oscillating electrons and protons themselves become sources of secondary scattered X-rays. Because the protons are far more massive than electrons, their oscillation amplitude, and hence their ability to reemit X-rays, is much reduced and is insignificant compared to scattering by electrons. Because the atoms, and hence their electrons, are ordered in a crystal, at certain angles in 3-D space, their scattered radiation will constructively interfere, giving rise to diffraction reflections, while elsewhere they will destructively interfere, canceling each other out.

Each diffraction point (also called a "reflection" or, colloquially, a "spot") has three characteristics: its 3-D indices $(h, k, l)$, its intensity $(\mathrm{I}_{hkl})$, and its phase angle $(\alpha_{hkl})$. The intensity depends on how the scattering from different atoms in the same unit cell interfere with each other (this should not be confused with the constructive/destructive interference of identical atoms in different unit cells), and the phase angle relates the various diffraction points to each other and to the unit cell origin. The indices are easily determined once the unit cell parameters and crystal orientation have been ascertained. The intensity is easily measured by some kind of X-ray detector, be it a piece of film or an electronic detector. Unfortunately, all information about the phase angle is lost and its recovery is the central problem in X-ray crystallography.

Two essential relationships in crystallography are those that equate diffraction and electron density. The diffraction pattern is the Fourier transform of the electron density and vice versa. The electron density $(\rho)$ for each point in $x$, $y$, $z$ space can be expressed as the following 3-D Fourier transform (ignoring certain scaling and constant terms):

$$\rho(x, y, z) = \sum_{h=-\infty}^{+\infty} \sum_{k=-\infty}^{+\infty} \sum_{l=-\infty}^{+\infty} F_{hkl} \cos(2\pi(hx + ky + lz) - \alpha_{hkl}), \qquad (1)$$

where $F_{hkl}$ is the structure factor amplitude $(\sqrt{\mathrm{I}_{hkl}})$.

If the structure (electron density) is known, it is possible to calculate the diffraction pattern (structure amplitudes and phases):

$$a_{hkl} = \sum_{x=0}^{1} \sum_{y=0}^{1} \sum_{z=0}^{1} \rho(x, y, z) \cos(2\pi(hx + ky + lz)) \qquad (2)$$

$$b_{hkl} = \sum_{x=0}^{1} \sum_{y=0}^{1} \sum_{z=0}^{1} \rho(x, y, z) \sin(2\pi(hx + ky + lz)) \qquad (3)$$

$$F_{hkl} = \sqrt{a_{hkl}^2 + b_{hkl}^2}; \qquad \alpha_{hkl} = \tan^{-1}(b_{hkl}/a_{hkl}), \qquad (4) \text{ and } (5)$$

where $x$, $y$, $z$ are in fractional coordinates. Actually, it is computationally more efficient to loop over the atoms in the unit cell.

An important consequence of these equations is that every reflection contributes to each point in the electron density and every point in the electron density function (or every atom) contributes to each diffraction point. There is no one-to-one correspondence between a particular atom and a particular diffraction point. The contribution of an atom to the scattering of X-rays is directly proportional to the number of electrons it has, and conversely the higher its electron density will be in the final electron density map. Hydrogen atoms, with their single electron, are too weak to be seen, and their positions can only be inferred from stereochemical considerations.

### 3. Resolution

To investigate the structure of something, the probe that is used must be at least as fine as the detail one would like to discern. Since we are interested in atomic detail, we need to use a probe that is as small as the distance between atoms, 1.0–2.5 Å. This corresponds to the X-ray region of the electromagnetic spectrum. Resolution, as the word implies, is the ability to resolve two points separated by some distance from each other. The higher the diffraction angle from a crystal is, the higher the resolution will be, the more data that can be collected, and the "sharper" our view of the molecule will be. Confusion sometimes arises becauses resolution is a reciprocal term: high resolution is expressed as a small number (e.g., 2.0 Å), whereas low resolution is expressed as a larger number (e.g., 3.5 Å). In practice, resolutions of 3.0–2.0 Å are usually achieved for biological macromolecules, whereas resolutions below 1 Å are routinely achieved in classical "small molecule" crystallography. This means that macromolecular crystallography seldom achieves true atomic resolution because the individual atoms themselves are not resolved. However, below around 2.3 Å resolution, it does become possible to resolve atoms separated by a third atom, e.g., the two $\gamma$-carbon atoms of a valine side chain (Fig. 1).

### 4. Crystallization

As mentioned earlier, the heart of every X-ray crystallographic investigation is a crystal. Producing a crystal of a biological macromolecule is usually the most critical hurdle in the investigation. Despite many years of experience, there are still few hard and fast rules. One is that the sample should be as pure as possible. The general approach to crystallization of proteins is to start with a solution of protein and to slowly change the conditions so that saturation is exceeded. This can be brought about by changing any of many parameters, the most important ones being ionic strength, pH, temperature of the solution, and the addition of such things as ions, organic molecules, cofactors, and inhibitors to the solution. The trick is to approach supersaturation slowly so that only a few nucleation sites are created which, with time, will grow larger. Too rapid changes often result in a shower of crystals, all of which are too small to be of any use. Figure 2 illustrates one of the more popular ways of bringing about slow saturation. Typically, a protein (or nucleic acid) crystal needs to be at least 0.2–0.3 mm in all three

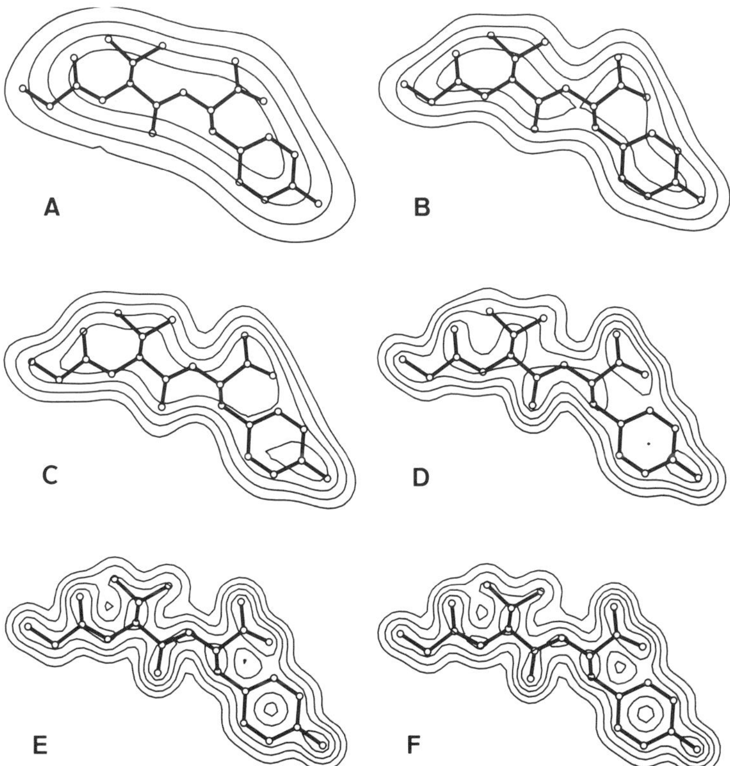


**FIGURE 1** Effect of resolution on the quality of an electron density map. Electron density for the tripeptide Gly–Val–Tyr contoured at 20, 40, 60, and 80% of $\rho_{max}$ at resolutions of (A) 5.0 Å, (B) 3.0 Å, (C) 2.5 Å, (D) 2.0 Å, (E) 1.5 Å, and (F) 1.0 Å. Note that beginning with the 2.0-Å map one begins to observe "bumps" for most of the individual atoms and the beginning of a hole in the tyrosine ring.

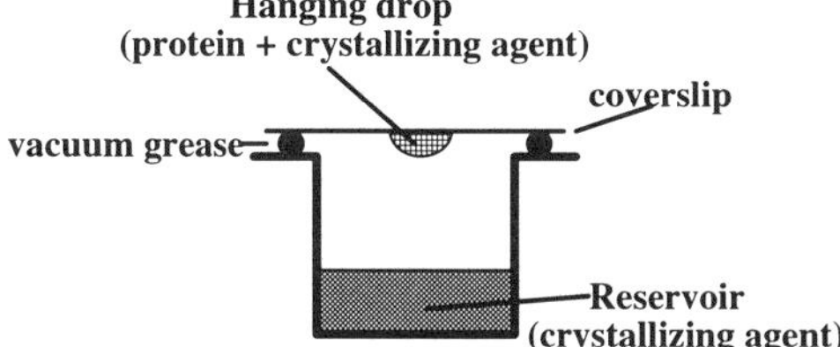


**FIGURE 2** "Hanging drop" crystallization experiment. A drop of protein solution below the saturation point is suspended above a reservoir above the saturation point. Through vapor diffusion, the two solutions slowly reach equilibrium, driving up the concentration in the drop beyond the protein saturation point.

dimensions to diffract strongly enough to achieve the resolution needed for a successful 3-D structure clarification. Alexander McPherson's book "Preparation and Analysis of Protein Crystals" (1988) and "Crystallization of Nucleic Acids and Proteins" by Ducruix and Giegé (1992) are probably two of the best publications dealing with protein and nucleic acid crystallization (respectively) written to date.

## 5. Data Collection

Crystals of biological macromolecules are quite different from small molecule crystals in that they contain a large amount of solvent. Typically around 50% of the volume of a crystal is solvent. This results in very fragile crystals which need to be kept in a moist environment to prevent their drying out. The large solvent content has the advantage that the molecules are still in an environment that is not too far removed from a physiological one. While this is good for the biochemist, for the crystallographer it means that the crystals are far less well ordered than those of small molecules, with the subsequent loss of diffraction and resolution.

The crystal is mounted in a thin-walled quartz capillary and set on the X-ray camera. Because of their large crystallographic unit cells, biological macromolecules give rise to a very large number of diffraction reflections, usually between a few thousand for small proteins and hundreds of thousands for large protein complexes and viruses. In addition, biological macromolecules deteriorate under intense X-rays, hence it is imperative to collect data as quickly, yet as accurately, as possible. Diffractometers, which measure one reflection at a time, are the standard devices for classical small molecule X-ray crystallography, but are impractical for macromolecular crystallography. They have the great advantage, however, that they are almost completely automated. Film detectors allow a large number of data to be collected simultaneously, but films are extremely tedious to digitize and process. Electronic area detectors combine the automation and ease of use of the diffractometer with the large sampling area of film detectors and are the ideal device for collecting diffraction data from large molecules. Recent developments include the use of charge coupled devices (CCDs) and film-like image plates in which absorbed X-rays cause excitation of rare earth atoms in the plate to meta-stable states. These can then be excited with a laser of appropriate wavelength and their decay to the ground state is accompanied by a loss of a photon of known wavelength, which can then be measured. These devices, coupled to automatic scanners, allow a large number of quite accurate data to be measured because of their large areas and high signal-to-noise characteristics. Their one disadvantage is their relatively slow readback times.

The collected data must then be corrected for known systematic errors such as background radiation, decay of the crystal over time, absorption of the diffracted beam as it passes through the crystal, and polarization of the

X-ray beam. Reflections that should have identical intensities because of symmetry are averaged together to get the best possible estimation of their true value. The agreement of these measurements, called $R_{merge}$, gives an indication of the accuracy, or at least the reproducibility, of the data. For macromolecular crystals, $R_{merge}$ is usually around 5–10%. Due to counting statistics, strong reflections are much more accurately measured than weak ones.

### 6. Structure Solution

In order to calculate an electron density map, both the structure factor amplitude ($F_{hkl} = \sqrt{I_{hkl}}$) and phase angle ($\alpha_{hkl}$) of each reflection is needed. While the intensities can be measured directly, all information about the phase angle is lost. There are three ways to derive these phases, depending on how much is known about the structure under investigation. In the easiest case, the structure is already approximately known in the same crystal form (isomorphous crystals). This might be the case of an enzyme/small molecule inhibitor complex vs the enzyme alone, or a mutant protein vs the wild type. In such a case, since most of the atoms are already in the correct position (or very nearly so), one assumes that the phases of the two structures are also very similar.

If the protein is related to another protein whose structure is known, but crystallizes in a different crystal form, then it is possible to use the known structure to solve the unknown one. This is called the molecular replacement method and has become a very powerful tool for solving protein structures as more and more structures from more and more different protein families become available. The proteins have to be quite similar in their three-dimensional structures for the method to be effective, implying a sequence identity of approximately 40% or higher. Since the two protein structures are similar, their Fourier transforms will also be similar. The diffraction pattern of the unknown can be measured and that from the known structure can be calculated. The calculated transform can then be rotated and compared to the measured transform (diffraction) of the unknown structure until a maximum overlap is found, implying that this is the orientation of the protein in the crystal. After determining the proper orientation (rotation) of the protein, one must then ascertain the proper location in the unit cell (translation). There are various ways of doing this, but they all boil down to moving the properly oriented molecule around in the unit cell until good agreement with the observed diffraction is found. Phases are then calculated from this starting model and corrections for the proper sequence are then made.

The most difficult case is when the protein is unrelated to any other protein for which a three-dimensional structure exists. To solve such a *de novo* structure, electron-dense ("heavy") atoms (such as mercury, platinum, and lead) are soaked into the protein crystals. In the ideal case, these atoms will bind with a high occupancy to a few identical sites on each protein molecule in the crystal without disrupting the protein structure itself. This is called the isomorphous replacement method. Because these heavy atoms are very electron dense relative to the protein atoms, even just a few of them will cause measurable changes in the diffraction intensities. By analyzing these changes, it is possible to locate where these few heavy atoms are in the crystal. From the calculated scattering of the heavy atoms, plus the measured intensities of the native protein and of the heavy atom derivatized protein, plus some fancy mathematical footwork, it is possible to determine the phase angles for each reflection for the native protein. Actually, because the angle is derived from an arc cosine, two possible phase angles per reflection are determined, one right and one wrong. To distinguish which one is the right one, a second, different heavy atom derivative must be made and solved, again yielding two possible phase angles per reflection, one of which will match one of the first two angles. This then is the correct angle. Naturally, real life is much more complicated. Each reflection is affected differently by the introduction of the heavy atom. Some will have large changes, making determination of its phase angle easy, whereas others will show small or even no change, imparting little or no information about its phase angle. Measurement errors also make things difficult since the amplitude of the changes for a good derivative is around 15% of the diffraction intensity, whereas the accuracy of the measurement itself is around 5%. Usually three or four derivatives (or even more if the derivatives are not so good) must be made before one has enough well-determined phase angles to calculate a first interpretable electron density map. Because of the need for many isomorphous derivatives, the method is called the multiple isomorphous replacement (MIR) method.

## 7. Fourier Synthesis and Map Interpretation

Once phase angles for each reflection have been determined, they are combined with the measured structure factor amplitude as in Eq. (1) to calculate the electron density in three dimensions. The electron density is then examined on a computer graphics system. If the resolution is sufficiently high and the phases sufficiently good, one should begin to recognize structural features such as $\alpha$-helices and $\beta$-strands. Long tubes of density representing the main chain should have blobs of density branching off denoting side chains. Although it is seldom possible to recognize side chains from their density, they can be classified as small, medium, long, or aromatic. One can then try to correlate a stretch of such side chains with the sequence, if it is known. The direction of the main chain may be determined by the location of bumps along the main chain tube caused by the peptide oxygen atoms. It is very seldom that the entire main chain can be traced continuously from a first electron density map because of weak ("smeared out") electron density at exposed, flexible loops. The ease of interpretation, or "fitting the electron density," is strongly dependent on the resolution of the data and especially on the quality of the phases. Slowly, steadily, the protein is built up.

## 8. Structure Refinement

Once the electron density has been fitted with a structure, Eqs. (2)–(5) can be used to calculate what the diffraction of the model should look like. The structure factor amplitudes of this calculated diffraction can then be compared to those actually measured. This is usually expressed as the crystallographic reliability factor or simply R factor ($R_f$) and is defined:

$$R_f = \frac{\Sigma\,|F_o - F_c|}{\Sigma\,F_o}, \tag{6}$$

where $F_o$ are the observed (measured) structure factor amplitudes and $F_c$ are the ones calculated from the model, summed over all measured reflections. A perfect fit of the calculated structure factors to the observed gives an $R_f$ of zero, whereas a completely random structure gives an $R_f$ of 0.59. A fairly good starting model built into a MIR electron density map usually gives an $R_f$ around 0.4–0.5, whereas a good starting structure from molecular replacement is usually between 0.35 and 0.45, depending on how similar the trial (search) structure is to the unknown structure.

The goal of refinement is to improve the agreement of the calculated structure factors with the observed. The method of choice is least squares in which the quantity $\Sigma(F_o - F_c)^2$ is minimized. In linear least squares this is very straightforward, and parameters which give the absolute minimum can be calculated directly. In crystallography the relationship between the parameters ($x$, $y$, $z$) and the quantity minimized (structure factors) is not linear, but, as seen from Eqs. (1)–(5), sine and cosine functions. This requires linearization of the functions via a Taylor series expansion with the result that one cannot directly calculate the best parameters, but only shifts toward the best parameters. This also implies that a fairly good starting model is needed in order to even start refinement.

One of the problems of macromolecular crystallographic refinement is the rather low number of observations (data) per parameter refined. One way around this under determinacy problem comes from the fact that the positions of the atoms are not totally independent. Atoms are bonded together, some groups of atoms are known to be coplanar, and so on. This a priori stereochemical information can be incorporated as observations (data) and added to the quantity to be minimized. One now minimizes $\Sigma(F_o - F_c)^2 + \Sigma(Q_m - Q_i)^2$, where $Q_m$ is some model stereochemical quantity, like a bond distance or bond angle, and $Q_i$ is its ideal value determined from small molecule X-ray crystallography. In this case the stereochemistry is not rigidly constrained, but is restricted to lie near the ideal values, hence this method is called restrained least squares refinement.

Refinement works well if the atoms lie close to their correct positions. For larger errors, the structure must be corrected by hand, usually on a computer graphics system. After all questionable parts of the structure have been thus corrected, least squares refinement is performed on the updated model until local convergence is achieved and the model is again examined for errors. This cyclic operation is repeated until there is no evidence of wrong structure. Molecular dynamics can also be used to sample a much larger section of conformational space, thereby increasing the radius of convergence for refinement. Using classical restrained least squares procedures, 10 cycles of refinement are needed to reach the local minimum, before manual intervention is required. Usually 5 to 10 cycles of automatic refinement/manual correction are required before the global minimum is reached. If molecular dynamics are included, usually one cycle of automated refinement is required to reach the local minimum, although the individual cycles are much longer. In such a case, less manual intervention is necessary, and usually only for very large errors in the original structure. As in most refinements, about 90% of the structure is correctly built with 10% of the total effort, with the remaining 90% of the effort being expended on the last 10% of the structure: poorly defined loops and side chains.

During refinement, water molecules that are hydrogen bonded to the protein in the crystal lattice will also appear in the electron density map. These are also included as part of the model and are refined along with the protein atoms. Since the model tries to represent the electron density of the structure in the crystal, one needs to account for the fact that not all atoms are rigidly locked into place in the crystal lattice. The electron density that is seen on the computer graphics is actually the average structure of all the molecules in the crystal averaged over the data collection time (usually a few days). There may be both static and dynamic disorders which have the effect of smearing out the electron density of an atom. This is taken into account by an isotropic temperature factor (B factor), which theoretically describes thermal vibration within the crystal, but for the rather open lattices of macromolecules, also compensates for disorder. This then gives actually four parameters per atom to be refined: $x$, $y$, $z$ and B. The temperature factors show which parts of the crystal are fairly rigid and which parts are "floppy." There are often even areas where the electron density is so weak, i.e., the structure is so disordered, that nothing can be fitted. This is found most frequently for long, flexible surface side chains and at the termini and exposed loops of the protein and all that can be said about these areas is that there is no fixed structure.

## 9. Analysis of Results

At the end of an X-ray crystallographic investigation, one has a number of parameters that attempt to describe the system under investigation, the most important of which is a list of coordinates that describes the 3-D structure of the molecule. The overall quality of the coordinates is usually measured by three quantities: the resolution, the crystallographic R factor, and some measure of the stereochemical soundness of the structure. Resolution has already been discussed. Suffice it to say that, in general, the higher the resolution, the better (more precise) the coordinates. Roughly speaking, at resolutions lower than about 3.5 Å, it is difficult to even trace the main chain unambiguously. Between 3.5 and 2.5 Å, the folding of the protein is usually pretty certain, secondary structural elements ($\alpha$-helices and $\beta$-strands) are clear, loops less so. Aromatic side chains are well placed, but other side chains are less distinct. Only a few, very tightly bound solvent molecules can be found and individual temperature factors cannot be refined. Between 2.5 and 2.0 Å, the orientations of side chains become less ambiguous, more solvent molecules are found, individual temperature factors can be refined, and, because the peptide oxygen atoms are well resolved, positioning of the peptide plane is less dubious. At resolutions better than 2.0 Å, the coordinates and temperature factors become more precise, and discrete conformations for side chains may become visible.

The crystallographic reliability factor [R factor, $R_f$, Eq. (6)] measures the agreement between observed structure factors and those calculated from the model. A well-refined structure has a R factor of 0.15 to 0.20 (often expressed as a percent, 15–20%). Low resolution structures (lower than roughly 2.5 Å) seldom get below 0.20 because the lack of high resolution data does not allow very accurate positioning of the atoms. The R factor can be "fudged" by discarding weak reflections or by refining more parameters than are justified by the resolution. For example, refinement of individual temperature factors or the addition of excessive solvent molecules when the data do not extend beyond 2.5 Å resolution is probably not justified.

A correct structure will have good stereochemistry as well as a low R factor. The contribution of these two terms can be adjusted by playing with their relative weights. A proper balance should roughly give the following root mean square deviations from ideal for various stereochemical restraints: $<0.015$ Å for bond distances, $<3.0°$ for bond angles, $<0.015$ Å for planar atoms out of the plane, and no incorrect chiral centers. Another very important stereochemical restraint is close (van der Waals) contacts, but unfortunately, the atoms are either given generous (small) radii with very high restraints once they are in contact (giving few, small violations) or very large radii and weaker restraints (giving many, seemingly larger violations), so it is difficult to judge their meaning without more information. If the model has consistently significantly worse stereochemical deviations than those listed earlier, that is an indication that the structure is not as well refined as the R factor might at first suggest.

## B. Nuclear Magnetic Resonance (NMR)

From the late 1950s until the early to mid-1980s, X-ray crystallography was the only method available for determining the full 3-D structure of a biological macromolecule. Since the mid-1980s, techniques based on NMR spectroscopy have also allowed the full elucidation of 3-D structures of biological macromolecules to atomic detail. Because NMR examines molecules in solution instead of in the crystalline state and because NMR allows investigations into the dynamics of structure whereas the crystal structure is a much more static view, the two methods complement each other extremely well. Again, what follows is a qualitative description of NMR phenomena. A basic introduction to general NMR can be found in almost any book on physical methods in chemistry or biochemistry. A deeper, lucid (nonmathematical) description can be found in "Modern NMR Techniques for Chemistry Research" by Andrew Derome (1987). Well-written texts on NMR of biological macromolecules are "NMR of Proteins and Nucleic Acids" by Kurt Wüthrich (1986) and "Nuclear Magnetic Resonance Spectroscopy in Biochemistry" by Justin Roberts and Oleg Jardetzky (1985), although due to the rapid advancement in the field, many of the more recently developed techniques currently in use are not covered. The mathematics of NMR are treated more fully in "Principles of Nuclear Magnetic Resonance in One and Two Dimensions" by Ernst *et al.* (1987). A review has been published by Clore and Gronenborn (1991).

NMR spectroscopy takes advantage of the fact that some nuclei have protons with unpaired spins. This spinning charge generates a magnetic field that will either align with (parallel to) or against (antiparallel to) an applied external magnetic field. Absorption of energy supplied by electromagnetic radiation at the resonance frequency can cause the nucleus to reverse its orientation ("flip"). The energy separation of these two states is very small compared to, for example, that between the ground and an excited electronic state or even rotational and translational states. This results in there being only a small population excess in the lower energy state at room temperature because of the thermal Boltzmann distribution, and consequently NMR is plagued by the need to measure very weak signals and the reduction of noise that accompany such subtle signals. The energy (wavelength) of electromagnetic radiation characteristic for this orientation "flip" is in the radio wavelength area of the electromagnetic spectrum and is dependent on the strength of the external field and the nucleus under investigation. Fortunately, not only the type of nucleus, but also its local microenvironment, will affect the magnetic field experienced by this nucleus and hence the energy necessary to "flip" orientation. It is this property that allows NMR to probe local atomic environments. Much more important, from the structural chemist's point of view, is the effect of neighboring spin centers on a nucleus of interest. It is the identification of these interactions that allow the eventual solution of the 3-D structure of the molecule under investigation.

Although a number of NMR-sensitive nuclei can be investigated (e.g., $^{1}H$, $^{13}C$, $^{15}N$, $^{19}F$, $^{31}P$), the sensitivity, high natural abundance, and ubiquity in biological macromolecules make $^{1}H$, or proton, NMR the most important type, although developments in molecular biology now allow uniform labeling of proteins with $^{13}C$ and/or $^{15}N$. Proton NMR will be used as an example in the following discussion, although the principles apply equally well to the other nuclei.

### 1. Chemical Shifts

Perhaps the most basic contribution of structure to the resonance of a particular nucleus is the local electronic environment. The negatively charged electrons also generate magnetic fields which shield the nucleus. In particular, electronegative atoms (e.g., oxygen or nitrogen) or delocalized electrons in aromatic rings will cause a shift of the resonance frequency for neighboring spin systems. Since the absolute magnitude of this chemical shift is also dependent on the strength of the external field, such shifts are usually reported as parts per million (ppm) relative to this field so that spectra from different experimental conditions can be directly compared. It is also this dependence on the external magnetic field that leads to the development of increasingly more powerful NMR spectrometers. Higher field NMR spectrometers have higher resolution, higher sensitivity, and give rise to stronger dipole relaxation and consequently better signals. Chemical shifts are important in the interpretation of NMR spectra because they spread out the resonances and allow identification by type, e.g., protons bonded to peptide (amide) nitrogen atoms will be grouped together away from those bonded to, for example, aliphatic carbon atoms.

More detailed information can be gained from the effect that nearby nuclei have on each other. Protons can affect neighboring protons by spin–spin interactions either through bonds or through space.

### 2. Spin–Spin Coupling

Through bond interaction, called variably spin–spin coupling, J-coupling, scalar coupling, or coherence transfer, occurs because the spin state of a proton affects the spin state of its valence electron. The spin state of this electron affects the spin state of the valence electron of the bonded atom with which it is paired, which in turn affects the spin state of the nucleus of the bonded atom. It is this J-coupling through bonds that gives rise to the familiar line splitting in NMR spectra. These interactions provide connectivity information because they occur only between protons on covalently bonded atoms. In addition, the J-coupling constant between two such protons is dependent on the electronic orbital overlap of the hydrogen atoms and the heteronuclear atoms to which they are bonded. Theoretically then, the J-coupling constant between, for example, one of the C5′ protons and the phosphorus atom in a nucleic acid allows determination of the $\beta$ torsion angle (Fig. 3). Sugar pucker angles in nucleic acids should also be amenable to investigation by examination of their proton J-couplings. Unfortunately, spectral overlap in biological macromolecules usually makes accurate determination of the J-coupling constants ambiguous, although with modern pulse sequences (see below) on $^{13}C$-enriched samples one can measure J-coupling constants that are even smaller than the line widths.

### 3. Nuclear Overhauser Effect

Another type of spin–spin interaction occurs via a dipole relaxation coupled between nuclei through space. Cross relaxation of the nuclei results in an exchange of their magnetization. The strength of this interaction falls off roughly with $1/r^6$, where $r$ is the separation between the two nuclei. Because of this rapid falloff with distance, only protons separated by less than about 5 Å can be observed. The change in magnetization of one nucleus due to perturbation of the magnetization of a nearby nucleus is termed the nuclear Overhauser effect (or enhancement, NOE) and provides information about which protons are near to each other in space but not necessarily in sequence. Because of the dependence of the NOE on the precise spatial arrangement of the nuclei (and not just the distance between them), the strength of the NOE between any two protons can only be calculated by solving a set of coupled differential equations incorporating knowledge of the 3-D structure of the macromolecule of interest, a clear case of putting

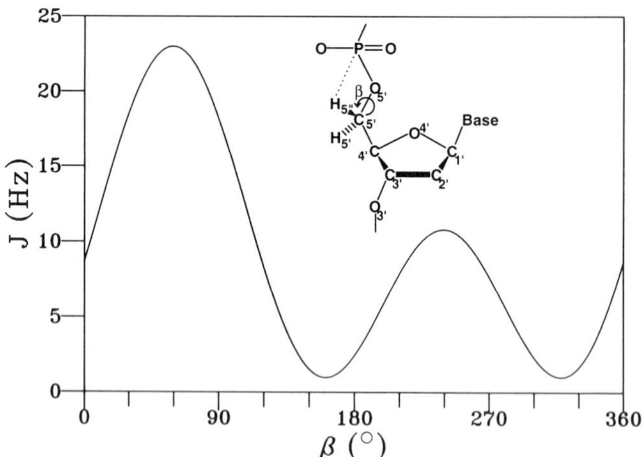


**FIGURE 3** Calculated J-coupling constant between $H_{5''}$ and the phosphorous atom of a nucleic acid as a function of the backbone $\beta$ conformational angle. Adapted from Blommers (1990).

the cart before the horse. In practice, NOEs are usually classified as strong ($r < \sim 2.7$ Å), medium ($r < \sim 3.3$ Å), or weak ($r < \sim 5.0$ Å).

There are then two fundamentally different measurements available from NMR experiments: J-coupling constants between protons close together in bonded sequence and NOE measurements between protons close together in 3-D space (Fig. 4). Because of the large number of protons in even small biological macromolecules, with much spectral overlap, the major effort in macromolecular NMR spectroscopy is separating out the many observed interactions and identifying the specific protons in the molecule giving rise to them.

### 4. Proton Exchange

NMR experiments examine molecules in the solvated state, generally in an appropriately buffered aqueous solution. Clearly the intensity of the water proton resonance (concentration $\sim 110\ M$) is much stronger than those of the molecule of interest (concentration 1–5 m$M$). For some experiments the sample is lyophilized and dissolved in $D_2O$ (D = deuterium, $^2H$ which is far less NMR sensitive than $^1H$). For observing the proton resonances of the exchangeable protons, e.g., the NH amide resonances for proteins or NH imino resonances for nucleic acids, pulse experiments must be performed where the water resonance is suppressed either by irradiation or by selective excitation. The latter technique allows one to measure exchange rates which gives qualitative information about its exposure to solvent. This is especially useful for the backbone amide protons in proteins. Since proton exchange is acid and base catalyzed, the pH of the solution will also influence the rate of exchange. In particular, slightly acidic conditions retard proton exchange of the backbone amide protons in proteins.

Through space
Interaction
(NOE)

Through bond
Interaction
(J-couplings)

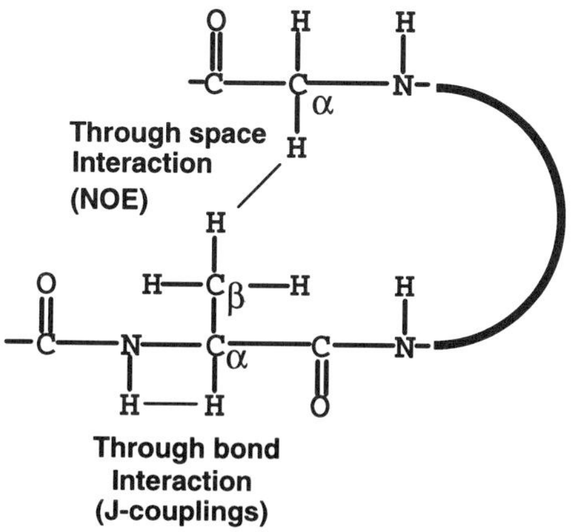

**FIGURE 4** Interactions observed in NMR spectroscopy. J-couplings can be detected for protons separated by (usually) not more than three chemical bonds. The nuclear Overhauser effect (NOE) is detected for protons separated in space by $< \sim 5.0$ Å.

## 5. Sample Considerations

One of the greatest advantages of NMR spectroscopy over X-ray crystallography is that the biomolecule under investigation is prepared in solution rather than in the crystalline state. This has the advantage that the molecule is in a more physiological environment and there is no need to prepare well-ordered single crystals. Nevertheless, because of the inherent weakness of the NMR signal, the protein or nucleic acid concentration must be relatively high, generally not less than 1 m*M* (roughly 10 mg/ml for a molecule with a molecular weight of 10,000). It is important that the protein (or nucleic acid) does not aggregate at this concentration. The typical volume of an NMR sample is 0.6 ml. In addition, the relaxation times (necessary for NOE measurements) of the excited states are correlated with the tumbling times of the molecule, which in turn are correlated with the size (radius) of the molecule. With current spectrometers, this places an upper limit of about 30,000 molecular weight (250–300 amino acid residues) for molecules which can be investigated.

## 6. Experimental

As mentioned earlier, the signal from an NMR experiment is exceedingly weak and repeated scans are necessary to improve the signal-to-noise ratio. All frequencies are excited simultaneously, and the signals arising from the decay of all resonances are measured simultaneously. This complex signal is then deconvoluted by the use of Fourier transforms into a set of frequencies and their corresponding amplitudes.

There are many different types of NMR experiments, but they typically adhere to the following general procedure: The sample solution is placed in a strong, uniform magnetic field. The unpaired spin centers orient themselves either parallel to or antiparallel to this field. Actually, the spin centers precess about the axis of the external field $B_0$, by convention called the z axis of a Cartesian coordinate system (Fig. 5). The precessing motion is similar to what happens when a spinning mass is placed in a gravitational field (a gyroscope). A pulse of radio frequency electromagnetic radiation creating a fluctuating magnetic field $B_1$ is then applied perpendicular to the direction of the external field, i.e., in the *xy*-plane (for example, along x) for a short time, causing temporary partial alignment along this axis. The amount of "flipping" onto this axis for a particular nucleus depends on the length of the pulse, and pulses are usually timed to reorient the nuclei 90° or 180° from $B_0$. The spin–spin relaxation of the nuclei back to a uniform distribution around the z axis in the *xy*-plane is accomplished by free induction decay (FID) and is measured as the acquisition time $t_2$. In addition, spin–lattice relaxation ($T_1$) also occurs which brings the spin back to thermal equilibrium along the z axis. This is the basic arrangement for a classical 1-D NMR spectra. Because of the hundreds of protons in biological molecules of inter-

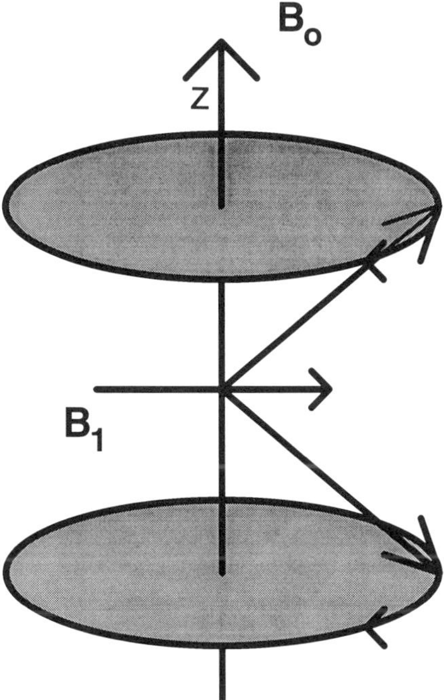


**FIGURE 5** Possible alignments of an unpaired spin (represented by arrows) with an external magnetic field. The spin can either be "up" or "down." The angle of the axis of the spin from perfect alignment with (or against) the magnetic field depends on the strength of the external field, the spin state of the nucleus of interest, and the local microenvironment. The spin axis precesses around the z-axis (gray disks). At any one moment, a large population of similar spin systems will be randomly distributed around the z-axis for a net magnetization parallel or antiparallel to the external field.

est, the piling up of resonance peaks (spectral overlap) makes such a spectra practically useless for 3-D structure solution.

### 7. Multidimensional NMR

In a 2-D NMR experiment, more pulses are applied. Incrementing the delays between them allows the labeling of frequencies in the "artificial" time domain $t_1$. The pulses are applied in such a fashion that coherences are selected for detection during the acquisition time $t_2$. A 2-D Fourier transform of the result produces a spectrum where the off diagonal signals correspond to the detected magnetization transfers between the spins. Correlated spectroscopy (COSY) uses a pulse sequence that gives rise to cross peaks corresponding to the covalent structure (J-couplings), while a different set of pulses is used in NOE spectroscopy (NOESY).

The resonances are first assigned by looking at the characteristic chemical shifts and cross peak patterns for the individual amino acid or nucleic acid types. Off diagonal signals can then be sequentially assigned by a combination of the identification of spin patterns in COSY-type experiments with the cross peaks observed in the NOESY spectrum.

Higher order NMR experiments are also possible. Heteronuclear 3-D or 4-D NMR requires the use of NMR-sensitive heteroatoms, especially $^{13}C$ and $^{15}N$. This adds considerably to the expense of the experiments (especially for $^{13}C$), as the protein must be uniformly labeled with these isotopes, usually by expressing the protein in an organism (bacteria, yeast) grown in minimal media with $^{15}NH_4Cl$ and/or [$^{13}C$]glucose as the sole nitrogen and/or carbon source. In a 3-D or 4-D NMR experiment, large through bond heteronuclear couplings are exploited. In these experiments, proton–proton cross peaks that are piled up in the 2-D spectra are spread out along the $^{13}C$ or $^{15}N$ frequency. This has allowed assignment of proton resonances for proteins of up to 200 amino acids.

### 8. Generation of the 3-D Structure

Once all resonances have been assigned, one has a list of rough interproton distances from the intensity of the NOE signal between interacting protons and some dihedral angles from J-coupling constants. The most popular method of generating a 3-D structure from these distances and dihedral angle restraints is the distance geometry method which also makes use of a priori knowledge of macromolecular structure such as bond lengths and angles, planar groups, chirality, and van der Waals contacts. One begins with two distance matrices (**U** and **L**) whose elements are either the upper or the lower limits derived from both the NMR distance restraints and from stereochemical constraints. A third matrix **D** is constructed whose elements are randomly selected to lie between those in the upper and lower limit matrices **U** and **L**. Actually, only the first element is truly randomly selected since subsequent elements are somewhat dependent on the selection of the first element. A 3-D structure is then generated which satisfies the distance restraints of matrix **D**. The structure is then optimized using some linear minimization algorithm like molecular dynamics. This is then repeated using different random matrices **D** (i.e., different first selections) so that one ends up with an ensemble of automatically and randomly generated structures which best satisfy the distance restraints. Because this procedure is automatic and does not require a starting structure, it avoids incorporating any possible bias from the investigator, but it has been shown not to sample possible conformational space in a truly random fashion, meaning some possible structure which equally well satisfy the distance restraints might not be generated. The entire ensemble of structures derived from an NMR experiment are usually presented. Parts of the structure that are rigid tend to overlap quite well between the various structures, whereas "floppy" parts of the structure show a greater discrepancy. However, this discrepancy may also be due to a lack of NOEs for this part of the structure because they could not be separated from overlapping peaks. Presentation of the structure in this fashion then provides not only the structure itself, but also a feeling for its dynamics.

### 9. Refinement of the Structure

As in a crystallographically derived structure, the first model can usually be improved by refinement. In particular, the first NMR-derived models tend to be high energy structures (eclipsed torsion angles, unfulfilled hydrogen bonds and salt bridges, etc.). The structure is then refined by calculating the NOEs from the structures by solving a set of coupled differential equations and fitting against the measured NOE intensities. Molecular dynamics refinement incorporating such NOE restraints is usually used to generate structures that are minimized with regard to stereochemical energy as well as with regard to the NMR data in a very similar fashion as was described for crystallographically derived structures. One important difference is that this must be repeated for the many structures derived from the NMR data.

### 10. Interpretation of Results

At the end of an NMR investigation, one has an ensemble of refined structures, all of which more or less equally well satisfy the measured NOE distance restraints and have reasonable stereochemical parameters. This can also be expressed as a single "average" structure and root mean square deviations (RMSDs) of the individual atoms derived from the multiple structures. Clearly the quality of the final structure depends on both the number and quality of the NOE measurements. Larger structures need more than just a proportionally large number of interactions to define the structure accurately since the total number of interactions increases more rapidly than the number of measurable short range ($<5$ Å) interactions. Parts of the structure which are flexible give rise to fewer NOE restraints and consequently more different structures will fit the restraints, resulting in larger RMSDs for those atoms. There is no analogous crystallographic reliability factor in NMR because of the difficulty in measuring accurate NOE values. Although it is possible to measure a NOE intensity quite precisely, it actually represents those from a possibly large number of different conformations. Because of the $1/r^6$ dependence of the NOE intensity with distance, the measured NOE intensity will be biased in an unpredictable way toward those conformations which bring the interacting protons closest together. The quality of the structure is then measured by the RMSDs of the atomic positions (usually divided into main chain and side chain atoms), the overall stereochemical soundness (van der Waals violations, ($\phi$, $\psi$) conformation angle analysis, side chain torsion angle analysis, etc.), and the number of unfulfilled NOE restraints ("violations").

## C. Results of Experimental Structures

Experimentally derived structures form the factual foundation on which our knowledge and understanding of biological macromolecular structures are built. Small molecule X-ray crystallography has given us very accurate measurements of the basic structural quantities, bond lengths, bond angles, group planarity, and chirality of the building blocks of macromolecules, the amino acids and nucleosides. In addition, they show what sort of variances can occur for these quantities.

Knowledge of the structures of the basic building blocks led to predictions about how polymers of these could arrange themselves in 3-D. The $\alpha$-helix and $\beta$-strands of proteins were predicted long before the first experimental structures were determined in the late 1950s and early 1960s. The double-helix structure of DNA was predicted with the help of fiber diffraction data long before the first 3-D structures were experimentally determined to atomic resolution. Predictions about interactions of macromolecules with other macromolecules or with small molecules have also raced ahead of experimental proof.

### 1. Protein–Protein Interactions

The first protein structure solved crystallographically was hemoglobin, a complex of four separate polypeptide chains which already told us something about protein–protein interactions. Since then a large number of multisubunit proteins, enzyme–protein inhibitor complexes, and antibody–antigen complexes have been examined. What has come out of these investigations is that contrary to first conjecture, it seems that the many weak nonpolar interactions are more important than the few strong ones (salt bridge or hydrogen-bonding). Complementarity of surfaces and sequestering of hydrophobic groups from the aqueous environment seem to be the driving force behind protein–protein association. A crude rule of thumb seems to be that at least 600 Å$^2$ of surface area per polypeptide becomes "buried" (sequestered) at a protein–protein interface.

An example of a protein–protein interaction would be transforming growth factor-$\beta$2 (TGF-$\beta$2), a cytokine whose crystal structure was solved simultaneously in two different laboratories (Daopin *et al.*, 1992; Schlunegger and Grütter, 1992). The active protein is a dimer formed by two identical 112 amino acid chains (Fig. 6). The interactions are extensive and include a covalent disulfide bond between equivalent cysteine residues in each chain. By examining the solvent accessible surface of an isolated monomer and of a dimerized monomer, one can examine the interactions between the monomers in the dimer. Approximately 950 Å$^2$ surface area per monomer becomes solvent inaccessible ("buried") on dimer formation. Almost 75% (700 Å$^2$) comes from hydrophobic side chain atoms, with main chain and side chain hydrophilic interactions accounting almost equally for the rest

**FIGURE 6** Schematic drawing of the dimer of transforming growth factor-β2 (TGF-β2) with one monomer drawn plain and one striped. The monomers are covalently linked by a disulfide bond between equivalent cysteine residues. Numerous side chain hydrophobic interactions occur along the interface between the monomers, especially between α-helices and the concave side of the opposing β-sheet.

(Fig. 7). TGF-β2 cannot exist as a monomer, presumably because of the large hydrophobic surface which would become exposed to solvent.

### 2. Protein–Nucleic Acid Interactions

Unlike the interactions between proteins, those between proteins and nucleic acids seem rather restricted. Only a few basic protein constructions

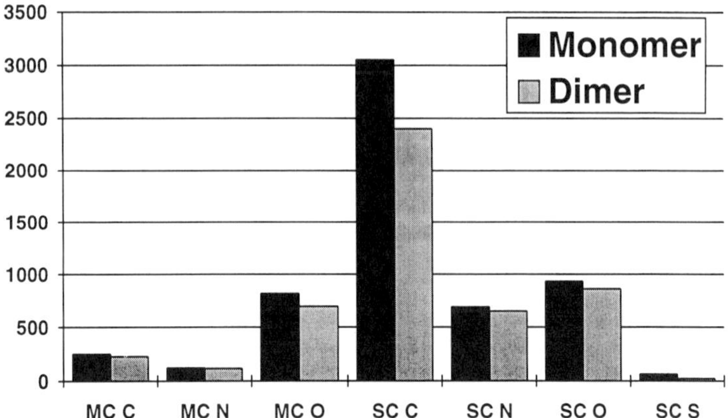


**FIGURE 7** Comparison of the solvent accessible surface area (in Å²) as a function of atom type in the dimer and isolated monomer of TGF-$\beta$2. MC C refers to main chain carbon atoms, SC O to side chain oxygen atoms, etc. Hydrophobic side chain interactions account for approximately 75% of the total buried solvent accessible surface area in the formation of the dimer.

(isolated helix, helix-turn-helix motif, zinc finger) allow good interaction with DNA and RNA, probably due to the very restricted helical structures of these target molecules. Proteins that interact with DNA tend to be symmetrical dimers which recognize palindromic DNA sequences. Nonspecific interactions with the sugar-phosphate backbone are followed by sequence-specific recognition achieved by a few specific hydrogen bonds between polar protein side chains (especially arginine residues) and the DNA/RNA bases.

A typical example would be that of the phage 434 *cro* repressor with synthetic 434 operator DNA (Anderson *et al.*, 1987). The *cro* repressor is a dimer and possesses the helix-turn-helix DNA-binding motif with many positively charged lysine and arginine residues interacting with the DNA sugar-phosphate backbone and with the bases (Fig. 8). In the dimer of the *cro* repressor, about 1150 Å² of solvent accessible surface area is lost upon binding to DNA. In contrast to TGF-$\beta$2 dimer formation, hydrophilic side chain interactions account for about half of this area (580 Å²) with hydrophobic side chain interactions accounting for only 40% of the total and main chain interactions contributing only 10% (Fig. 9).

### 3. Protein–Small Molecule Interactions

Unlike the association between proteins, in which relatively flat surfaces come together, small molecules tend to be completely engulfed by their macromolecular hosts. The interaction areas are also roughly 600 Å² or more, but the interactions themselves tend to be more specific: more hydrogen bond formation and better surface complementarity.

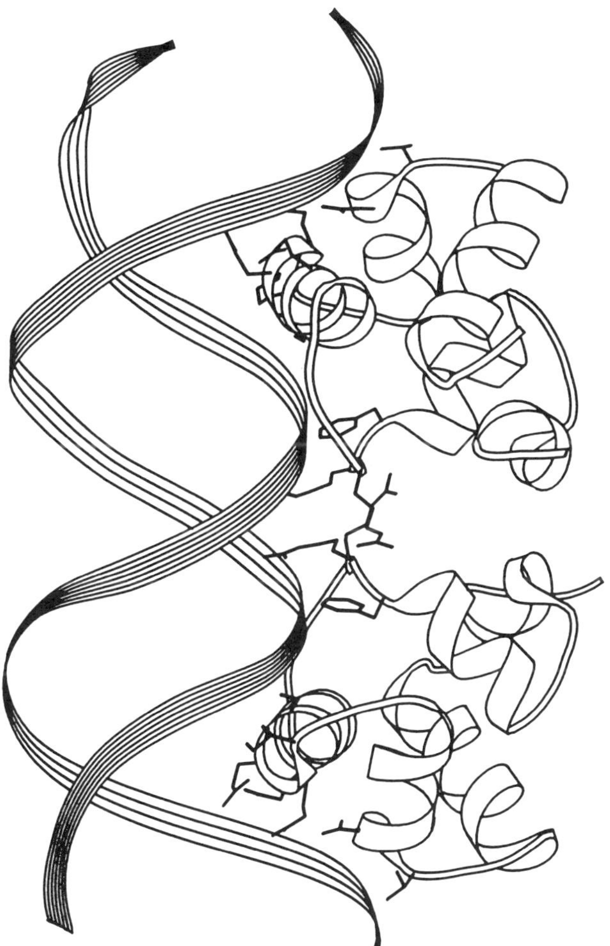

**FIGURE 8** Schematic drawing of the dimeric *cro* repressor bound to its cognate DNA with some of the more important interacting side chains drawn in.

An example of current interest would be the complex between HIV proteinase and the transition-state analog inhibitor CGP-53437 (Grütter and Priestle, unpublished results). The HIV proteinase is a dimeric aspartic acid proteinase which possesses two long "flaps" which close down on the protein substrate (Fig. 10). This is also a water molecule mediating interaction between the substrate/inhibitor and these flaps, which has itself been incorporated into the design of some HIV proteinase inhibitors (Lam *et al.*, 1994). Removing CGP-53437 makes about 500 Å$^2$ more of the surface of HIV proteinase dimer solvent accessible (Fig. 11), but this is not the full story since without substrate or inhibitor, both "flaps" open up and become

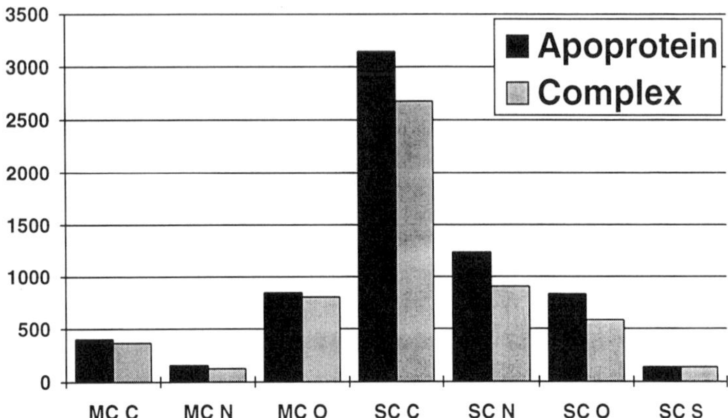


**FIGURE 9** Comparison of the solvent accessible surface area (in Å$^2$) as a function of atom type for one monomer of the *cro* repressor alone and complexed with DNA. MC C refers to main chain carbon atoms, SC O to side chain oxygen atoms, etc. Side chain interactions, mostly hydrophilic, account for approximately 90% of the total buried solvent accessible surface area in the formation of the complex.

floppy. Unlike the TGF-β2 and *cro*/DNA example where main chain interactions make a very minor contribution, in the HIV proteinase/CGP-53437 complex they account for over one-third of the buried surface (170 Å$^2$). This is typical of peptide-like inhibitors which bind to the enzyme forming main chain hydrogen bonds reminiscent of β-sheets. Specificity is then provided by side chain interactions, as attested to by the large hydrophobic side chain contribution (220 Å$^2$). Hydrophilic side chain interactions account for about one-fifth of the total (110 Å$^2$) buried surface area.

### 4. Macromolecule–Solvent Interactions

Both X-ray crystallography and NMR spectroscopy allow investigation into how a macromolecule interacts with its aqueous environment. Crystals of macromolecules are usually 30–70% solvent and a large number of solvent molecules can usually be located. Although the solution out of which proteins crystallize often have very high salt concentrations (1–2 *M* ammonium sulfate, sodium chloride, etc.), very few salt ions have been found in crystal structures. The water molecules surrounding proteins are held in place by hydrogen bonds, predominantly to main chain carbonyl oxygen and amide hydrogen atoms. Solvent molecules often play a critical role in the protein structure, forming hydrogen bonds to many protein atoms widely separated along the sequence. These molecules often have temperature factors as low as those of the main chain protein atoms, implying tight bonding and high occupancy. Not surprisingly, the temperature factors of solvent molecules quickly increase as one moves into the second or even third solvent layer, until there is a merging with the bulk (dynamic) solvent.

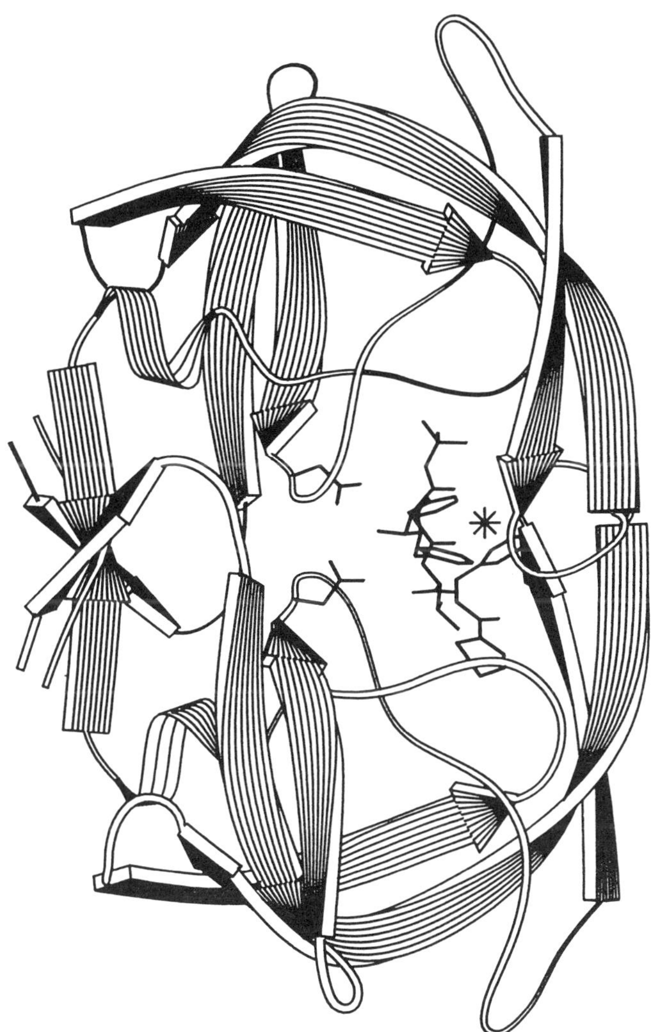

**FIGURE 10** Schematic drawing of HIV proteinase complexed with the inhibitor CGP-53437. Also shown as a star is a water molecule that bridges the inhibitor and the proteinase.

NMR investigations have shown that some side chain protons are rapidly exchanged with those from the surrounding solvent. The rate of main chain amide proton exchange with solvent protons has been correlated with both solvent accessibility and the degree of hydrogen bonding to other protein atoms.

In the active site of an enzyme, the solvent structure is important for understanding binding since these molecules must be first be displaced. If the solvent structure is known, important water molecules may even be

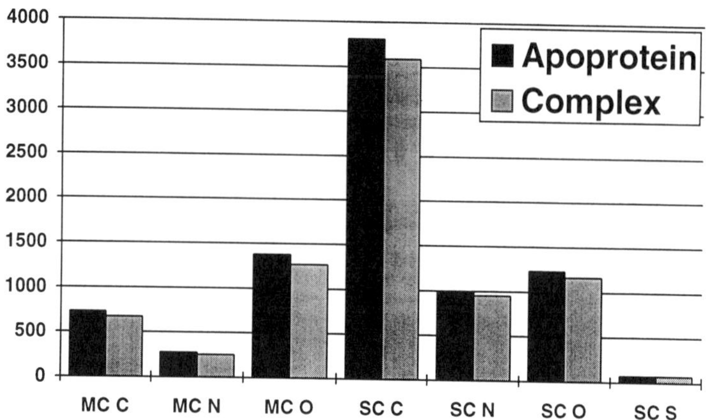


**FIGURE 11** Comparison of the solvent accessible surface area (in Å$^2$) as a function of atom type in free and inhibited HIV proteinase. MC C refers to main chain carbon atoms, SC O to side chain oxygen atoms, etc. Main chain hydrogen bond formation accounts for about one-third of the total buried solvent accessible surface area in the formation of the complex.

incorporated into a potential inhibitor structure. NMR spectroscopy can provide even more quantitative and dynamic information about the interactions between macromolecules and solvent molecules since the solvent also contributes to the local microenvironment of external protein protons.

### 5. Experimental Binding Constants

Binding constants are experimentally determined by Scatchard analysis. Reversible binding can be expressed as

$$\mathrm{P} + n\mathrm{L} \overset{K_a}{\longleftrightarrow} \mathrm{PL}n,$$

where P is protein, L is ligand, $n$ is the number of binding sites on the protein for the ligand, PL$n$ is the protein–ligand complex, and $K_a$ is the binding (affinity) constant of the protein for the ligand. Defining a variable r as the average number of ligand molecules bound to each protein molecule at a fixed concentration of [P] and [L]:

$$\mathrm{r} = \frac{[\mathrm{PL}n]}{[\mathrm{P}]}$$

the Scatchard equation relates these by:

$$\frac{\mathrm{r}}{[\mathrm{L}]} = nK_a - \mathrm{r}K_a.$$

A plot of r/[L] vs r has a slope of $-K_a$ and a y intercept of $nK_a$ (Fig. 12). If there is interaction between multiple sites on the protein (positive or

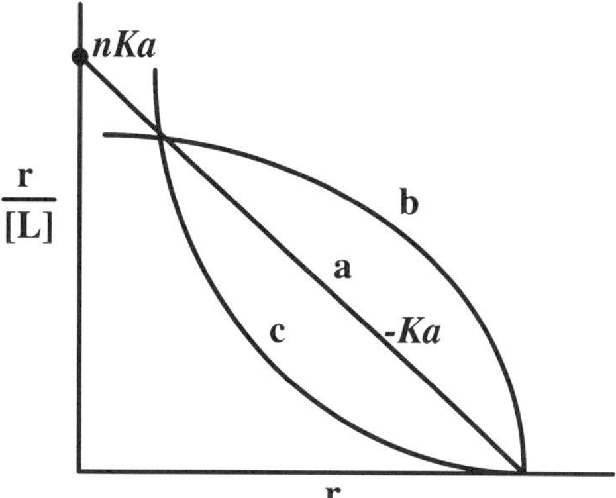


**FIGURE 12** Scatchard plots of r vs r/[L], where r = [PL*n*]/[P]. Plot a is for multiple independent sites; the slope is the negative of the binding (affinity) constant $K_a$ and from the y-intercept the number of binding sites ($n$) can be determined. Plot b is for multiple binding sites showing positive cooperativity and plot c shows negative cooperativity.

negative cooperativity), then the line will not be straight. To avoid having the variable r on both axes, a double reciprocal plot of 1/r vs 1/[L] can be made:

$$\frac{1}{\mathrm{r}} = \frac{1}{n} + \frac{1}{n[\mathrm{L}]K_a}.$$

Experimentally, the protein concentration is usually held fixed and the amount of bound vs unbound ligand is measured as a function of varying ligand concentration. For small ligands, the most common technique is equilibrium dialysis in which a semipermeable membrane separates two chambers. Protein is in one chamber and ligand is introduced into the other. The ligand moves freely through the membrane which is impermeable to the protein. Upon reaching equilibrium, the difference in concentration of ligand between the two chambers is due to binding to protein in the one chamber. Other techniques such as ultrafiltration, polarography, distribution analysis, and absorption spectroscopy may also be used. When investigating macromolecular interactions, it is usually necessary to immobilize one of the molecules, introduce the other, and again measure the amount of binding as a function of ligand concentration.

## D. Use of Experimental Structures in Protein Modeling

The most basic data gathered from experimental structures are general principles of chemical structures. From small molecule crystallography, accurate values for atomic bond lengths and angles, contact distances, and conformational preferences have been determined. These have been translated into the force constants and structural restraints inherent to all energy minimization and molecular dynamics routines and form the basis for the structural databases discussed in Section II. From macromolecular crystallography and NMR spectroscopy we have learned how proteins fold and how they interact with other macromolecules and small molecules, as discussed in the previous section. Analysis of protein foldings has even allowed a new type of modeling, that of the target protein itself, in cases where an experimental structure is unavailable. There are currently well over 1300 protein structures available in the Brookhaven Protein Data Bank (PDB), albeit many are not unique (the most extreme case being the 100-plus structures which exist for lysozyme and its muteins). Analysis of the structures of related proteins not only indicates how such modeling can be performed, but also allows at least an intuition about the accuracy and therefore the usefulness of such models.

One of the most obvious features of a macromolecular structure is that it gives the modeler a scaffold onto which, or more usually, into which, modeling can be carried out. Quantitative structure–activity relationship (QSAR) analysis among active compounds may suggest the important pharmacophores, but only analysis of these compounds interacting with the target macromolecule can explain why these groups are important for function. This forms the basis of so-called "rational drug design" and is becoming an increasingly important tool as more complicated systems are investigated and larger molecules are needed to guarantee tight binding and high specificity.

Now that X-ray crystallographers and/or NMR spectroscopists are becoming active members of drug design teams in the larger pharmaceutical companies, there is the chance of using experimental structures to test, validate, and/or improve modeling methods. The proposed mode of binding of a new molecular entity suggested by the modeler can now often actually be checked experimentally. A correct prediction validates the assumptions and methods used, whereas an unexpected binding mode allows revision and improvement of the prediction methods, and becomes an excellent starting point for further design.

## II. THREE-DIMENSIONAL STRUCTURE DATABASES

This section is intended to describe what chemical databases are, how they operate, the variety of their contents, and a number of ways that they may be used. It is not intended to provide design details for the construction of databases or of database software; entire volumes may be devoted to this arcane task. Nor is it intended to provide detailed operating instructions for a specific software product. Likewise, reaction databases and their literature—although important to the synthetic chemist—are ignored. This section provides guidance in the selection of already constructed databases and search software ("what information is available?"), in the expectation of functionality ("what can I do with it?"), and ideas about how these may be used in the design of small organic compounds ("why would I want to use it?")

The majority of software described is commercially available. Although research and proprietary efforts are described, an emphasis is maintained on commercial products because of their widespread availability, research utility, commercial success, and well-characterized performance.

The literature of chemical databases is vast, and the research community is active. Several books and review articles provide additional background for chemical databases in general (Warr and Suhr, 1992; Ash *et al.*, 1985) and for 3-D databases in particular (Willett, 1991; Martin, 1992; Martin *et al.*, 1990). At least one peer-reviewed journal is devoted exclusively to chemical information and related topics in computer science.[1] Library classification schemes conveniently shelve books dealing with chemical databases in only a few locations.[2] Several congresses and symposia are held annually which address progress in the field.[3]

## A. Components of a Chemical Database

Chemical databases are databases containing encoded chemical structures along with molecular and atomic data. Because the term database is often used in inappropriate contexts, it will be defined here and distinguished from other forms of data compilation.

In the most general sense, a database (or database system) is formally composed of three components: data in a searchable form, a generic search engine, and a query language appropriate to the domain of the data. Since few databases exist which are not computerized, the first two components become electronic manifestations of data stored on disk or in memory, and a software program is designed around the representation of this data.

Although these components may be discussed separately, of necessity they are designed as an integral unit; each is intimately dependent on the structure of the others. Search software must be able to translate and apply

[1] *Journal of Chemical Information and Computer Science* (ISSN 0095-2338); published bimonthly by the American Chemical Society, Washington, D.C.

[2] Additional reference material may be found clustered under the following subject headings as classified by the Library of Congress: "Chemistry—Information services" (QD8.3), "Chemistry—Mathematics" (QD39.3), and "Molecular structure—Databases" (QD461). Related topics include: "Chemistry, Pharmaceutical—Data processing" (RS401), "Information storage and retrieval systems—Chemistry" (Z699.5.C5), and "Information storage and retrieval systems—Chemistry, Pharmaceutical" (RS421).

[3] The ACS Chemical Information section (CINF) meets every 6 months; numerous database software vendors hold users' meetings annually or more frequently.

any query that may be phrased in its query language. The query language must be capable of translation by the search software and must refer to data attributes that are present in the database. The data must be encoded in such a way as to permit efficient storage and rapid pattern matching at search time. The three components are thus interdependent.

### 1. Structural Data in Searchable Format

Chemical structure data must unambiguously define the atomic connectivity within the molecule and, in many cases, the precise 3-D coordinates of all atoms. Additional data are often present, although its nature is highly dependent on the subdomain which the data represent: structures determined analytically by X-ray diffraction will include crystal-packing interactions; structures computed with molecular orbital methods may include partial electronic charges; and structures associated with a corporate synthetic program may lack 3-D coordinates, but may include such mundane information as inventory and sample location.

Chemical representation within computerized databases is a Byzantine and highly specialized area of research and, luckily for most scientific users, is buried invisibly within any particular database system. The most common representation of chemical connectivity relies on the data structures and algorithms of graph theory; the graph must be transformed and stored in a searchable form that preserves atom identity, connectivity, and stereochemistry, yet permits easy dissection of substructural subgraphs. One common method for this encoding is SEMA (stereochemically extended Morgan algorithm: Wipke and Dyott, 1974).

Structural data in any particular database tend to be compiled from within a particular subdomain of chemistry or biochemistry. Thus, seldom does one see databases containing both proteins and small organic molecules, for these reasons: (a) lack of special interest groups; (b) difficult issues of representation for different classes of compounds; and (c) software design and functionality. Three familiar collections of data, which will be cited frequently, illustrate these reasons. The Brookhaven PDB (Bernstein *et al.*, 1977; Abola *et al.*, 1987) is predominately designed for the encoding of crystal structures of macromolecular polymers, usually proteins or nucleic acids, and accommodates small organic molecules only marginally. The Cambridge Structural Database (CSD; Allen *et al.*, 1991a) is designed for the encoding of crystal structures of small organic and organometallic compounds with a wide variety of bonding patterns and shelters few peptides. And lastly, the commercial and corporate small molecule databases compiled worldwide are frequently designed under the limitations of atom- and bond-count maxima presented by the software used to construct and search these libraries efficiently. The limitations alluded to here will become clearer in subsequent sections.

Another characteristic of a database is that the data are indexed or keyed in a variety of ways appropriate to the domain; such keys facilitate extremely rapid searches. Without keys or screens, it may be possible to examine only hundreds or thousands of structures in a file; this represents a practical limit. The identity of keys in a database is usually a closely guarded trade secret of the vendor of the search engine. Common 2-D keys for chemical databases include distinctive substructural fragments, atomic number of composite atoms, and gross structural features such as multiple fragments or residues. Common 3-D keys are distances between noncarbon heteroatoms and distance ranges for atom pairs. The selection of 3-D key classes is an active area of current research, particularly in the laboratory of Willett (Poirrette *et al.*, 1991, 1993).

### 2. Graphical Query Language

Chemical databases are intended to be used; they are not static libraries, archived forever with no interest in retrieval. It is necessary to pose questions about the contents of a database and to retrieve example structures. These questions must be phrased in a chemical language that is syntactically simple, but semantically rich. Usually, this query language is built on top of standard graphical chemical nomenclature, augmented by features appropriate to the subdomain of chemistry represented in the database, and by methods of associating these features with the graphical connectivity.

Why must a chemical query language be graphical? In part, because the state of the art allows it and because users of chemical databases have become less tolerant of "primitive" query interfaces. However, the best reason is that chemical structures are by their very nature graphical; the diagrammatic shorthand with which we are familiar represents a powerful language that is easy to learn, simple to use, terse, visually compact, unambiguous, and incredibly rich. It is literally the "language of chemistry" (Hoffmann and Laszlo, 1991).

Each chemical or biochemical subdomain has defined structural and meta-structural (descriptive) features appropriate to its data, and these features must be an integral part of the query semantics. Thus, when working with small flexible organic compounds serving as leads for future drugs, one must have the ability to position substructural fragments relative to one another in 3-D space; to define computed features from these fragments such as centroids, planes, and normal lines; and to define intramolecular constraints such as distances or torsion angles. When working with structures derived from crystalline arrays, one must be able to define intermolecular features of interest, such as molecular packing derived from crystal symmetry rules and consequent intermolecular hydrogen bonds. When working with macromolecular polymers such as proteins, one must have the ability to define motifs such as helices and helical axes, loops and sheets and their geometric interrelationship, perhaps independently from primary sequence data.

## 3. Generic Search Engine

The chemical structure data in a database are organized and indexed for efficient searching by the search engine. This software is responsible for four tasks: (a) the translation of the user's query into the encoded "space" of the data; (b) preliminary screening of the query against database keys; (c) pattern matching of the query across the applicable subdomain of the data; and (d) presentation of the resultant "hits." These are four very different tasks.

The data representation underlying an interactively constructed query diagram may be, though seldom is, appropriate for direct use when scanning the database structures. The nature of the query determines the disposition of the underlying data. On the one hand, a small organic molecule query may be SEMA-encoded for a rapid identity search; this encoding followed by searching a 250,000 compound database may take only a few seconds. Alternatively, the query may be analyzed for its composite substructural and meta-structural features in preparation for a substructural search; this analysis is usually accomplished in under a few seconds.

The vast majority of database searches are not simple identity checks, but substructural or similarity searches. The query features discovered by analysis in the preceding step are applied as filters against the key indexes for the database. This results in an intersection of keys and the identification of a subset of the database across which subsequent operations must traverse. Depending on the complexity of the query, this key search may reduce the scope of the search from a domain of 250,000 compounds to a mere few thousand, in the process consuming tens of seconds or less of elapsed time.

The most time-consuming portion of a small molecule substructural search is involved with the atom-to-atom mapping of the query against each compound entry in the key-screened database subset; the more successful the application of the key search, the fewer compounds that must be manipulated individually. The pattern-matching algorithms associated with 2-D chemical substructural searching rely on the theories of subgraph isomorphism (Ullmann, 1976). For small molecule rigid 3-D searching, Brint and Willett (1987) have shown the Ullmann algorithm to be best; for protein motifs, they prefer to first use Lesk (1979), then Ullmann. The algorithms associated with 3-D feature mapping, especially with conformational flexible searching (CFS), are generally tightly guarded by software vendors; however, some of the underlying theory is derived from the work of Shenkin *et al.* (1987) on "random tweak" (originally used in the computational fusion of different loop regions of heterologous proteins during immunoglobulin modeling).

Depending on the nature of the query and the software, there may be one additional step. Search engines for 3-D models may treat the 3-D coordinate data as immutable or, as in the case of packages supporting on-the-fly conformationally flexible searching, simply as one arbitrary conformer of a potentially very flexible structure. In the latter case, another function of the search software is the chemically meaningful perturbation of the coordinate data in the database to facilitate pattern matching of the query (see Section II,C,5). This manipulation must be done on a per-model basis, and is computationally intensive.

The complexity of the query and the speed of the underlying hardware will affect the search performance radically, as will the individual software. A typical 2-D substructural search across a screened subset database of a few thousand entries with MACCS (see Section II,D,1) might range from 30 sec to a few minutes on contemporary mainframe architectures; these performance statistics are the subject of much intervendor competitive pressures and are continually improving. A typical rigid 3-D substructural search across this same few thousand entry subset might range from 2 to 4 min. With current software, if this same 3-D substructural search were conducted by use of an on-the-fly conformationally flexible algorithm, it would be three to five times less efficient; current research in 3-D key selection for flexible queries is anticipated to be able to reduce these times in the near future.

Once the search has finished and the resultant "hits" have been identified, the final function is their presentation to the user. While commonly the search engine is responsible, separate specialized modules may be involved. The two most commonly exercised options are interactive browsing and printed reports, and they are commonly used in precisely that order. When browsing, it is useful to understand how the query was fitted to the molecule currently being displayed; most software provides an overlay and colored highlight capability for this visualization. When browsing lists produced by conformationally flexible searches, the 3-D model may be displayed in the conformation which was computed to satisfy the query geometry; however, the density of such a display often dictates that such browsing be done in a molecular modeling package with more sophisticated interactive visualization and manipulation.

#### 4. What Is Not a Database?

Not all high quality collections of structural data may formally be called a database, based on the criteria listed earlier. The classic case in point is the Brookhaven PDB, which explicitly calls its product a "data bank," recognizing the distinction. The PDB structural data is encoded in a format designed to encourage standardized interchange, but this format is not efficiently searchable. No generic search engine is provided with the PDB, although numerous vendors and research groups have attempted to fill this gap. The PDB is discussed in depth in Section II,F.

## B. Data and Sources of Data

A chemical database is useful only in proportion to the provenance, quality, and relevance of the data in the database.

Most small-molecule chemical databases store representations of both a 2-D diagram and one or more 3-D structural models. When present, the 2-D domain is sometimes separated logically from the 3-D domain and may be searched separately without reference to any 3-D features. Macromolecular databases seldom store a separate 2-D diagram because of its complexity and uselessness; instead they retain a shorthand notation for the polymeric structure, usually peptide or nucleotide sequences. The remainder of this subsection deals only with the sources of 3-D structural models; it will be assumed that when appropriate, a 2-D diagram has been provided.

It is common to classify databases along two dimensions, with one dimension defining the source of data and the other dimension defining the content, in the sense of chemical domain. For example, a large portion of the CSD may be defined by the intersection between experimental X-ray diffraction (as the source) and organometallic complexes (as the content). Individual examples of these categories are detailed below.

The primary categorical dichotomy of importance to computational chemists or molecular modelers is the distinction between structural data that is derived experimentally and that which is computed *de novo*. That phraseology is important because the distinction does not lie solely in the realm of computation: the computational techniques of fast Fourier transform, distance geometry, molecular mechanics, and dynamics are essential to the accurate determination of molecular structure. The critical factor is the use of experimentally derived information to constrain these techniques, and indeed whether such data exist at all for the project of interest.

Comprehensive reviews of individual crystallographic and molecular databases may be found in Warr and Suhr (1992) and International Union of Crystallography (1987). The databases discussed in this section have been selected to be of interest to medicinal chemists and molecular modelers and to be searchable at least for structural data. Specific sizes of databases cited are for releases current in April 1994 and for 6 months prior. With only a few exceptions, copies of all databases may be acquired or licensed for local installation and searching.

### 1. Analytical/Experimental Data Sources

Databases containing experimentally derived structures are most commonly viewed as libraries or archives of reference material, with well-defined origins in techniques developed over many decades, the *sine qua non* of data sources. Due primarily to the historically slow accretion of data from these sources, there are only a few of these databases available today, and they are still of moderate size.

There are three primary sources of experimental 3-D structural data—X-ray diffraction, NMR spectroscopy, and neutron diffraction—and one source of slowly growing importance—electron diffraction.

***X-ray crystallography*** The methods of experimental structure determination through the use of X-ray diffraction have been detailed earlier in this chapter. At least three large commercial, research, or public-domain databases or collections exist whose content has been determined primarily by X-ray diffraction. The most relevant to the medicinal chemist or molecular modeler are the Brookhaven PDB (Bernstein *et al.*, 1977; Abola *et al.*, 1987) and the CSD (Allen *et al.*, 1991a), both of which are profiled in depth later in this chapter.

One of the largest databases derived from evaluated X-ray diffraction data is the NBS Crystal Data collection, produced by the NIST Office of Standard Reference Data (NBS Crystal Data Center, 1982). The content overlaps with that of CSD (organics and organometallics) and extends into inorganics; however, it may only be searched on-line and does not at this time support structural searching, so is of little interest to the medicinal chemist.

***NMR spectroscopy*** The methods of experimental structure determination through the use of NMR spectroscopy have also been detailed earlier in this chapter; these are especially useful when the solution structures are desired or when the molecule of interest cannot be crystallized sufficiently for X-ray analysis. The major role in structure determination played by the computational methods of distance geometry, molecular mechanics, and dynamics must be reiterated here: experimentally derived NOE distance constraints "ground these techniques in reality."

No commercial or public domain databases exist that are devoted solely to structures derived from NMR spectroscopy. However, a growing subset of the entries in the Brookhaven PDB are so derived; examples include entry 4AIT (tendamistat, an $\alpha$-amylase inhibitor) and entry 5HIR (hirudin, an anti-coagulant), just 2 of over 100 such entries in a recent release.

***Neutron diffraction*** Only one commercial database contains structures derived from neutron diffraction—the Cambridge CSD—and only 0.7% (830) of the entries were derived in that way.

***Electron diffraction*** Very few databases contain structures derived from electron diffraction. The rare exceptions are data for certain membrane-associated proteins which do not crystallize, but form pseudo-crystalline stacks. For example, the Brookhaven PDB contains entry 1BRD, a structure for bacteriorhodopsin, derived solely from electron cryomicroscopy studies; this is the single entry in PDB so derived.

## 2. Computed Models

Databases containing 3-D structures derived from *de novo* computational modeling are to some extent controversial in the field of molecular modeling, not having yet gained complete acceptance. This controversy has traditionally been rooted in the preference of the computational chemist for fewer numbers of high-quality, experimentally derived structures. If such data are not available, preference has then been given to structures derived from quantum mechanical methods or molecular mechanics, especially the latter if the molecule's accessible conformational space has been extensively sampled.

Minimal crossover exists between the domain of computational chemistry and that of chemical data librarians and archivists. Interest in the application of contemporary molecular modeling techniques across large databases is only of recent origin and has, in part, been driven by the use of targeted screening as a lead-finding mechanism in pharmaceutical research.

***Quantum and molecular mechanics*** With one possible exception, there are no commercial or public domain databases whose 3-D models are primarily based on modeling by the methods of either quantum or molecular mechanics. Although these methods are valuable and well-proven (witness the extensive coverage elsewhere in this volume), they present several difficulties to a researcher who wants to model a collection of thousands of compounds.

First, even for molecules with 25 heavy atoms, these methods are slow, requiring anywhere from a number of seconds (molecular mechanics minimization from a crude starting geometry) to a number of hours (for high level ab initio calculations); for larger molecules, the time scales quadratically. The computational investment required for thousands of compounds is prohibitive. Second, these methods frequently involve manual manipulation and interactive visualization for satisfactory results, thus are difficult to automate completely. And finally, because of the complexity of the conformational behavior of a given molecule, these methods (molecular mechanics or dynamics in particular) may explore hundreds or thousands of conformers in the process of investigating the accessible conformational space of even the simplest organic molecules, thus passing the problem of combinatorial explosion onto the database manager. It is no wonder that these techniques are reserved for studies of smaller populations of compounds.

Although no formal database systems are based on these data, at least one data bank exists. Over the years, a data bank of molecular structures and properties calculated by the Gaussian program has been assembled (the "Gaussian archives"; see Appendix).

***Automated model builders*** The consistent use of quantum and molecular mechanics as modeling tools are laudable goals, but are not always achievable, especially when the task is the production of models for hundreds of thousands of compounds, generally small organic molecules, that might exist in a corporate synthetic library. A viable alternative is the use of one of the automatic model builder software packages. This category of tool offers features of fully automatic, good-to-high quality model generation from a variety of 2-D connectivity table formats, in times that average under a second per compound.

The costs of such speed in model building are obvious: the quality of modeling would be inadequate for any in-depth molecular modeling study. However, when the resultant models are to be used primarily in 3-D databases, where typical queries may be posed with distance tolerances between 0.25 and 1.0 Å and valence angle tolerances of 5° to 10°, it becomes clear that the accuracy of molecular mechanics is seldom needed (Van Drie, 1993). Once one accepts the compromise of slightly inaccurate modeling of individual compounds against the coverage of many thousands of candidate molecules, the next step is the selection of a specific tool for model generation.

Software products to automatically and rapidly generate reasonable 3-D models from 2-D connectivity tables have come a long way since their advent in the mid-1980s. Several high-quality 3-D model builders are available commercially, including COBRA (Leach *et al.*, 1990; derived from WIZARD); CONCORD (Rusinko *et al.*, 1993); and model builders incorporated into Chemical Design's ChemDBS-3D (ChemModel: Davies *et al.*, 1991), Biosym's Insight (Converter), and BioCAD Catalyst (Teig *et al.*, 1993) products. The development of high-quality model builders is still a topic of ongoing research; it should be expected that many private corporations are engaged in similar, proprietary, research.

The quality of models generated by these automatic builders may be measured by any number of criteria, but most commonly used is RMS deviation from some standard of reference (CSD and PDB structures have served as reference), specifically examining issues of bond lengths, and of bond valence and torsion angles. Several studies have been conducted comparing the quality of (a) CONCORD-modeled structures with experimentally determined structures from the CSD (Hendrickson *et al.*, 1993), (b) ChemDBS-3D-modeled structures with CSD structures (Nicklaus *et al.*, 1993), and (c) models generated by many builders with ligand molecules extracted from PDB entries (Ricketts *et al.*, 1993). The evaluations are consistent: bond lengths are quite accurate and bond valence angles are well modeled, but torsion angles compare less favorably. As one might expect, torsion angles in constrained or rigid ring systems are well modeled, but in regions of high flexibility, the modeled torsion angles tend to adopt an arbitrary lower energy form which may not be exhibited in crystal forms. This merely implies that proper use of these models must account for conformational flexibility, either with extensive conformational analysis with molecular modeling or with appropriate flexible database search techniques.

From the viewpoint of being independent from any specific vendor's database environment, and as a stand-alone program with uses outside of database construction, CONCORD is the most frequently cited and most widely used approximate model builder today. CONCORD generates models in two phases. The first phase combines the logic of an "expert system" approach with a pseudo-minimization procedure based on a univariate strain function for certain "nontrivial" bond valence and torsion angles; it does not perform an actual energy minimization to generate rings or ring systems, whose bond lengths are assigned from an extensive table. The second phase is optional (and new in version 3.0.1); when requested, CONCORD performs a true molecular mechanics minimization for those rotatable dihedral angles exhibiting "close contacts," as defined in the first phase. There are atom-count and ring-size limits on conversion, and CONCORD only handles compounds composed of the common organic elements (H, C, N, O, F, Si, P, S, Cl, Br, and I). If explicit hydrogens or lone pairs are provided on input, they are retained; on request, CONCORD can generate and output coordinates for all hydrogen atoms.

### 3. Categories of 3-D Data: A Survey

Typically, the contents of individual databases are limited to specific subdomains in chemistry or biochemistry. The following general categories survey the breadth of data available.

***Peptides/proteins*** The premiere "database" for the molecular structures of proteins and large peptides is the Brookhaven PDB. As is detailed in Section II,F, 92% (2143) of the PDB entries are proteins, and this active collection is continuing to grow at an increasing rate. These structures are primarily derived from X-ray diffraction and NMR spectroscopy. Two-dimensional diagrams are not provided with this data, although sequence information is available.

Although not primarily devoted to this type of chemistry—and the formal atom and bond storage architecture precludes compatibility even with small proteins—about 1% (1122) of the entries in the Cambridge Structure Database are peptides and are stored with sequence data to aid in retrieval. The vast majority of these peptide structures were derived from X-ray diffraction.

There are sure to be a number of proprietary 3-D databases of computed models of small peptides scattered throughout the corporate research world. When combined with conformationally flexible 3-D searching, such data offer an interesting and rapid survey tool for suggesting lead compounds that are particularly accessible synthetically.

***DNA/RNA/RNP complexes*** Those few structures that have been determined and published for nucleic acids are found in the Brookhaven PDB.

Some 8% of its contents are devoted to this chemistry (174 entries), and most of the structures were derived from X-ray diffraction. As described earlier, 2-D diagrams are not provided, but sequence information is.

One academic group is creating a new database devoted entirely to nucleic acid structures (the NDB of Berman *et al.*, 1992). It is managed with a relational database manager (not unlike BIPED: Islam and Sternberg, 1989; see below for more details) and contains 3-D coordinate data based primarily on PDB records of nucleic acids (although CSD and "other sources" are included), along with sequence information and derivative geometric data. The project is still young and its search semantics are incomplete.

***Organics/organometallics*** The premiere database for the evaluated structures of small organic and organometallic compounds is the Cambridge Structure Database, which now surpasses 120,000 entries. Over the years, this growing database has become the repository of "interesting" crystallographic problems, resulting in a high percentage (over 43%) of organometallics ($\pi$-bonding complexes in particular) and metal complexes. CSD structures are derived from X-ray and neutron diffraction. A subset of the CSD is now available in the proprietary binary format of MDLI (San Leandro, CA).

Three commercial 3-D model databases are marketed by MDLI in their proprietary binary format; all models were computed by CONCORD. ACD-3D, the Available Chemicals Directory, augmented from FCD over the past few years, contains 125,999 2-D entries and 107,612 3-D models for commercially available organic compounds. MDDR-3D, the MACCS-II Drug Data Report, contains 47,602 2-D entries and 42,805 3-D models for biologically active compounds developed worldwide; this database is derived from Drug Data Report.[4] CMC-3D, Comprehensive Medicinal Chemistry, contains 6141 2-D entries and 5374 3-D models for commercially available drugs, derived from Hansch *et al.* (1990). Each database is updated in its entirety every 6 months, occasionally incorporating the enhanced features of the most current release of CONCORD.

Three commercial 3-D model databases are marketed by Chapman & Hall (England) in the format proprietary to ChemDBS-3D (Chemical Design); all models were computed by the builder ChemModel which is incorporated into ChemDBS-3D. Dictionary of Drugs contains 3-D models for about 12,000 compounds, covering the contents of British and American Pharmacopoeia and compounds undergoing clinical trials. Dictionary of Natural Products contains 3-D models for about 50,000 compounds from the Chapman & Hall printed Dictionaries of Alkaloids, Antibiotics, and Terpenoids. Dictionary of Fine Chemicals contains 3-D models for about

[4] *Drug Data Report* (ISSN 0379-4121); published monthly by J. R. Prous Science Publishers, Barcelona.

120,000 organic compounds available from commercial sources. Each database is available on a yearly subscription basis.

The Derwent Standard Drug File is available from Chemical Design as a 3-D database in ChemDBS-3D format. The models are constructed by ChemModel. This database contains structural and chemical information on over 31,000 compounds.

The Chemical Abstracts Service (Columbus, Ohio) maintains the world's largest and most comprehensive collection of chemical databases. The primary CAS structure registry file (2-D) contains upwards of 14 million entries covering organics and a wide variety of other chemical moieties. This registry file also contains over 4.6 million 3-D models generated by CONCORD in a research effort to investigate the practicality of searching a 3-D database of this size (Fisanick *et al.*, 1993). Although these 3-D models cannot be directly searched on-line at present, they may be retrieved via STN. CAS offers the CAST-3D structure templates file as an indirect alternative to access these 3-D models; this file is a 370,000 model subset identified as being "rigid" (limited conformational flexibility), supplied in the MDLI SDfile interchange format, and provided to customers for registration into their own local 3-D databases. User-defined subsets of CAST-3D are also available.

A 3-D subset of the NCI database (the National Cancer Institute Drug Information Service, currently over 450,000 entries; see Milne and Miller, 1986) is available for use with Catalyst software (BioCAD). The source database is a repository of small organic and organometallic compounds solicited from a variety of sources by the NCI since 1955 for testing in a variety of antitumor screening assays. Conformational analysis and modeling of about 96,000 entries were performed by the Catalyst builder at an average of 8.3 models per compound. A CONCORD-modeled version of the entire database was prepared in-house by NCI (Hendrickson *et al.*, 1993) but is not generally available; a subset is in preparation for more general distribution.

***Corporate synthetic programs*** Many chemical and pharmaceutical companies have had synthetic programs, some extending as far back as 1900. The compounds in these programs have been recorded both for patent protection and as a reference library or archive—a form of institutional memory. Since the mid-1980s, many of these collections have been computerized for rapid access by synthetic and medicinal chemists, and are now embodied in large chemical database systems. At least in the pharmaceutical industry, these collections are an exceedingly diverse *gemisch*, primarily small organics (as would be placed in classes 1–59 of the CSD classification system; see Table I), or their salt forms. They are important both in their sheer quantity and for their very diversity.

**TABLE I** Census of the Cambridge Structural Database, April 1994

| Basic classes[a] | Generic coverage | % of database |
|---|---|---|
| 1–12 | Simple aliphatics | 5.0 |
| 13–23 | Monocyclic hydrocarbons | 4.9 |
| 24–31 | Polycyclic hydrocarbons | 4.2 |
| 32–42 | Heterocycles | 16.1 |
| 43–59 | Natural products | 11.3 |
| 60–61 | Molecular complexes, clathrates | 2.7 |
| 62–70 | Main group compounds | 12.0 |
| 71–75 | Organometallics (Metal—C and $\pi$-bonds) | 17.5 |
| 76–86 | Metal complexes | 26.1 |

[a] A numeric chemical classification scheme in use since the inception of CSD. Since a compound may be assigned to multiple chemical classes (e.g., as with a steroidal alkaloid), an arbitrary precedence sequence has been devised to unambiguously classify all compounds into one and only one "basic" class (denoted by data field *BCLASS). Thus while steroids belong to class 51 and alkaloids to class 58, the precedence rule decides that a steroidal alkaloid has basic class 58; class 51 is assigned as a cross-reference class and may be discovered by searching on data field *CLASS.

With the advent of rapid model builders and 3-D searching software, many corporations have been or are in the process of building 3-D databases from these corporate archives, turning large libraries of 2-D diagrams into valuable mines of 3-D information. The sizes of most proprietary databases are not generally available, but one of the more active Ciba-Geigy Pharmaceutical research 3-D databases contains about 250,000 2-D entries and over 465,000 3-D models derived from CONCORD modeling, a rich source for "database mining." Pharmacophore-based searches across corporate databases allow the identification of critical lead compounds that may be obtained directly from the stockroom shelf, and which may be input to automated rapid screening systems.

Chemical databases may be used as a library across which searches for new ideas may be conducted; this applies to all the sources mentioned earlier, but is particularly apt for corporate databases since their content is more easily directed. Contents are not simply static reference compounds, but, when searched and thought about, become conceptual catalysts for chemical design. Particularly important here are databases constructed from models of compounds that have not yet been synthesized, but could be. Examples include (a) explicit models of diastereomers and tautomers of compounds from a corporate database; (b) 3-D databases of flexible and cyclic peptides; and (c) data collected from experiments exploring the complementary chemical graph space available to ligands in a macromolecular binding site.

## C. Queries and Sources of Queries

This section defines the common modes by which a database can be searched, detailing the various semantic elements that are permitted in a query. These search modes are discussed in the context of small molecule databases; special mention will be made where they are applicable to macromolecules.

An often neglected aspect of chemical database use is the derivation of the queries themselves. Models for pharmacophores do not appear full blown from the head of Zeus; construction and formulation of 3-D queries is not an intuitive art. Discussed in this section are methods for the assembly of queries from molecular modeling data and other sources.

### 1. Query Syntax: What Kinds of Questions May Be Posed?

Starting from the graphical representation of a 2-D diagram or a 3-D model, what type of questions may be posed, and what kind of information is returned? How many ways can one formulate a chemically meaningful query?

The kinds of questions one normally asks may be categorized roughly into either 2-D or 3-D modes. Common 2-D search modes include substructure, query enhancements or "fuzzy" search, R-group or Markush, and similarity. Common 3-D search modes include 3-D feature (both rigid and flexible) and 3-D similarity. The use of each of these different search modes is described in the following sections, although emphasis is placed on 3-D queries. Many database search engines allow most of these query modes to be combined; this is important to the precise definition of a pharmacophore or ligand.

The performance of any specific mode of searching depends on the complexity of the query and the size of the database. In general, the different search modes rank as follows: substructure search speed is roughly equivalent to similarity search, which is faster than Markush searching, which is faster than rigid 3-D searching, which is faster than CFS.

First, several points of terminology. The "query" is the operand on which the database search engine functions. The query may have one or more "features." The result of a successful search is the identification of one or more compounds from the database which satisfy the query; this result is often referred to as the "hit list." In some contexts, the hit list is a list of identification names or numbers; in other contexts, the hit list is actually an ordered collection of all the compound structures to which the list refers.

### 2. 2-D Substructure Searching

Substructure searching involves the identification of all molecules in a chemical database that contain a user-defined partial structure (Ash *et al.*,

1985). The query identifies this partial structure exactly. Substructure searching will always identify structures which are equal in size or larger than the query.

Most search engines offer a number of "fuzzy" query features, for the precise control of more broadly defined queries. The following list gives an overview of some of the possibilities: atom lists ["at this position may be found any of these three atoms (C, N, O)"]; variable bond types ("this bond may be either single or aromatic"); chain bond ("no ring must contain this bond"); variable hydrogen count ("this atom may have attached one or two hydrogens"); substituent count ("this atom may only be substituted as shown"); no closed ring ("this fragment may not be part of a larger ring system"); or no cross link ("these two atoms may not be linked by any length chain or ring"). Since the syntax by which these features are illustrated differs for each vendor's products, not all these features are illustrated. A few are seen in Fig. 13.

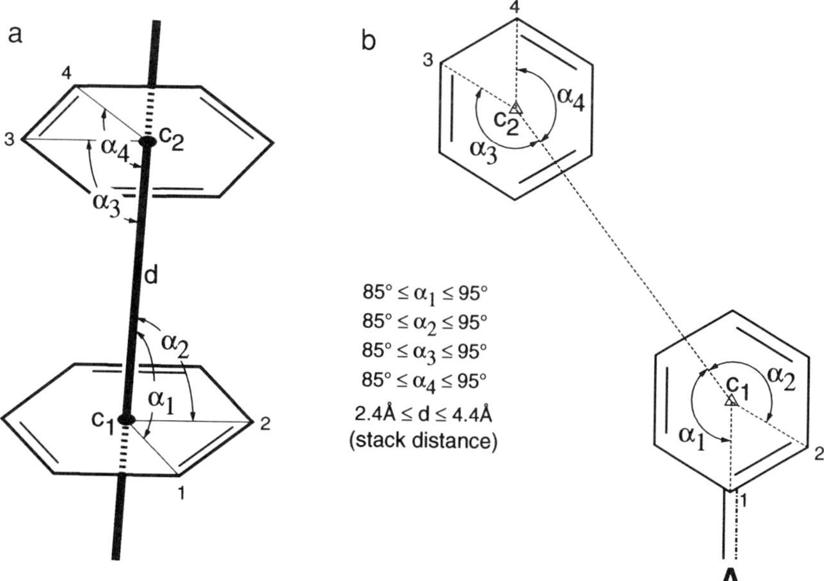


**FIGURE 13** Example of 3-D query formulation designed to extract models which exhibit $\pi$-stacking of a pair of benzenoid rings. A centroid is defined for each ring system. A single distance is defined between these centroids "c1" and "c2" such that the $\pi$-stacking distance range becomes constrained. Four angle constraints are defined, two for each ring, to constrain the centroid-to-centroid vector to be perpendicular to the ring, with minor variations. The query features of "single or aromatic bond" and "any nonhydrogen atom" are used to anchor the 3-D constraints to the substructure fragment and to prevent false hits.

### 3. R-group (Markush) Searching

Markush searching is a powerful enhancement of substructure searching and involves the identification of all molecules in a chemical database that contain one or more combinations of a user-defined substituent pattern. The query defines each substituent exactly, the substitution pattern on a substrate substructure, and (optionally) any logic associated with their occurrence in a matching structure. As with substructure searching, Markush searching always identifies structures that are equal in size or larger than the query.

### 4. 2-D Similarity Searching

Similarity searching involves the identification of all molecules in a chemical database that are loosely related to the query structure. The query is exactly comprised of an entire structure; seldom is the query a substructure or collection of fragments. To the user, the measure of similarity is often an integer or percentage which denotes "how similar" a target compound must be for it to be returned on the hit list. For the MACCS search engine, the similarity coefficient is based on the 2-D search keys as defined for the query and the database entry, as applied with a variant of the Dice algorithm. If $N_q$ is the number of keys set in the query structure, $N_t$ is the number of keys set in the target structure, and $N_c$ is the number of keys which are set in common for both the query and target structure, then:

$$\text{similarity coefficient} = N_c/(N_q + N_t),$$

$$\text{subsimilarity coefficient} = N_c/N_q, \text{ and}$$

$$\text{supersimilarity coefficient} = N_c/N_t.$$

The QUEST-3D search engine offers a nearly identical similarity coefficient, based as well on the classification bit screens. Some search engines will weight keys unequally (e.g., MACCS). Most similarity coefficients are reported as percentage values.

Many search engines will offer a choice of three variants for 2-D similarity searching: normal, subsimilarity, and supersimilarity. Subsimilarity compares the number of common key settings to the keys set in the query only. Hits from a subsimilarity search are typically structures that are larger than the query and contain a substructure similar to the query. Supersimilarity compares the number of common key settings to the keys set in the target structure. Hits from a supersimilarity search are typically smaller than the query and contain features similar to those in the query.

From a theoretical viewpoint, the concept of chemical similarity in databases is a complex topic. The use of database keys to establish similarity is not the only method, nor often the most useful; devising different schemes by which to define and measure molecular similarity has been a topic of research for some time. Other methods have been implemented which provide a different "view" of the database; three particularly interesting methods have been published from the laboratory of Venkataraghavan (Carhart *et al.*, 1985; Nilakantan *et al.*, 1987, 1990). A comprehensive review by Barnard (1993) of developments in searching is available.

One way in which a similarity search may be put to use has to do with the clustering of a large proprietary database to assist in the balanced sampling of the collection in a targeted screening process. However useful targeted screening may be in leading finding, one problem is determining if an adequate sampling of chemical diversity has occurred. Clustering of an entire collection (Hodes and Feldman, 1991) is time-consuming—the example cited is the NCI database with over 400,000 entries—but the tradeoff is a better understanding of the collection and a reduced duplication of effort during screening.

### 5. 3-D Feature Searching

3-D searching differs from 2-D searching in that the connectivity table is no longer the sole arbiter of search matches: the relative positions of atoms in 3-D coordinate space, and of derived features anchored to these atoms, now become of primary interest. When one is designing a compound to mimic the bioactive conformation of a substrate or agonist, the chemical connectivity is of less importance than is the precise placement of particular substituents in space so that they may promote the interaction of the ligand with its host macromolecule. Development of these pharmacophore models, or of compounds which can adopt the conformation of these models, is a major step in the drug design process. Tools such as 3-D searching, which can apply a pharmacophore model as a query across a large "space of chemical possibilities," can provide enormous assistance to this process. Experience has shown that 3-D searching often returns structures in very different classes of compounds as compared with the compound from which the query originated.

***Definitions of 3-D features*** At the root of all 3-D features are the atomic coordinates of the atoms to which these features are anchored. The term "atom" is used to mean not only heavy atoms, but hydrogens (which may be sprouted if not already present) and lone electron pairs (for some search engines only). Three-dimensional features are generally not defined relative to bonds, although features can be defined which will mimic a bond's vector.

Commonly defined 3-D features include both geometric objects and constraints, defined by reference to atoms or other features in the query. Derived objects may include centroids, points, lines, planes, normals, and "exclusion spheres." Constraints usually include distances, angles, and dihedrals, and have numeric values or ranges assigned to them. Lines and planes may often have RMS "best-fit" constraints attached to them.

Two additional constraint concepts have proved useful. The first is the "fixed atom," a constraint defined by four or more atoms in a "submodel" whose coordinates are fixed relative to one another as a single entity, generally in association with user-defined tolerances. The second is the "transform," a constraint calculated at search time as a composite algebraic function of one or more other constraints (usually distances, angles, or dihedrals). Fixed atoms are presently only supported by MACCS, and transform features only by QUEST-3D.

To some extent, the 3-D features described can be assembled together in a nested hierarchical fashion; this provides substantial syntactic power for the precise specification of a complex 3-D object. The depth to which these definitions may be nested is dependent on the particular search engine.

***Rigid 3-D searching*** Rigid 3-D searching involves the identification of all models in a chemical database that contain a user-defined set of geometric features. The query defines the exact substructural features which are to serve as anchors, the individual geometric features, and any constraint values associated with these features which must be satisfied. "Rigid" applies to the interpretation of models in the target database; registered conformations are examined unperturbed. As currently implemented, rigid 3-D searching requires that all query features be present in any prospective target.

Figure 13 shows a simple 3-D query in the syntax of MDLI's ISIS which is designed to extract all models which exhibit $\pi$-stacking of a pair of benzenoid rings. This query contains seven 3-D features and two "fuzzy" query features. The four angle constraints are required to prevent planar naphthalene-like systems from satisfying the query and to position the two rings properly in space (as illustrated in Fig. 13a). Although there are other ways to phrase this query, the angle features were chosen both for their computational efficiency at search time and because they are acceptable to all 3-D search engines. When executed against the MDDR-3D database, this query retrieved five hits, all of which exhibited $\pi$-stacking in the registered computed conformation.

Rigid 3-D searching assumes that the registered conformations are relevant to the query and present an accurate sampling of the compound's conformational space. This assumption is seldom exactly true. Techniques of query formulation have been developed to alleviate this problem somewhat; the options described by Güner *et al.* (1992, 1990) are good approaches to creating a flexible query, relying on extensive knowledge of the set of compounds from which the query has been derived.

Rigid 3-D search is one mode supported by macromolecular search engines such as PROTEP (Tripos Associates, 1992). Under the "complete search" mode, a structural pattern and its tolerances are defined and used as a query. This is directed against a protein under the assumption that the protein is a rigid body.

***Conformationally flexible 3-D searching*** Conformational flexible searching is a variant of 3-D searching devised to address one of the major problems of rigid searching: the flexibility of small molecules. How can one ensure that a 3-D query will execute successfully against a stored model present in a realistic, low energy conformation, when this conformation may be subtly and arbitrarily distinct from the conformation demanded by the query?

In the past, this problem was dealt with in one of two ways: (1) by loading multiple 3-D models of a compound into the database, and using a more or less standard query targeted at the pharmacophore or active configuration; or (2) by placing a single conformer in the database and designing a query with sufficiently wide tolerances to accommodate some portion of the conformational space available to the hypothetical target (Güner *et al.*, 1992, 1990). Although rigorous, the first approach is problematic: on the one hand, the conformational space available to a small organic molecule may not be sampled adequately with only a few conformers, thus requiring tens to hundreds of models per molecule; on the other hand, search performance will suffer in direct proportion to the completeness of sampling. Although the second approach works for databases of all types, there are some problems: decreased search performance due to wide tolerances, the level of conceptual complexity involved in constructing such a flexible query, and a virtual certainty that at least some models will still be missed. Neither approach is satisfactory.

The culmination of several research programs has been the introduction of several database search engines with the capability of perturbing, during the search, the individual 3-D structures stored in the database to discover if any might be able to adopt the configuration presented by the query. This is conformational flexible searching (CFS) and it addresses all of the problems cited earlier, at the costs of computational complexity and reduced search-time performance. Several commercial systems now exist that include CFS algorithms. In approximate chronological order of introduction, these include Chemical Design's ChemDBS-3D system (1991), BioCAD's Catalyst (1992), Tripos' SYBYL/3D Unity (1993), and MDLI's MACCS and ISIS products (1994).

Both the ChemDBS-3D and Catalyst systems address the CFS problem by collecting rotational isomeric states at a specified density for each compound, effectively performing a rapid analysis of conformational space at registration time, and either storing discrete coordinates for multiple conformers or rapid shortcuts for their generation at search time. For database generation, both can generate multiple conformers for flexible rings only if they exist in a fragment library or lookup table. In addition, Catalyst employs a proprietary algorithm which allows a molecule's conformational space to be sampled at uniform "intervals" to any arbitrarily specified conformer density.

Both SYBYL/3D Unity and the MDLI MACCS and ISIS systems address the CFS problem by requiring only a single registered conformer, performing sophisticated feature identification and keying at registration time, and then adopting search algorithms which perform a directed torsional search of conformational space at search time. Both vendors offer search options for the perturbation of flexible rings.

Concern about the performance of CFS algorithms is tripartite. First, how fast is CFS, in sheer performance? Not unexpectedly, current CFS algorithms suffer in performance relative to their 2-D counterparts, taking between 4 and 20 times longer to complete over databases of the same size. Second, it should be realized that neither registration-time nor search-time CFS algorithms are intended to comprehensively sample the entirety of conformational space. And third, how efficient is CFS in discovering valid active compounds that would have been missed by rigid or flexible query searches? As an example, the $\pi$-stacking query mentioned earlier, when executed against MDDR-3D under CFS, returned 27 hits, 22 more than the rigid 3-D search. On several occasions, rigid 3-D searches have been shown in comparison with flexible searching to collect only a fraction of the relevant target structures in a database: Haraki *et al.* (1990) found values ranging from 5 to 80%, and Davies and Upton (1990) found values around 25%. Curiously, one interpretation of the study by Haraki *et al.* (1990) is that while CFS searches are more efficient than rigid searches in total hits, they are less efficient when judged by the "noise" (the percentage of hits known *not* to be active). The trade-off is between exhaustivity and specificity.

## 6. 3-D Similarity Searching

As with the 2-D mode, 3-D similarity searching involves the identification of all models in a chemical database that are loosely related to the query structure. The query comprises the complete structure, and geometry, to which similar hits are sought. However desirable it might be, in the context of existing database systems the definition of "similar" is not taken to mean "similar shape," at least by the tenants of molecular shape recognition. Instead, 3-D similarity is defined by analogy to the 2-D version of similarity and uses the 3-D search keys as defined for the query and the database entry, as applied with a variant of the Dice algorithm. The use of 3-D feature keys as a similarity metric is a compromise; true 3-D shape similarity in chemical databases is an active research topic.

Three-dimensional similarity searching is another mode supported by macromolecular search engines such as PROTEP (Tripos). Under the "partial search" mode, a template protein (and its motifs) is identified, geometric tolerances and similarity measures provided, and used as a query. This is directed against a collection of proteins with the assumption that not all motifs need be present for the target to be considered a hit; this judgment is performed by clique detection under the guidance of the similarity measure provided.

## 7. Derivation of Pharmacophore Models

The concept that a receptor recognizes electronically similar regions on molecules which are substructurally different but geometrically similar was proposed by Kier (1971). He borrowed Ehrlich's phrase "pharmacophore" and provided its current meaning as the three-dimensional arrangement of functional groups necessary for a molecule to exert its biological effect. This section will cite a number of methods by which 3-D queries may be formulated; the emphasis is that these templates can be computed deterministically. The review by Martin (1992, Table 1) provides additional suggestions about query formulation.

*Queries from published pharmacophores* Proposals for pharmacophore models have appeared in the medicinal chemistry literature for several decades. In most cases, these models may be used directly in the formulation of 3-D queries. Translation of hydrogen bond donor and acceptor character can be accomplished by use of Markush structures, and vectors can be defined with explicit hydrogens or lone pairs. A detailed review and exhaustive compilation of these sources would be a valuable asset to the domain of 3-D searching.

*Queries from single compound modeling* Often a single active lead compound, which happens to be moderately rigid, can serve as the basis of a useful pharmacophore model; the structure can be obtained by small molecule crystallography or solution NMR spectroscopy. Even for a flexible compound, the bioactive (bound) conformation can often be derived from X-ray complexes or NMR studies using transfer NOE.

In the absence of extensive structure–activity data, requirements for specific functional groups will be uncertain, as will be the importance or accuracy of their relative orientations in space; the pharmacophore hypothesis will therefore rely on the chemical intuition of the researcher. This is sufficient data, however, to perform a 3-D similarity search (preferably with a system supporting shape similarity or CFS) to obtain ideas for design or targeted screening efforts. Selective deletion of the supporting skeleton can yield a set of fragments which may be used for a "fix all atom" 3-D submodel search, with the same intention.

*Queries from ensemble molecular modeling data* Ensembles of ligands or macromolecules may be used as the basis for a pharmacophore model.

The active analog approach of Marshall *et al.* (1979) can be used profitably for the development, through superposition of models, of the identity and geometry of functional groups strongly implicated in the biological activity of a lead series. Likewise, the "ensemble approach" based on distance geometry (Sheridan *et al.*, 1986) can also be used to deduce a pharmacophore. The aggregate model may be directly translated into a 3-D query by judicious editing and measurement of geometric parameters in any molecular modeling package, and transfer of the minimal fragment geometry into a query formulation front end.

A similar approach is useful for macromolecular ensembles. Multiple conformations of an active agonist, obtained by the techniques of NMR, molecular dynamics, or conformational analysis, are first overlaid to form an aggregate model. Important functional groups are identified by other techniques (directed single point mutation, affinity studies, etc.), and the backbone skeleton and unimportant residues are deleted. The resulting "clouds" of fragments represent the space which may be occupied by those groups deemed important to biological activity. A statistical analysis is performed on the spatial arrangement of these "clouds," the centroid of each "cloud" determined, and the mean and standard deviation of the intercentroid distances obtained. The same analysis may be performed on intercentroid angles and dihedrals. Finally, the representative functional groups from a single structure are extracted, and the average geometric values are applied to these substructural fragments as 3-D query constraints.

***Queries from ligand binding sites*** The binding site of a receptor or enzyme, as determined by any analytical technique, is a prime candidate for pharmacophore derivation.

In the absence of a bound ligand, an extra-radius molecular surface ("Lee and Richards surface") of the site may be prepared, and complementary regions of potential ligand binding are identified and visualized (Bohacek and McMartin, 1992). Likewise, the GRID method of Goodford (1985) and Boobbyer *et al.* (1989), which calculates the optimal interaction energy between a variety of chemical probe groups and a target macromolecule, can be used to generate energy contours and visualize "hot spots" for probe binding or repulsion. The chemical nature of these regions is then translated into substructural fragments or Markush structures, and their geometry is embedded in distance and angular constraints. One problem with this technique is that macromolecular flexibility often cannot be accounted for in the absence of a bound ligand; it is therefore limited to those cases where the enzyme or receptor is suspected or known to exhibit limited flexibility in the vicinity of the binding site.

The ideal case is where one has a structure illustrating the complexed ligand. The advantage of having structural data for one or more ligands complexed with their binding site is that the binding geometry with receptor sites is clear; these vectors can be formulated into the 3-D query with reasonable confidence of their accuracy. Conformational analysis of the receptor site and the variability with which each residue may be located can be converted into tolerance on a per-residue basis.

## D. Search Engines

This section describes in detail the capabilities of two popular chemical database search software packages and briefly summarizes the specialized capabilities of one macromolecular search system. The systems chosen are representative of a much larger collection of commercial and academic packages, many of whose features have been described elsewhere in this section. See Appendix for complete details.

### 1. Commercial Products

Each of the following three products has been developed for a specific niche, and each has one or more specialized features that are not shared. Most of these features are dependent on the nature of the data being searched.

*MDLI MACCS* The MACCS program (Molecular Access System) from MDLI began as a database creation, searching and reporting package for 2-D structure databases, and has grown over the years to acquire additional modular functionality and interfaces to other nonstructural databases. MACCS is a "mainframe" package, with versions for VAX/VMS, VM/CMS, and MVS/TSO. Buried within MACCS is a chemistry-knowledgeable application-specific programming language with which simple forms or sophisticated systems can be built. In 1990, the database format was extended, providing 3-D data storage and rigid 3-D searching (Christie *et al.*, 1990b). In early 1994, the first release was made of a CFS algorithm already tested in MDLI's ISIS product.

Interactive query formulation and searching occur through a terminal-based, mouse-driven graphical user interface; reports may be directed to the screen for interactive browsing or routed through custom forms to paper. Queries may be superimposed on the molecules in either form. The terminal-based environment does not support high-quality molecular visualization and manipulation.

The 2-D domain is limited to chemical structures with 255 atoms, 255 bonds, and 128 stereo centers; MACCS is not intended to store and manipulate macromolecules. Although some performance tests indicate that storing greater than 350,000 compounds may result in inefficient searches (Hicks and Jochum, 1990), there is no formal storage limit. Several "fuzzy" substructural query options are supported, including atom lists, variable bond types, variable connectors, and R-groups (Markush patterns). Preliminary 2-D screening uses 960 substructural and atom-centered keys, 166 of which are known to the user and may be accessed explicitly.

Entries in the 3-D domain must map isomorphically with a registered 2-D structure (except for hydrogens and lone pairs). Per-model, per-atom, and per-atom-pair data are supported. Hydrogens and lone pairs may be sprouted during searches when so dictated by the query. The following 3-D query features are supported for rigid searches: point, line, plane, centroid, normal, distance, angle, dihedral, or exclusion sphere. Both angles and dihedrals may be proper or improper (i.e., may be defined by points or atoms that are not contiguously bonded). Three-dimensional screening uses 1886 keys based on distances and angles between selected features (described by Christie *et al.*, 1990b), none of which may be accessed explicitly by the user.

Similarity searching, available only in the 2-D domain, is simply based on key ratios, following a modified Dice algorithm. Specialized search capabilities include a recent implementation of run-time CFS and radically enhanced search performance across the 2-D structure domain.

***CCDC QUEST-3D*** The search and analysis software associated with the CSD started out as the BIBSER, CONNSER, and GEOM programs (Allen *et al.*, 1979), and developed into QUEST and GSTAT (Allen *et al.*, 1991a); the statistical analysis of molecular geometry was always a critical portion of the software suite. Starting with version 4, a graphical user interface version of QUEST became available, which supplanted the user-unfriendly CONNSER syntax in many cases. Two major features in the version 5 release—nearly complete 2-D to 3-D matching records and rigid 3-D feature searching in QUEST-3D—support a higher level of 2-D and 3-D search integration than previously possible. GSTAT is still necessary for postsearch statistical analysis, although a new graphical interface, VISTA, has been released. The CSD system runs on a variety of platforms, from "mainframes" to window-based UNIX workstations.

Interactive query formulation and searching occur through a mouse-driven graphical user interface (GUI). Search results may be directed to the screen for interactive browsing, but several intermediate data files serve to communicate with GSTAT and VISTA; one may browse, viewing either or both of the 2-D diagram and the 3-D unit cell (packed or asymmetric unit). When browsing, queries may be superimposed on the diagrams or the 3-D models. The standard GUI does not support high-quality molecular visualization and manipulation; plotting is the domain of the PLUTO program. Statistical analysis of the values adopted by query constraint parameters is an integral part of the summary reports generated at the close of each search.

Three modes of searching CSD are provided: 1-D, 2-D, and 3-D. These modes are described in Table II: 1-D information consists of several per-

**TABLE II** Data Fields in the Cambridge Structural Database[a]

| Data field | Description |
|---|---|
| REFCODE | Unique identification code for each entry; consists of six alphabetic characters which "define" the chemical compound and a possible further two numeric digits which "trace the publication history." |
| 2-D diagram | Based in the FCON format, multiple fields comprising residue association and net charge, atom identity (or special group symbol) and formal electronic charge, and connectivity with bond order. |
| 3-D model | Based in the FDAT format, multiple fields comprising crystallographic unit cell parameters, symmetry matrix, atomic radii, 2-D to 3-D matching information, atom identity and coordinates, and "crystal connectivity." |
| Citation | Multiple fields (AUTHOR, SURNAME, CODEN, VOLUME, PAGE, YEAR) defining the original reference; the underlying fields are separately searchable. |
| COMPOUND | Chemical name originally assigned by author or systematic name. |
| SYNONYM | Alternate name in about 7% of cases (tend to be drug molecules). |
| BCLASS | Basic chemical class; a numeric code from 1 to 86 classifying the primary residue in the entry (see footnote for Table I). |
| CLASS | Chemical class; multiple classification codes may be assigned if the primary residue in the entry carries multiple membership (see footnote for Table I). |
| ADATE | Accession date; date of first inclusion in CSD. |
| MDATE | Date of most recent modification. |
| Identification numbers | Cross references to other indexes: CAS (Chemical Abstract Services registry number, 12% of entries); MSD (published in Molecular Structures and Dimensions, 34% of entries); and NBS (National Institute for Standards and Technology identification number, formerly the National Bureau of Standards, 99% of entries). |
| SPACEGROUP | Space-group symbol. |
| SPGN | Space-group number. |
| RFACTOR | Measure of precision of structure determination. |

[a] This summary of data records is derived from Allen *et al.* (1992) and is not intended to be exhaustive; only representative fields and those deemed of interest to medicinal chemists have been detailed. In some cases, several logically related data fields have been combined.

entry bibliographic, identification, and text/numeric data fields; 2-D and 3-D information is as defined. A query may be formulated from any combination of these modes and may, in addition, contain explicit "bit screens," 2-D, and/or 3-D constraints.

The CSD contains nearly 700 bit screens which are derived from the entries themselves, serve as search performance enhancers in QUEST-3D, and are accessible for user queries. Many more bit screens are used for 3-D constraint filtering, but these are not user-accessible. The underlying 3-D features that are keyed include intramolecular distances (bonded and nonbonded, atoms, and aromatic centroids), proper torsion angles, and adjacent torsion angles (a pair of proper torsions sharing two edges); details are provided in Allen *et al.* (1992; Vol. 1, Section 7.16).

Two-dimensional constraints allow "fuzzy" specification of both atom-centered and bond-centered environments. Examples include atom lists, element groups, variable hydrogen and connected atom counts, variable bond types, and prohibitions on interresidue, cyclic, or direct linking. Three-dimensional constraints involve the definition of various geometric objects such as points (dummy atoms and centroids), vectors, planes, normals, distances, angles, dihedrals, and inclusion and exclusion spheres. Various special-purpose ring puckering and directionality parameters are also supported. Similarity searching is supported only in 2-D and is based on the chemical connectivity bit screens as manipulated by the algorithms derived by Willett *et al.* (1986).

The nature of the data, a fully defined crystallographic unit cell, permits certain specialized search capabilities from within QUEST-3D dealing with intermolecular interactions. In particular, one may search for single or multiple nonbonded contacts, in which van der Waals contacts are only one category; special parameters exist to permit lone-pair directionality to play a role in queries. The various symmetry operators already in place also support searches which automatically handle structural inversion. One extremely powerful special feature, unrelated to the nature of the data, is the ability to define derived numeric and 3-D features as query elements. These "composite geometrical parameters" are achieved with the TRANSFORM operator, which allows construction of unary and binary equations whose parameters are the values of already defined 3-D constraints. Thus one may search on the absolute value of the cosine of a dihedral or perform simple arithmetic.

***Tripos PROTEP*** PROTEP (Protein Topographic Exploration Programs) is designed to perform 3-D similarity searches for secondary structure and substructural motifs in proteins, independent of the constraints of primary sequence (Tripos Associates, 1992). This suite of programs grew out of the work of Artymiuk and others in the laboratory of Willett (Mitchell *et al.*, 1990; Artymiuk *et al.*, 1992; Grindley *et al.*, 1993) and relies strongly on modifications of the subgraph isomorphism algorithm to work on representations of previously identified motifs. Protein motifs are identified with the techniques of Kabsch and Sander (1983); the motif records in the original PDB files are ignored. The PROTEP modules are part of the Tripos SYBYL molecular modeling suite, having been introduced in 1992; PROTEP only runs in a UNIX workstation environment.

The collection of data that is searched is a local copy of the Brookhaven PDB, in native format. PROTEP is one example of a program that has evolved as a search engine for PDB to fill a special niche.

Interactive query formulation and searching occur through a command line-based window; formulation via GUI is not yet supported. Searches through the entire PDB may take 10–30 min, depending on the nature of the query. Search results may be browsed manually using the default visualization and manipulation routines of SYBYL; query and target motifs may be displayed with colored-filled arrows, superimposed on the C$\alpha$ trace or complete protein structure.

PROTEP is designed to find purely structural trends in sets of proteins. Searches may be for "partial match" or "complete match," which correspond roughly to 3-D similarity search and substructural search, respectively. In a complete match search, a pattern is defined (for example, an assemblage of $\beta$-sheets in a $\beta$-barrel, a helix bundle), then the geometric constraints which must be satisfied are defined (intersheet distances and angles). In a partial match search, an entire protein structure from the PDB can be used as a template, the minimum matching number of motifs can be defined (as a measure of similarity), and then the geometric constraints which must be satisfied can be defined. In either type of search, the primary sequence order may be ignored, as desired.

### 2. Integration with Molecular Modeling Packages

One important factor in selecting a chemical database search engine is how well it is integrated with, or interfaces with, the current variety of molecular modeling software packages. The major issues are the interaction between the search engine and the model visualization software for the tasks of (a) query formulation, and (b) hit list visualization and manipulation. A search engine that is integrated by a vendor into a suite of modeling programs may provide more facile browsing of the resultant list of models, but the trade-off may be the need to use software with features unsuitable to the task at hand, e.g., incomplete query semantics.

Here is a brief evaluation of several commercial database search engines along these lines. When reference is made to "viewing the resultant molecules," note that in relevant cases, the registered 3-D conformation has been torsionally transformed to conform to the query, and the query has been superimposed on the model.

Both ChemDBS-3D and the SYBYL/3D Unity and PROTEP systems have integrated the 3-D searching into a wide variety of commercially successful molecular modeling modules, and consequently the query and browsing process is simple. The SYBYL system in particular has an elegant "Molecular Spreadsheet" module at its core providing excellent process and data organization. Catalyst is primarily a query formulation and statistical evaluation package, with integrated 3-D searching, browsing, and visualization; on earlier releases, browsing the results database was quite arduous, but this is to be enhanced in future releases. In all the aforementioned packages, when operated on an appropriate workstation, visualization and interactive manipulation is of high quality. Neither ISIS/3D nor MACCS-II/3D are integrated with molecular modeling packages; the explicit intent is to specialize on database construction and searching and to work with other vendors of modeling software to achieve interfaces via standard file formats or interprocess communication, depending on the platform. Hence, although browsing is rapid, visualization is of low quality (Tektronix graphics), and "transparent" exchange with modeling software presents difficulties.

### E. Successes of Three-Dimensional Database Searching

That 2-D and 3-D database searching is a valuable part of the drug design process has become apparent since the 1980s as a series of successes have accumulated and been published. Many more success stories remain hidden behind the wall concealing corporate proprietary research. This section cites and summarizes some noteworthy examples.

Sheridan *et al.* (1989) discussed three examples in their summary: the CNS pharmacophore and the ACE inhibitor pharmacophore, both of which were based on published SAR and proposed structural features, and the L-tryptophan repressor pharmacophore, which was developed from the known structure of a protein–ligand complex, PDB entry 1WRP. Christie *et al.* (1990a) cited three examples, but the one of import is the cardiotonic pharmacophore based on Milrinone. Although both of these studies comprise "retrospective" analysis, several valuable design insights were achieved in addition to the retrieval of known active compounds. Several other "success stories" are reported in Martin *et al.* (1990), Willett (1991; Chapter 6), and the review by Martin (1992).

The area of inhibitor design for the HIV-1 protease is one in which the value of 3-D searching has been particularly evident. Both Bures *et al.* (1990) and DesJarlais *et al.* (1990) discuss their work with this system. DesJarlais *et al.* (1988) used DOCK; although DOCK is not a complete chemical database system, its use is illustrative. Bures *et al.* (1990) used then-available protease–inhibitor complex structures with A-74704, devised a series of 3-D queries, and then used ALADDIN (Van Drie *et al.*, 1989) to conduct 3-D searches across both corporate and commercial 3-D databases. Several hits thus identified were assayed and were discovered to exhibit micromolar activity, even without further refinement.

A group from DuPont Merck has cited the critical use of 3-D searching, also in support of HIV-1 protease inhibitor design (Lam *et al.*, 1994). The problem was similar to that of Bures *et al.* (1990), basically the redesign of a known peptidomimetic inhibitor, a C2-symmetric diol, to reduce its molecular weight and to incorporate a mimic for a structural water molecule uniquely found in retroviral protease–inhibitor complexes. A 3-D pharmacophore was developed based on computer modeling of the diol-HIV interaction and was used as a query for MACCS against the CSD. One of the resultant hits only met the initial search criteria and suggested the use of a six-membered ring as scaffolding, but also included an oxygen that matched the structural water. No activities were reported for this molecule, but it then served as the starting point for the development of a seven-membered cyclic urea which exhibits subnanomolar activity and is highly bioavailable. The design success in both displacing and mimicking the structural water was confirmed by X-ray crystallographic studies.

## F. In Depth: The Brookhaven PDB

As an international repository, the Brookhaven PDB accepts depositions of atomic coordinates, bibliographic citations, and primary sequence and secondary structure information, as well as crystallographic structure factors and 2-D NMR experimental data, on biological macromolecules. Currently, this includes proteins, RNA, DNA, viruses, and carbohydrates. There is no restriction on the origin of these coordinate sets, but most are the result of X-ray diffraction or NMR spectroscopy. As an organization, the PDB was established at the Brookhaven National Laboratory in Upton, New York in 1971, with a charter to collect, standardize, and distribute atomic coordinates and other data from macromolecular studies. Many journals now require the submission of coordinates to PDB as a prerequisite for acceptance for publication; as a service, the PDB will hold coordinate sets private until their release is sanctioned by the submitting laboratory. The compilation is maintained in large part by U.S. government research funds, and thus macromolecular structures are in the public domain. Nominal charges are in place to cover media costs and shipping. The PDB release schedule is quarterly. In addition to the main depository at Brookhaven, subsidiary depositories in Cambridge (United Kingdom) and Tokyo were later established to serve Europe and Asia, respectively.

The PDB release of January 1994 contained over 2327 fully annotated structure entries. Table III provides statistics of coverage for this release across various macromolecular classes.

The PDB file format is governed by a published standard that is slowly changing to reflect the nature and quantity of data submitted. The current format is that of ASCII-encoded text partitioned into typed 80-character records (Protein Data Bank, 1989). Table IV describes the variety of data types that are supported by most PDB files. Only PDB files directly supplied by Brookhaven are assured of having all of these data properly formatted; the PDB "standard" is manipulated by many third-party software developers as an interchange format, and extremely wide variations may be seen in

**TABLE III** Census of the Brookhaven PDB, January 1994

| Count | Description |
|---|---|
| 2327 | Fully annotated atomic coordinate entries[a] (605 new full-release entries since previous release) |
| 2143 | Proteins, enzymes, and viruses[b] |
| 156 | DNAs |
| 9 | RNAs |
| 9 | tRNAs |
| 10 | Carbohydrates |
| 2013 | X-ray experimental entries |
| 286 | NMR experimental entries[c] |
| 1 | Electron diffraction experimental entry |
| 27 | Theoretical models |
| 353 | Structure factor entries |

[a] The total size of the atomic coordinate entry database is 678 Mbytes, uncompressed.

[b] About 32% of the entries contain macromolecules complexed with ligands (entry 1STP is streptavidin complexed with biotin; entry 1TLP is thermolysin with phosphoramidon); a few ligands are present as separate PDB entries in the same coordinate space as their macromolecular hosts (see entries 6LYZ and 9LYZ for lysozyme and a guest trisaccharide) so that they may be assembled by any molecular modeling package.

[c] PDB entries identified as NMR experimental entries are different in a number of subtle ways from the classical X-ray-derived PDB entry. The major difference is that they commonly consist of a pair of entries: one with a single best-fit computed conformation, and another—often characterized by explicit mention of NMR and MULTIPLE structures in the COMPND record—with multiple catenated conformers (each delimited by MODEL/ENDMDL records). The release of an entire ensemble of computed conformations is extremely useful but has been known to cause problems with some structure conversion programs that were not designed to handle multiconformer PDB files.

both the canonical PDB record structure and overall PDB format! One legitimate format variation is in support of entries with coordinates derived by solution NMR: whereas the standard X-ray-derived PDB entry contains a single structure or conformer, the majority of NMR-derived PDB entries are ensemble collections of multiple conformers. This provides access to the individual computed structures, but the format has proven troublesome to some third-party format translators.

The atomic coordinates stored in the PDB are orthogonalized from the original crystallographic coordinates. The PDB convention for orthogonalizing nonorthogonal crystal axes ($\vec{a}$, $\vec{b}$, $\vec{c}$) to orthogonal Cartesian coordinates ($\vec{x}$, $\vec{y}$, $\vec{z}$) is that:

1. $\vec{x}$ is parallel to $\vec{a}$
2. $\vec{z}$ is parallel to $\vec{a} \times \vec{b}$
3. $\vec{y}$ is parallel to $\vec{z} \times \vec{x}$.

**TABLE IV** Data Fields in the Brookhaven PDB[a]

| Record type | Description |
|---|---|
| HEADER | Unique PDB entry identification code consisting of one numeric and three alphanumeric characters, functional classification of the macromolecule, and date entered into the Data Bank. |
| COMPND | Name of molecule and identifying information; may include Enzyme Commission number for enzymes. |
| SOURCE | Systematic name of species, organ, tissue, and mutant from which the molecule has been obtained, where applicable. |
| EXPDTA | Source of experimental data; now being used to indicate NMR or electron diffraction as alternate sources to the traditional X-ray diffraction. |
| AUTHOR | Names of contributors. |
| JRNL | Literature citation that defines coordinate set. |
| REMARK | General remarks; some of these records are standardized for specific types of comment information, such as reference, resolution, refinement, or correction. |
| SEQRES | Residue sequence: identifies chain (if multiple) and amino acid or nucleotide, as appropriate, with three-character mnemonic. |
| FTNOTE | Footnotes relating to specific atoms or residues. |
| HET | Identification of nonstandard groups or residues. |
| FORMUL | Chemical formulae of nonstandard groups. |
| HELIX | Identification of helical substructures. |
| SHEET | Identification of sheet substructures. |
| TURN | Identification of hairpin turns. |
| SSBOND | Specification of disulfide bonds. |
| SITE | Identification of groups comprising the various sites. |
| CRYST1 | Unit cell parameters and space group designation. |
| ORIGX | Transformation from orthogonal Angstrom coordinates to submitted coordinates; three records: ORIGX1 through ORIGX3. |
| SCALE | Transformation from orthogonal Angstrom coordinates to fractional crystallographic coordinates; three records: SCALE1 through SCALE3. |
| MTRIX | Transformations expressing noncrystallographic symmetry; three records: MTRIX1 through MTRIX3; multiple sets may be present. |
| TVECT | Translation vector for infinite covalently connected structures; only relevant for about nine structures. |
| MODEL | Brackets the beginning of records for a single model; only found in NMR multiconformer files, paired with ENDMDL. |
| ATOM | Atomic coordinate records for "standard" groups. |
| HETATM | Atomic coordinate records for "nonstandard" groups. |
| ANISOU | Anisotropic temperature factors; rarely provided. |
| TER | Chain terminator record. |
| CONECT | Connectivity records. |
| ENDMDL | Brackets the end of records for a single model; only found in NMR multiconformer files, paired with MODEL. |
| MASTER | Penultimate master control record with check sums of total number of records in the file for selected record types. |
| END | End-of-entry record. |

[a] This partial summary of data records is derived from Protein Data Bank (1989) and lists records in their approximate order of inclusion in a typical PDB file. This list is incomplete due to the elimination of obvious, rarely used, or obsolete records (SIGATM is only present in a single entry, 4RXN; other records ignored include REVDAT, OBSLTE, and SPRSDE).

Deriving the orthogonal coordinates ($x_o$, $y_o$, $z_o$) from those along the crystallographic axes ($x_c$, $y_c$, $z_c$) is described by the matrix:

$$\begin{vmatrix} x_o \\ y_o \\ z_o \end{vmatrix} = \begin{vmatrix} 1 & \cos\gamma & \cos\beta \\ 0 & \sin\gamma & \dfrac{\cos\alpha - \cos\beta * \cos\gamma}{\sin\gamma} \\ 0 & 0 & \dfrac{\text{vol}}{\sin\gamma} \end{vmatrix} \begin{vmatrix} x_c \\ y_c \\ z_c \end{vmatrix} \quad (7)$$

$$\text{vol} = \sqrt{1 - \cos^2\alpha - \cos^2\beta - \cos^2\gamma + 2\cos\alpha * \cos\beta * \cos\gamma_w}$$

here and $\alpha$, $\beta$, and $\gamma$ are the unit cell angles. Clearly, for unit cells with $\alpha = \beta = \gamma = 90°$, no transformation is necessary.

No generic search engine is provided with the PDB, although numerous vendors and research groups have attempted to fill this gap with systems of astonishing variety. Some notable examples include the PROTEP program (Tripos); the IDITIS relational database (Gardner and Thornton, 1991); encoding the Kabsch and Sander algorithm (1983) in PROLOG and storing a subset of the PDB within the PROLOG system (Barton and Rawlings, 1990); and the BIPED relational database project (Islam and Sternberg, 1989), part of the ISIS effort undertaken by Birkbeck College and Leeds University. Likewise, no PDB-specific graphical query language exists targeted directly to the PDB molecular domain and its characteristic motifs and substructural features (although a semigraphic syntax was developed as an interface to PROTEP).

In most of the efforts to provide a search environment for the PDB contents, the motifs defined by the submitting authors in the PDB records HELIX, SHEET, and TURN are seldom used exclusively. Instead, the research efforts recalculate these features using published algorithms, e.g., the classic Kabsch and Sander (1983) work on secondary structure pattern recognition. The rationale for recalculation is the lack of unambiguous standards defining each motif in the PDB submission; which characteristics may be required as part of an $\alpha$-helix or $\beta$-sheet are different for each observer, especially if the assignment is by eye and not by algorithm. Use of well-defined and well-characterized secondary structure assignments as provided by Kabsch and Sander (1983) goes a long way to relieve this problem.

The files and supporting material of the PDB may be obtained quarterly in "full releases." The "pre-release" mechanism used briefly during 1993 is no longer in effect. Standard release medium has been computer tape, but more modern techniques are now also available. New submissions may be uploaded, and files may be retrieved from the PDB, via Internet and other Internet tools such as Gopher or ftp (see Appendix); software and data support from the PDB are also available through this same mechanism. Use of the Internet enhances tremendously the utility and convenience of these data. The PDB files have been released on CD-ROM by the journal *Protein Science* (White, 1993); the CD-ROM also includes the MAGE and PREKIN structure display and manipulation software (Richardson and Richardson, 1992).

Research over many years has relied on the data in the PDB. The macromolecular structures have been put to any number of diverse uses, including these examples. The PDB has become a reference source for protein substructures in assisting in homology modeling and X-ray structure determination; this has been described by Jones and Thirup (1986), Claessens *et al.* (1989), and Jones *et al.* (1991), which includes the X-ray map-fitting programs FRODO and "O," and by Shenkin *et al.* (1987). The PDB provided conformations of bound protein–ligand complexes for pharmacophore development and 3-D database searching; in Sheridan *et al.* (1989), the L-tryptophan repressor pharmacophore was developed from the known structure, 1WRP. It is interesting to note that about 32% of the PDB entries contain complexes of macromolecular hosts and guests, making this a prime source for studying protein/ligand interactions; these entries are characterized by the canonical use of the word COMPLEX in the COMPND record.

Only a few problems and shortcomings are worth citing, and most concern the PDB format itself. The common problem of partial occupancy in crystalline but mobile structures is dealt with, but not in a totally satisfactory manner. As mentioned earlier, the minor change to accommodate the ensemble of multiple conformers calculated from solution NMR data has presented some problems with third-party software. A partial solution taken by the PDB has been the provision of two files for these entries: one with a single computed conformer (perhaps the best-fit) in canonical PDB format, and the other a multiconformer partially compliant PDB file. Another problem with many PDB-like formats is the inability to distinguish between $\alpha$-carbons of proteins and calcium ions, both written as "CA." The obvious conclusion is that the PDB format has become obsolete and needs to be replaced.

The Crystallographic Information File (CIF; Hall *et al.*, 1991) archive file has been adopted as the future standard format for the PDB. The CIF format follows the STAR (self-defining text archive and retrieval) format (Hall, 1991). Although the CIF has been fully defined for small molecule crystallography, extensions necessary for macromolecular crystallographic data have not yet been established. An example of a small molecule entry in the CIF format is given in Table V.

## G. In Depth: The Cambridge Structural Database

The Cambridge Structural Database is a large database of high-quality evaluated structures of organic and organometallic compounds as determined by X-ray and neutron diffraction (Allen *et al.*, 1991a). The CSD has

**TABLE V** Example of a Crystallographic Information File (CIF)[a]

```
_chemical_name_systematic
;
2-aminopropanoic acid
;
_chemical_formula_sum           'C3 H7 N O2'
_cell_length_a                  6.137(1)
_cell_length_b                  12.784(2)
_cell_length_c                  18.485(1)
_cell_angle_alpha               90
_cell_angle_beta                90
_cell_angle_gamma               90
_cell_volume                    1450.2(3)
_cell_formula_units_Z                4
_symmetry_space_group_name_H-M       'P21 21 21'
loop_
_symmetry_equiv_pos_as_xyz
     +x,+y,+z  1/2-x,-y,1-2+z
     1/2+x,1/2-y,-z  -x,1/2+y,1/2-z
loop_
_atom_site_label
_atom_site_Cartn_x
 atom site Cartn y
_atom_site_Cartn_z
_atom_site_U_iso_or_equiv
_atom_site_thermal_displace_type

C1        1.075(3)  3.828(2)  4.811(2)  .065(1)   Uani
C2        1.906(4)  2.570(4)  5.018(2)  .056(2)   Uani
C3        3.057(6)  2.488(3)  4.025(3)  .084(1)   Uani
N2        1.049(3)  1.378(3)  4.883(2)  .041(1)   Uani
O1        0.326(5)  3.942(2)  3.843(1)  .047(1)   Uani
#...........................................data omitted for brevity
H31       3.94(1)   2.13(1)   4.54(1)   .17000    Uiso
H32       3.24(1)   3.48(1)   3.62(1)   .15000    Uiso
H33       2.78(1)   1.80(2)   3.23(1)   .28000    Uiso
_publ_cont_author

     Prof. Mark F. Schneider
     Dept. of Chemistry
     Texas State University
     Hendricks, TX 78462
     U.S.A.
```

[a] Data are identified by a data name (e.g., _chemical_name_systematic) followed by the data itself as *text*, *number*, or *character*. Large blocks of text are set off by lines beginning with semicolons (;). A long list of data of the same type and format is simplified by use of a loop_ statement followed by the data labels and then the data itself. Loops are ended by the appearance of a new data label. A full description of the CIF can be found in Hall *et al.* (1991).

been compiled by the Cambridge Crystallographic Data Centre (CCDC), a small nonprofit organization in existence since 1967 whose charter is the combing of the X-ray literature (Allen *et al.*, 1979) and the collection of structures into biannual CSD releases. Coverage of literature data is fully retrospective to 1935. Although the original published structures are in the public domain, several tasks (quality control, indexing, 2-D substructure, and 3-D feature keying) constitute value added by CCDC, resulting in a small one-time charge for the database. Search software is licensed following the practice of many software vendors, with periodic maintenance fees; separate price schedules apply for academic or commercial customers.

The CSD release of April 1994 contained over 120,480 structural entries relating to nearly 107,850 unique chemical compounds. Of these, 89% had 3-D atomic coordinates present and 83% had partial or complete 2-D to 3-D matching records. Table I provides statistics of coverage for this release across various compound classes.

The CSD format is a set of proprietary binary files which contain indexes, keys, and data for both 2-D and 3-D aspects of crystallographic data. Three common interchange formats have been published and are still in use by the QUEST, GSTAT, and PLUTO software and other third-party software; these are the FDAT, FCON, and FBIB "records" which describe, respectively, the 3-D crystallographic entry, the 2-D structural connectivity, and 1-D bibliographic data. Table II describes the variety of data types which comprise a complete binary record; for a description of the interchange formats, see Allen *et al.* (1992; Vol. 3). A subset of the CSD is now available from CCDC in MDLI's proprietary binary format containing only molecule (not crystallographic) data; this version, covering about 75% of the CSD, represents nearly all those entries for which 2-D to 3-D matching records exist. The intermediate files from this effort, in SDfile interchange format, may be used to construct 3-D molecule databases under other database managers.

The compound identification code (REFCODE, or reference code) explicitly carries information that allows tracking of publication history and consequent revisions of the data. The REFCODE format is described in Table II; the suffix numeric characters allow different crystal forms or different studies by different laboratories. Stereoisomers of a compound are assigned different REFCODEs; this is also the case for deuterated analogs of the normal hydrogen species.

Since both the software and data are licensed for local use, access is through the workstations or computers at the installed site. There is no centralized facility maintained for user access at, nor on-line access mechanism supported by, CCDC. Telephone support is available, but a more convenient mechanism for both software support and database questions is to use the Internet (see Appendix).

Research over many decades has relied on the quality of the data in the CSD. Many different types of experiments can be carried out with these data, and it has frequently been used not only as a reference data set for questions which probe not only the nature of the chemical entities contained therein, but also as a "standard" collection from which subsets are frequently developed, which are then searched or manipulated further. The following is a brief and very incomplete sampling of some important and unusual uses of the CSD.

Allen *et al.* (1983) summarized the broad nature of CSD use at that date; studies on intramolecular interactions included mean atomic and molecular geometries, and substituent and hybridization effects; studies on intermolecular interactions included hydrogen bonding, and the inference of reaction pathways. Taylor and Kennard (1984) discuss their classic work on hydrogen bonding. Dunne *et al.* (1991) discuss steric effects in determining phosphine geometry. Lawrence and Davis (1992) used the CSD as the source for a collection of small molecules to bind with a known protein site, in procedure development. Over the years, the Willett laboratory (Brint and Willett, 1987; Poirrette *et al.*, 1991, 1993) has used the CSD as a tool against which to check algorithm and screen development for novel 3-D searching techniques. Allen *et al.* (1991b, and earlier citations) and Allen and Johnson (1991) have reported on the development and testing of algorithms for the automated conformational analysis of structures in a database. Lipkowitz and Peterson (1993) analyzed benzene and cyclohexane structures from the CSD in showing that the benzene ring is not very rigid. Finally, crystal structures serve as an important source of data which may be used to calibrate the functional parameters of empirical (molecular mechanics) methods.

Only a few problems and shortcomings are worth citing in the CSD. The terminal-based graphical user interface maintains the look and feel of mid-1980s software; but as users move to windows-based, UNIX platforms this becomes less of a concern. Even though a proprietary format, the binary ASER file may be manipulated if a nondisclosure agreement is signed; however, on the VAX, this file is DCX compressed, and the lack of a CCDC-supported, decompression-mediating, programmatic interface to the CSD inhibits the access of the crystallographic data by in-house-developed, VAX-based, analytical software. The present inability to support the user creation of proprietary databases in CSD format, so that QUEST and GSTAT may be applied to unpublished data, may soon be alleviated with the release of their builder program.

## III. IUPAC CONVENTIONS FOR PEPTIDES AND NUCLEIC ACIDS

### A. Peptides

#### 1. Amino Acids

The full IUPAC-IUB (International Union of Pure and Applied Chemistry–International Union of Biochemistry) conventions for nomenclature and

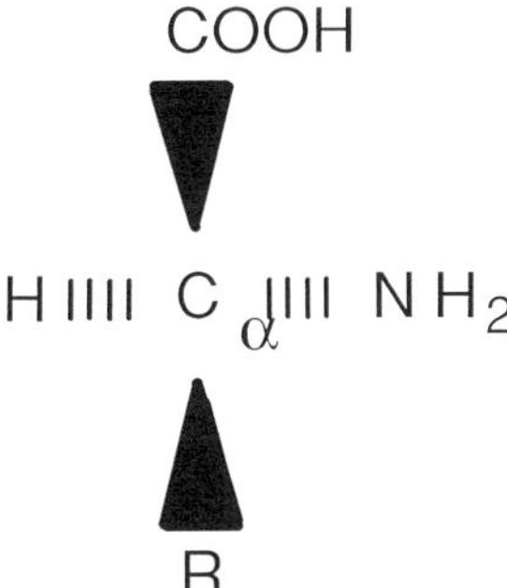


**FIGURE 14** Stereochemistry around the $\alpha$-carbon of the L-amino acids.

symbolism for amino acids and peptides can be found in *J. Biol. Chem.* (1984, Vol. 260, pp. 14–42). The names and structures of the 20 commonly occurring amino acids with their three- and one-letter codes are given in Table VI.

The chemical groups common to all the amino acids are an amino group, a carboxylic acid group, and a carbon atom connecting them to which is also bonded a hydrogen atom and a varying R-group. These common atoms are linked together by a condensation reaction between the carboxylic acid group of one amino acid with the amino group of the next, forming the main chain backbone of the polypeptide with the various R-groups designated as side chains. The central carbon atom, the $\alpha$-carbon (or $C\alpha$) in all naturally occurring amino acids, except glycine, is chiral and all have the same chirality, namely "S" or L- in relationship to L- and D-glyceraldehyde (Fig. 14). In addition to the $\alpha$-carbon, both threonine and isoleucine have chiral $\beta$-carbons. Only one chirality is found in nature, that shown in Fig. 15.

### 2. Atom Nomenclature

Two systems exist for identifying the atoms in the amino acids. The IUPAC system is based on numbering the carbon atoms from the carboxylic

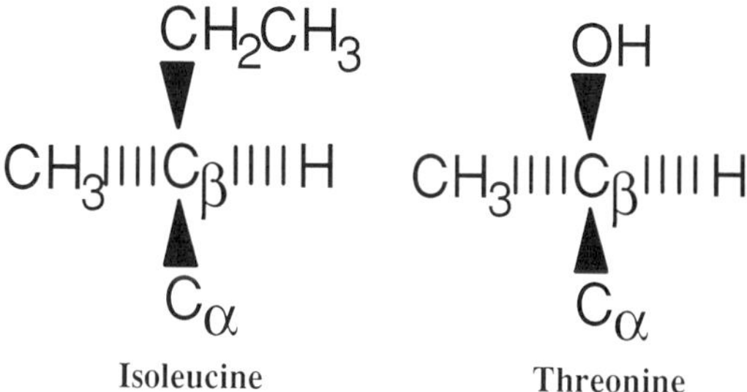


**FIGURE 15** Stereochemistry around the $\beta$-carbon of the naturally occurring diastereoisomers of isoleucine and threonine.

## TABLE VI Nomenclature and Structure for Amino Acids

$\overset{1}{HOOC}-\underset{\alpha}{\overset{2}{C}H}(NH_2)-\underset{\beta}{\overset{3}{C}H_3}$

Alanine (Ala, A)

$\overset{1}{HOOC}-\underset{\alpha}{\overset{2}{C}H}(NH_2)-\underset{\beta}{\overset{3}{C}H_2}-\underset{\gamma}{\overset{4}{C}H_2}-\underset{\delta}{\overset{5}{C}H_2}-\underset{\varepsilon}{NH}-\underset{\zeta}{C}(=\underset{\eta_1}{NH})-\underset{\eta_2}{NH_2}$

Arginine (Arg, R)[a]

$\overset{1}{HOOC}-\underset{\alpha}{\overset{2}{C}H}(NH_2)-\underset{\beta}{\overset{3}{C}H_2}-\underset{\gamma}{\overset{4}{C}}(=O\delta_1)-\underset{\delta_2}{NH_2}$

Asparagine (Asn, N)

$\overset{1}{HOOC}-\underset{\alpha}{\overset{2}{C}H}(NH_2)-\underset{\beta}{\overset{3}{C}H_2}-\underset{\gamma}{\overset{4}{C}}(=O\delta_1)-\underset{\delta_2}{OH}$

Aspartic acid (Asp, D)[a]

$\overset{1}{HOOC}-\underset{\alpha}{\overset{2}{C}H}(NH_2)-\underset{\beta}{\overset{3}{C}H_2}-\underset{\gamma}{SH}$

Cysteine (Cys, C)[b]

$\overset{1}{HOOC}-\underset{\alpha}{\overset{2}{C}H}(NH_2)-\underset{\beta}{\overset{3}{C}H_2}-\underset{\gamma}{\overset{4}{C}H_2}-\underset{\delta}{C}(=O\varepsilon_1)-\underset{\varepsilon_2}{NH_2}$

Glutamine (Gln, Q)

$\overset{1}{HOOC}-\underset{\alpha}{\overset{2}{C}H}(NH_2)-\underset{\beta}{\overset{3}{C}H_2}-\underset{\gamma}{\overset{4}{C}H_2}-\underset{\delta}{\overset{5}{C}}(=O\varepsilon_1)-\underset{\varepsilon_2}{OH}$

Glutamic acid (Glu, E)[a]

$\overset{1}{HOOC}-\underset{\alpha}{\overset{2}{C}H_2}-NH_2$

Glycine (Gly, G)

$HOOC-\underset{\alpha}{\overset{\alpha}{C}H}(NH_2)-\underset{\beta}{\overset{\beta}{C}H_2}-$ imidazole ring ($\gamma$ 4, $\delta_2$ 5, $\varepsilon_2$ $\tau$NH, 2 $\varepsilon_1$, $\pi$ N $\delta_1$)

Histidine (His, H)[c]

$\overset{1}{HOOC}-\underset{\alpha}{\overset{2}{C}H}(NH_2)-\underset{\beta}{\overset{3}{C}H}(\underset{\gamma_2}{\overset{3'}{C}H_3})-\underset{\gamma_1}{\overset{4}{C}H_2}-\underset{\delta_1}{\overset{5}{C}H_3}$

Isoleucine (Ile, I)

$\overset{1}{HOOC}-\underset{\alpha}{\overset{2}{C}H}(NH_2)-\underset{\beta}{\overset{3}{C}H_2}-\underset{\gamma}{\overset{4}{C}H}(\underset{\delta_2}{\overset{5'}{C}H_3})-\underset{\delta_1}{\overset{5}{C}H_3}$

Leucine (Leu, L)

$\overset{1}{HOOC}-\underset{\alpha}{\overset{2}{C}H}(NH_2)-\underset{\beta}{\overset{3}{C}H_2}-\underset{\gamma}{\overset{4}{C}H_2}-\underset{\delta}{\overset{5}{C}H_2}-\underset{\varepsilon}{\overset{6}{C}H_2}-\underset{\zeta}{NH_2}$

Lysine (Lys, K)

$\overset{1}{HOOC}-\underset{\alpha}{\overset{2}{C}H}(NH_2)-\underset{\beta}{\overset{3}{C}H_2}-\underset{\gamma}{\overset{4}{C}H_2}-\underset{\delta}{S}-\underset{\varepsilon}{CH_3}$

Methionine (Met, M)

$HOOC-\underset{\alpha}{\overset{\alpha}{C}H}(NH_2)-\underset{\beta}{\overset{\beta}{C}H_2}-$ benzene ring ($\gamma$ 1, $\delta_2$ 2, $\varepsilon_2$ 3, $\zeta$ 4, $\varepsilon_1$ 5, $\delta_1$ 6)

Phenylalanine (Phe, F)[c]

Pyrrolidine ring: $\overset{\gamma}{C}H_2$ (4), $\beta$ $CH_2$ (3), $\delta$ $H_2C$ (5), HN (1), $\underset{\alpha}{C}H$ (2) $-COOH$

Proline (Pro, P)[c]

$\overset{1}{HOOC}-\underset{\alpha}{\overset{2}{C}H}(NH_2)-\underset{\beta}{\overset{3}{C}H_2}-\underset{\gamma}{OH}$

Serine (Ser, S)

$\overset{1}{HOOC}-\underset{\alpha}{\overset{2}{C}H}(NH_2)-\underset{\beta}{\overset{3}{C}H}(\underset{\gamma_2}{\overset{4}{C}H_3})-\underset{\gamma_1}{OH}$

Threonine (Thr, T)

$HOOC-\underset{\alpha}{\overset{\alpha}{C}H}(NH_2)-\underset{\beta}{\overset{\beta}{C}H_2}-$ indole ring ($\gamma$ 3, $\delta_1$ 2, N 1 H $\varepsilon_1$, $\varepsilon_2$ 7a, $\delta_2$ 3a, $\varepsilon_3$ 4, $\zeta_3$ 5, $\eta_2$ 6, $\zeta_2$ 7)

Tryptophan (Trp, W)[c]

$HOOC-\underset{\alpha}{\overset{\alpha}{C}H}(NH_2)-\underset{\beta}{\overset{\beta}{C}H_2}-$ benzene ring ($\gamma$ 1, $\delta_2$ 2, $\varepsilon_2$ 3, $\zeta$ 4, $\varepsilon_1$ 5, $\delta_1$ 6) $-\underset{\eta}{OH}$

Tyrosine (Tyr, Y)[c]

$\overset{1}{HOOC}-\underset{\alpha}{\overset{2}{C}H}(NH_2)-\underset{\beta}{\overset{3}{C}H}(\underset{\gamma_2}{\overset{4'}{C}H_3})-\underset{\gamma_1}{\overset{4}{C}H_3}$

Valine (Val, V)

**TABLE VI** *(continued)*

[a] The structure of the free amino acids given above do not take into account potential ionization and resonance forms of the molecules. For example, under physiological conditions the carboxylic acid of glutamic and aspartic acids is negatively charged and the two oxygen atoms are chemically indistinguishable.

[b] Two cysteine residues can also be covalently bound through oxidation of their S$\gamma$-sulfhydral groups, giving rise to a disulfide bond. These two residues together are considered one cystine residue, whereas each of the constituent cysteine moieties are referred to as half-cystine residues.

[c] In amino acids with ring systems, the ring numbering takes precedence in the IUPAC numbering scheme. In these cases, for the rest of the molecule the $\alpha$ and $\beta$ designation is used. The two ring nitrogen atoms of histidine cause special problems in the IUPAC system because organic chemists and biochemists have employed two different numbering schemes. Therefore, the imidazole nitrogen atoms are denoted by $\pi$ (*pros*, "near") and $\tau$ (*tele*, "far") according to their distance from the rest of the molecule in order to avoid confusion.

acid carbon. The system used more frequently by biochemists is based on Greek letters, which denote the distance of all nonhydrogen atoms from the $\alpha$-carbon. This system is used by the Brookhaven Protein Data Bank (Section II,F). Where branching occurs, subscripts further distinguish between atoms equidistant from the $\alpha$-carbon. In cases where atoms are identical because of resonance (e.g., the N$\eta$ atoms of arginine or C$\delta$ atoms of phenylalanine and tyrosine), the atom whose torsion angle is closest to zero is indicated by the subscript "1." Table VI gives the structures for the 20 commonly occurring amino acids showing both labeling systems. In cases where Greek letters cannot be used, e.g., in ASCII files, the Roman letters listed in Table VII are preferred.

### 3. Conformation of Peptides and Proteins

The full IUPAC-IUB conventions for the description of the conformation of polypeptide chains can be found in *J. Biol. Chem.* (1970, Vol. 245, pp. 6489–6497). As mentioned earlier, amino acids can be linked together giving rise to an amide bond, called the peptide bond. Sequences of amino

**TABLE VII** Greek Letters and Their Preferred Roman Equivalents

| | | | | | | | | |
|---|---|---|---|---|---|---|---|---|
| Alpha | $\alpha$ | A | Iota | $\iota$ | I | Rho | $\rho$ | R |
| Beta | $\beta$ | B | Kappa | $\kappa$ | K | Sigma | $\sigma$ | S |
| Gamma | $\gamma$ | G | Lambda | $\lambda$ | L | Tau | $\tau$ | T |
| Delta | $\delta$ | D | Mu | $\mu$ | M | Upsilon | $\upsilon$ | U |
| Epsilon | $\varepsilon$ | E | Nu | $\nu$ | N | Phi | $\phi$ | F |
| Zeta | $\zeta$ | Z | Xi | $\xi$ | X | Chi | $\chi$ | C |
| Eta | $\eta$ | H | Omicron | $o$ | O | Psi | $\psi$ | Y |
| Theta | $\theta$ | Q | Pi | $\pi$ | P | Omega | $\omega$ | W |

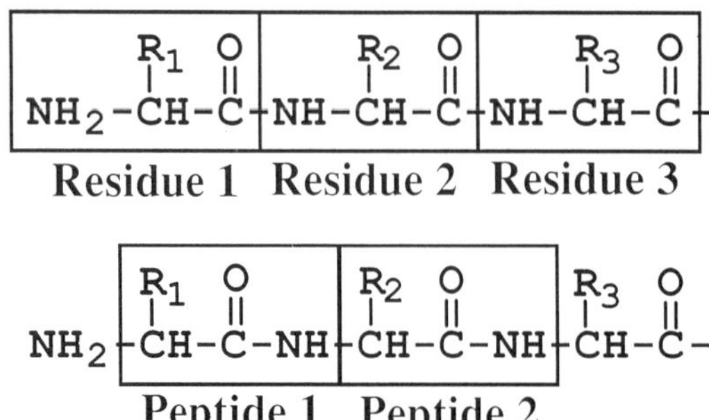


**FIGURE 16** Definition of residue and peptide units. A residue's peptide bond is that on the carboxyl side of its $\alpha$-carbon.

acids so joined are called peptides and large peptides (> ~20 amino acids) are called proteins. The individual amino acid groups in peptides are referred to as residues, with the peptide bond after the residue being considered as "belonging" to it (Fig. 16).

A torsion angle (also called a conformational or dihedral angle) can be defined for a system of four bonded atoms A—B—C—D by projecting the atoms onto a plane normal to the bond B—C. The angle between the projected bond A—B and C—D about B—C is the torsion angle. The torsion angle is considered positive when the upper bond needs to be rotated to the right (clockwise) to superimpose on the bottom bond and negative when it needs to be rotated to the left (counterclockwise) as illustrated in Fig. 17. It can also be thought of as the angle between the planes defined by the atoms A, B, C and B, C, D. Torsion angles are usually considered to run from $-180°$ to $+180°$ instead of from $0°$ to $360°$ so that relationships between enantiomeric conformations can easily be seen.

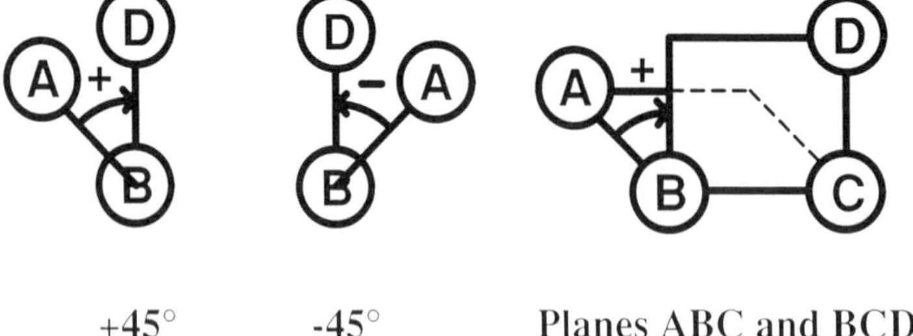


**FIGURE 17** Definition of torsion angles. The torsion angle defined by the four bonded atoms A—B—C—D is that along the bond B—C and is that formed by the vectors $\vec{BA}$ (front) and $\vec{CD}$ (behind) projected onto a plane perpendicular to the line $\overline{BC}$. It can also be thought of as the angle between the plane defined by atoms A, B, and C and that defined by atoms B, C, and D.

The conformation of the main chain of the *i*th peptide is defined by three torsion angles $\phi$ (phi), $\psi$ (psi), and $\omega$ (omega) defined as:

$$\phi\text{: } C_{i-1}\text{—}N_i\text{—}C\alpha_i\text{—}C_i \quad \psi\text{: } N_i\text{—}C\alpha_i\text{—}C_i\text{—}N_{i+1} \quad \omega\text{: } C\alpha_i\text{—}C_i\text{—}N_{i+1}\text{—}C\alpha_{i+1}$$

whereas side chain conformation angles are denoted by $\chi_1$ to $\chi_5$ and are defined by the atoms listed in Table VIII. Most side chain torsion angles tend to cluster around the gauche conformations +60° (+*gauche*, $g^+$), −60° (−*gauche*, $g^-$), and 180° (*trans*, *t*).

The $\omega$ conformation angle is that around the $C_i$—$N_{i+1}$ bond of the peptide. Because of its partial double bond character, the peptide bond is nearly planar with just two possible angles: $\omega \sim 0°$ and $\omega \sim 180°$. With respect to the two $\alpha$-carbons of the peptide bond, when $\omega \sim 180°$ (the usual case) the $\alpha$-carbons are *trans* to each other. When $\omega \sim 0°$ (rare) the $\alpha$-carbons are *cis* to each other (Fig. 18). This strong restriction on the $\omega$ conformation angle means that the folding of a peptide is determined by just the $\phi$ and $\psi$ angles. The few regular structures (repeating $\phi$, $\psi$ angles) that are recognized for proteins are listed in Table IX.

## B. Nucleic Acids

### 1. Nomenclature

The full IUPAC-IUB conventions of nomenclature and symbolism for nucleic acids can be found in *Eur. J. Biochem.* (1983, Vol. 131, pp. 9–15).

**TABLE VIII** Side Chain Conformational Torsion Angles

| Amino acid | $\chi_1$ $\chi_2$ $\chi_3$ $\chi_4$ $\chi_5$ |
|---|---|
| Arginine | N—Cα—Cβ—Cγ—Cδ—Nε—Cζ—$N\eta_1$ |
| Asparagine | N—Cα—Cβ—Cγ—$O\delta_1$ |
| Aspartic acid | N—Cα—Cβ—Cγ—$O\delta_1$ |
| Cysteine | N—Cα—Cβ—Sγ |
| Glutamine | N—Cα—Cβ—Cγ—Cδ—$O\varepsilon_1$ |
| Glutamic acid | N—Cα—Cβ—Cγ—Cδ—$O\varepsilon_1$ |
| Histidine | N—Cα—Cβ—Cγ—$N\delta_1$ |
| Isoleucine | N—Cα—Cβ—$C\gamma_1$—$C\delta_1$ |
| Leucine | N—Cα—Cβ—Cγ—$C\delta_1$ |
| Lysine | N—Cα—Cβ—Cγ—Cδ—Cε—Nζ |
| Methionine | N—Cα—Cβ—Cγ—Sδ—Cε |
| Phenylalanine | N—Cα—Cβ—Cγ—$C\delta_1$ |
| Proline | N—Cα—Cβ—Cγ—Cδ—N |
| Serine | N—Cα—Cβ—Oγ |
| Threonine | N—Cα—Cβ—$O\gamma_1$ |
| Tryptophan | N—Cα—Cβ—Cγ—$C\delta_1$ |
| Tyrosine | N—Cα—Cβ—Cγ—$C\delta_1$ |
| Valine | N—Cα—Cβ—$C\gamma_1$ |

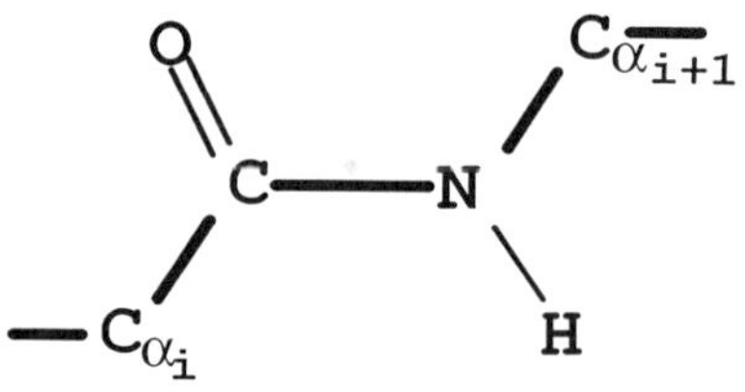

**normal *trans*-peptide (ω ~180° )**

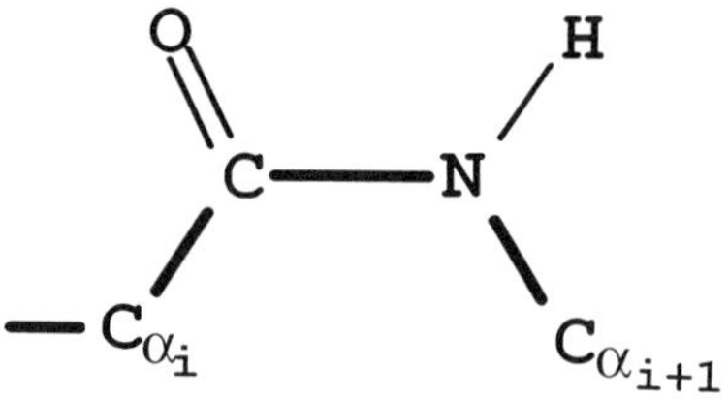

**rare *cis*-peptide bond (ω ~0° )**

**FIGURE 18** The C—N bond in the peptide has partial double bond character restricting its conformational angle ($\omega$) to roughly 0° or 180°, meaning that successive $\alpha$-carbons will be either *trans* or *cis* to each other. *cis*-peptides are rare (about 1% of all measured peptide bonds) and are almost exclusively found before proline residues.

As shown in Fig. 19, a nucleotide unit is composed of three distinct parts: a sugar ring (D-ribose or 2-deoxy-D-ribose), a phosphate group, and a purine or pyrimidine base. The sugar ring and phosphate group form the backbone and progress from the 3′ end of one sugar unit through the phosphate to the 5′ end of the next sugar unit.

There are five commonly occurring bases, three of which are found in both ribonucleic acids (RNA) and deoxyribonucleic acids (DNA) and one each found in only RNA or DNA. Their structures, three- and one-letter codes, and atom nomenclature are given in Fig. 20. Two types of sugars

**TABLE IX** Regular Structures of Peptides and Proteins

| Structure name | $\phi$ value | $\psi$ value |
|---|---|---|
| Fully extended[a] | 180° | 180° |
| Right-handed $\alpha$-helix | −57° | −47° |
| Left-handed $\alpha$-helix[a] | +57° | +47° |
| Parallel pleated sheet | −119° | +113° |
| Antiparallel pleated sheet | −139° | +135° |

[a] Structures seldom found in natural peptides/proteins.

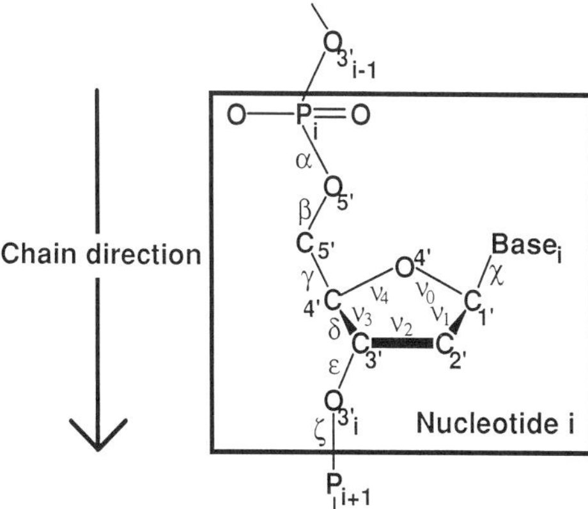


**FIGURE 19** A nucleotide unit consists of a purine or pyrimidine base, a sugar, and a phosphate unit bonded to the 5′ oxygen of the sugar unit. The sugar atoms are denoted by primed numbers. The backbone conformational angles are denoted by the Greek letters $\alpha$–$\zeta$, those of the sugar unit by $\nu_0$–$\nu_4$, and that for the planar base by $\chi$ (O4′—C1′—N9—C4 for purine bases and O4′—C1′—N1—C2 for pyrimidine bases).

Adenine (Ade, A)

Guanine (Gua, G)

Thymine (Thy, T)

Cytosine (Cyt, C)

Uracil (Ura, U)

**FIGURE 20** Structures, three- and one-letter codes, and atom names of the five commonly occurring purine and pyrimidine bases found in nucleic acids. The sugar attachment point in nucleotides is shown in parentheses.

are found in nucleic acids: D-ribose in RNA and 2′-deoxy-D-ribose in DNA (Fig. 21).

A nucleotide unit is defined as being composed of the base, sugar, and phosphate group esterified to the O5′ sugar atom. A nucleoside consists only of the base and sugar moieties. The preferred names for the free bases, ribonucleosides, deoxyribonucleosides, and their three- and one-letter codes are given in Table X. In the case of mixed nucleotides, prefixes "r" (for ribonucleotides) and "d" (for deoxyribonucleotides) should both be used.

## 2. Conformation of Nucleotides

***Backbone conformation*** The backbone conformation of a nucleotide is defined by six torsion angles/nucleotide along the sugar-phosphate backbone (Fig. 19):

$$\mathrm{O3'_{i-1}} \overset{\alpha}{—} \mathrm{P_i} \overset{\beta}{—} \mathrm{O5'_i} \overset{\gamma}{—} \mathrm{C5'_i} \overset{\delta}{—} \mathrm{C4'_i} \overset{\varepsilon}{—} \mathrm{C3'_i} \overset{\zeta}{—} \mathrm{O3'_i} — \mathrm{P_{i+1}} — \mathrm{O5'_{i+1}}$$

**ribonucleoside (sugar=D-ribose)**

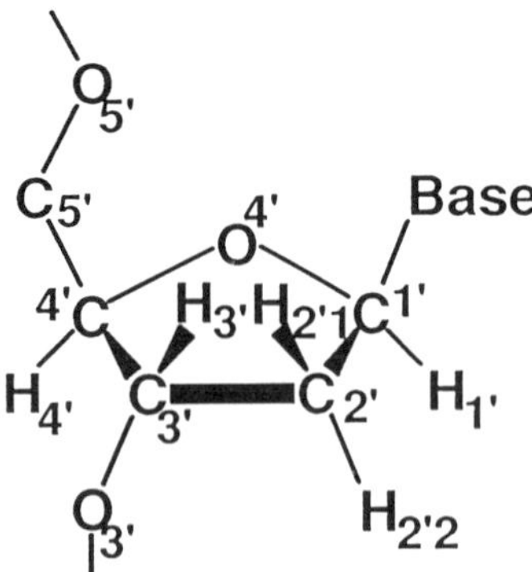


**deoxyribonucleoside (sugar=2-deoxy-D-ribose)**

**FIGURE 21** Structures and atom names of the sugar units found in RNA (D-ribose) and DNA (2-deoxy-D-ribose).

**TABLE X** Nucleoside Nomenclature

| Free base | Ribonucleoside | Deoxyribonucleoside[a] |
|---|---|---|
| Adenine (Ade, A) | Adenosine (Ado, A) | Deoxyadenosine (dAdo, dA) |
| Guanine (Gua, G) | Guanosine (Guo, G) | Deoxyguanosine (dGuo, dG) |
| Thymine (Thy, T) | Ribosylthymine (Thd, T)[b] | Thymidine (dThd, dT) |
| Cytosine (Cyt, C) | Cytidine (Cyd, C) | Deoxycytidine (dCyd, dC) |
| Uracil (Ura, U) | Uridine (Urd, U) | Deoxyuridine (dUrd, dU)[b] |

[a] Deoxyribonucleosides may be preceded by "2′-", but is assumed if not explicitly given.
[b] Uncommon in natural RNA and DNA.

***Sugar conformation*** Since the sugar ring is not planar, the torsion angles within the sugar ring are necessary to fully describe the system. The sugar ring torsion angles are called $\nu_0$–$\nu_4$ and are defined (Fig. 19):

$$\mathrm{C4'}\overset{}{—}\mathrm{O4'}\overset{\nu_0}{—}\mathrm{C1'}\overset{\nu_1}{—}\mathrm{C2'}\overset{\nu_2}{—}\mathrm{C3'}\overset{\nu_3}{—}\mathrm{C4'}\overset{\nu_4}{—}\mathrm{O4'}—\mathrm{C1'}$$

Note that the main chain conformation angle $\delta$ and the sugar ring torsion angle $\nu_3$ are both along the same bond (C3′—C4′), but are different, although clearly related, as their "end" atoms are different. Often sugar torsion angles are given in the Klyne–Prelog nomenclature (IUPAC Commission, 1976) which divides torsional angle space into large areas and is related to the torsion angle as shown in Table XI. In addition, the hemisphere $0° \pm 90°$ is often referred to as *syn*, whereas $180° \pm 90°$ is *anti*. The following torsion angles also have special names: 0° (*cis*), +60° (+*gauche*, $g^+$), 180° (*trans*, *t*), and −60° (−*gauche*, $g^-$).

In describing the sugar ring conformation, if four atoms lie in a plane, the conformation is described as envelope (E). If this is not the case, then the conformation is twist (T) and the reference plane is described by the three atoms nearest to the least squares plane defined by all five ring atoms.

**TABLE XI** Klyne–Prelog Torsion Angle Nomenclature

| Torsion angle | Full Klyne–Prelog name | K-P abbreviation |
|---|---|---|
| 0° to +30° | + Synperiplanar | +sp |
| +30° to +90° | + Synclinal | +sc |
| +90° to +150° | + Anticlinal | +ac |
| +150° to 180° | + Antiperiplanar | +ap |
| 180° to −150° | − Antiperiplanar | −ap |
| −150° to −90° | − Anticlinal | −ac |
| −90° to −30° | − Synclinal | −sc |
| −30° to 0° | − Synperiplanar | −sp |

Looking down onto this plane such that the ring atoms are sequentially numbered in a clockwise fashion, if an atom is up out of this plane (old nomenclature "*endo*"), its number is written as a superscript and precedes the letter (E or T). An atom that is down and out of the plane (old nomenclature "*exo*") is written as a subscript and follows the letter. Hence, C3′-*endo*/C2′-*exo* is now $^{3}T_{2}$. If both extraplanar atoms are equally displaced out of the plane in a twist conformation, both atom numbers are written preceding the letter T, e.g., ${}^{2}_{3}T$. Table XII gives the relationship of the envelope and twist (E/T) notation and the older *endo*/*exo* notation.

***Base conformation*** The conformation of the base relative to the sugar is defined by the $\chi$ torsion angle along the N-glycosidic bond (Fig. 19). The $\chi$ angle is defined by the four atoms O4′—C1′—N9—C4 for purine bases and O4′—C1′—N1—C2 for pyrimidine bases. For substituted groups in the sugar or base, torsion angles defining their conformation are given the letter $\eta$ followed by the number of the substituted atom. For example, if the O2′ of ribose is phosphorylated, the torsion angle defining the position of the phosphorous atom (C1′—C2′—O2′—P) is designated $\eta 2'$; if the amine of a cytosine base (N4) is methylated, then the torsion angle defining the position of the methyl group (N3—C4—N4—$CH_3$) is designated $\eta 4$.

***Helical parameters*** In describing the helix formed by a polynucleotide, the following characteristics should be given: $n$ is the number of residues/turn; $h$ is the translation/residue along the helical axis (in Å); $t$ is the angle of rotation/residue about the helical axis (360°/$n$); and $p$ is the pitch height of the helix ($n \times h$, in Å).

Relative to the helical axis and some reference point, the position of any atom can also be described by the parameters r, $\phi$, and z, where r is the radial distance from the helical axis, $\phi$ is the angle in the plane perpendic-

**TABLE XII** Sugar Pucker Notation

| *endo*/*exo* notation | E/T notation | *endo*/*exo* notation | E/T notation |
|---|---|---|---|
| C2′-*exo*/C3′-*endo* | ${}^{3}_{2}T$ | C2′-*endo*/C3′-*exo* | ${}^{2}_{3}T$ |
| C3′-*endo* | $^{3}E$ | C3′-*exo* | $_{3}E$ |
| C3′-*endo*/C4′-*exo* | ${}^{3}_{4}T$ | C3′-*exo*/C4′-*endo* | ${}^{4}_{3}T$ |
| C4′-*exo* | $_{4}E$ | C4′-*endo* | $^{4}E$ |
| C4′-*exo*/O4′-*endo* | ${}^{0}_{4}T$ | C4′-*endo*/O4′-*exo* | ${}^{4}_{0}T$ |
| O4′-*endo* | $^{0}E$ | O4′-*exo* | $_{0}E$ |
| O4′-*endo*/C1′-*exo* | ${}^{0}_{1}T$ | O4′-*exo*/C1′-*endo* | ${}^{1}_{0}T$ |
| C1′-*exo* | $_{1}E$ | C1′-*endo* | $^{1}E$ |
| C1′-*exo*/C2′-*endo* | ${}^{2}_{1}T$ | C1′-*endo*/C2′-*exo* | ${}^{1}_{2}T$ |
| C2′-*endo* | $^{2}E$ | C2′-*exo* | $_{2}E$ |

ular to the helical axis (in the x, y plane with z along the helical axis) between the reference point and the atom of interest, and z is the height along the helical axis from the reference point.

## ACKNOWLEDGMENTS

The authors thank Marcel Blommers, Sandra Cowan, Wayne Guida, and Joanne Williamson for discussion and critical review; and Binu Chaudhuri, Deborah Juterbock, and Laura Reimer for reference and literature search assistance.

## APPENDIX

Several academic research efforts and commercial products have been described in this chapter. Vendor information and literature references are collected together in this Appendix. Use of trademarked names throughout this chapter is for reference purposes only; no infringement is intended. Lack of mention of any particular product does not imply a judgment on quality or utility, merely the lack of space for a comprehensive summary. Other than the CSD and PDB, the following order of listing is alphabetical by vendor name. A large variety of structure file formats are represented here, and "utility programs" are often available to translate between formats.

Brookhaven PDB files (Bernstein *et al.*, 1977; Abola *et al.*, 1987) are available from Protein Data Bank, Chemistry Department, Brookhaven National Laboratory, Upton, New York 11973. Format document is Protein Data Bank (1989). The Brookhaven PDB organization can be accessed via the Internet in a variety of ways: Gopher and anonymous ftp protocols at address pdb.pdb.bnl.gov; fileserver at address fileserv@pdb.pdb.bnl.gov. Affiliated centers for PDB release exist around the world; most of them have Internet addresses. General questions about Internet access should be directed to address pdb@bnl.gov.

CSD (Allen *et al.*, 1991a) and QUEST search software at two version levels is available from the Cambridge Crystallographic Data Centre, 12 Union Road, Cambridge CB2 1EZ, United Kingdom. Manuals are Allen *et al.* (1992). The Cambridge Crystallographic Data Centre can be accessed via the Internet for software support and bug reports at address software@chemcrys.cam.ac.uk.

Catalyst (Sprague, 1991; Teig *et al.*, 1993) is available from BioCAD, 1390 Shorebird Way, Mountain View, California 94115. Insight and Converter are available from BIOSYM Technologies, Inc., 9685 Scranton Road, San Diego, California 92121-3752. CAST-3D is available from Chemical Abstracts Service, 2540 Olentangy River Road, P.O. Box 3012, Columbus, Ohio 43210. ChemDBS-3D (Murrall and Davies, 1990) and ChemModel

(Davies *et al.*, 1991) are available from Chemical Design, Ltd., Cromwell Park, Chipping Norton, Oxon OX7 5SR, United Kingdom. Aladdin (Van Drie *et al.*, 1989) is available from Daylight Chemical Information Systems, Inc., 18500 Von Karman Avenue, Irvine, California 92715. Gaussian-92 and the Gaussian archives are available from Gaussian, Inc., Carnegie Office Park, Pittsburgh, Pennsylvania 15106; information is available via Internet at address info@gaussian.com. MACCS, ISIS, and the 3-D modules for each (Moock *et al.*, 1995) are available from MDLI Information Systems, Inc., 14600 Catalina Street, San Leandro, California 94577. COBRA (Leach *et al.*, 1990) is available from Oxford Molecular, Magdalen Centre, Oxford Science Park, Sandford-on-Thames, Oxford OX4 4GA, United Kingdom. The QCPE may be contacted at Quantum Chemistry Program Exchange, Indiana University, Bloomington, Indiana 47405 and on the Internet at address qcpe@ucs.indiana.edu. CONCORD (Pearlman, 1987), PROTEP, and SYBYL/3D Unity are available from Tripos Associates, Inc., 1699 S. Hanley Road, St. Louis, Missouri 63144-2913. Reference for CONCORD manual is Rusinko *et al.* (1993). Reference for PROTEP manual is Tripos Associates (1992).

## REFERENCES


Abola, E. E., Bernstein, F. C., Bryant, S. H., Koetzle, T. F., and Weng, J. (1987). *In* "Crystallographic Databases: Information Content, Software Systems, Scientific Applications" (International Union of Crystallography, ed.), pp. 107–132. Data Commission of the International Union of Crystallography, Bonn, Cambridge, and Chester.

Allen, F. H., and Johnson, O. (1991). *Acta Cryst.* **B47**, 62–67.

Allen, F. H., Bellard, S., Brice, M. D., Cartwright, B. A., Doubleday, A., Higgs, H., Hummelink, T., Hummelink-Peters, B. G., Kennard, O., Motherwell, W. D. S., Rodgers, J. R., and Watson, D. G. (1979). *Acta Cryst.* **B35**, 2331–2339.

Allen, F. H., Kennard, O., and Taylor, R. (1983). *Acc. Chem. Res.* **16**, 146–153.

Allen, F. H., Davies, J. E., Galloy, J. J., Johnson, O., Kennard, O., Macrae, C. F., Mitchell, E. M., Mitchell, G. F., Smith, J. M., and Watson, D. G. (1991a). *J. Chem. Info. Comp. Sci.* **31**, 187–204.

Allen, F. H., Doyle, M. J., and Taylor, R. (1991b). *Acta Cryst.* **B47**, 50–61.

Allen, F. H., Johnson, O., Macrae, C. F., Smith, J. M., Motherwell, W. D. S., Galloy, J. J., Watson, D. G., Rowland, R. S., Edgington, P. R., Garner, S. E., Davies, J. E., and Mitchell, G. F. (1992). "Cambridge Structural Database System, User's Manuals," Vols. 1–4, Cambridge Crystallographic Data Centre, Cambridge, England.

Anderson, J. E., Ptashne, M., and Harrison, S. C. (1987). *Nature* **326**, 846–852.

Artymiuk, P. J., Grindley, H. M., Rice, D. W., Ujah, E. C., and Willett, P. (1992). *In* "Recent Advances in Chemical Information" (Collier, H., ed.), pp. 91–106. The Royal Society of Chemistry, Cambridge.

Ash, J. E., Chubb, P. A., Ward, S. E., Welford, S. M., and Willett, P. (1985). "Communication, Storage and Retrieval of Chemical Information." Ellis Horwood Ltd., Chichester.

Barnard, J. M. (1993). *J. Chem. Inf. Comput. Sci.* **33**, 532–538.

Barton, G. J., and Rawlings, C. J. (1990, published 1992). *Tetrahedron Comput. Methodol.* 3(6C), 739–756.

Berman, H. M., Olson, W. K., Beveridge, D. L., Westbrook, J., Gelgin, A., Demeny, T., Hsieh, S.-H., Srinivasan, A. R., and Schneider, B. (1992). *Biophys. J.* **63,** 751–759.

Bernstein, F. C., Koetzle, T. F., Williams, G. J. B., Meyer, E. F., Jr., Brice, M. D., Rodgers, J. R., Kennard, O., Shimanouchi, T., and Tasumi M. (1977). *J. Mol. Biol.* **112,** 535–542.

Blundell, T. L., and Johnson, L. N. (1976). "Protein Crystallography." Academic Press, New York/London.

Bohacek, R. S., and McMartin, C. (1992). *J. Med. Chem.* **35,** 1671–1684.

Bonchev, D., and Rouvray, D. H. (eds.) (1991). "Chemical Graph Theory: Introduction and Fundamentals." Gordon and Breach Science Press, Amsterdam.

Boobbyer, D. N. A., Goodford, P. J., McWhinnie, P. M., and Wade, R. C. (1989). *J. Med. Chem.* **32,** 1083–1094.

Brint, A. T., and Willett, P. (1987). *J. Mol. Graph.* **5,** 49–56.

Bures, M. G., Hutchins, C. W., Maus, M., Kohlbrenner, W., Kadam, S., and Erickson, J. W. (1990, published 1992). *Tetrahedron Comput. Methodol.* **3**(6C), 673–680.

Carhart, R. E., Smith, D. H., and Venkataraghavan, R. (1985). *J. Chem. Info. Comput. Sci.* **25,** 64–73.

Christie, B. D., Henry, D. R., Güner, O. F., and Moock, T. E. (1990a). *In* "Proceedings of the 14th International Online Information Meeting" (D. Raitt, ed.), pp. 137–161, Learned Information, Oxford.

Christie, B. D., Henry, D. R., Wipke, W. T., and Moock, T. E. (1990b, published 1992). *Tetrahedron Comput. Methodol.* **3**(6C), 653–664.

Claessens, M., Van Cutsem, E., Lasters, I., and Wodak, S. (1989). *Protein Eng.* **2,** 335–345.

Clore, G. M., and Gronenborn, A. M. (1991). *Prog. NMR Spectr.* **23,** 43–92.

Daopin, S., Piez, K. A., Ogawa, Y., and Davies, D. R. (1992). *Science* **257,** 369–373.

Davies, K., and Upton, R. (1990, published 1992). *Tetrahedron Comput. Methodol.* **3**(6C), 665–671.

Davies, K., Dunn, D., and Upton, R. (1991). Poster of the 5th Molecular Modeling Workshop, Darmstadt, Germany (May 1991).

Derome, A. D. (1987). "Modern NMR Techniques for Chemistry Research." Pergamon Press, Chicago.

DesJarlais, R. L., Sheridan, R. P., Seibel, G. L., Dixon, J. S., Kuntz, I. D., and Venkataraghavan, R. (1988). *J. Med. Chem.* **31,** 722–729.

DesJarlais, R. L., Seibel, G. L., Kuntz, I. D., Furth, P. S., Alvarez, J. C., Ortiz de Montellano, P. R., DeCamp, D. L., Babé, L. M., and Craik, C. S. (1990). *Proc. Natl. Acad. Sci. USA* **87,** 6644–6648.

Ducruix, A., and Giegé, R. (1992). "Crystallization of Nucleic Acids and Proteins, A Practical Approach." IRL Press at Oxford University Press, Oxford/New York.

Dunne, B. J., Morris, R. B., and Orpen, A. G. (1991). *J. Chem Soc. Dalton Trans.* 653–655.

Ernst, R. R., Bodenhausen, G., and Wokaun, A. (1987). "Principle of Nuclear Magnetic Resonance in One and Two Dimensions." Clarendon, Oxford.

Fisanick, W., Cross, K. P., Forman, J. C., and Rusinko, A., III (1993). *J. Chem. Info. Comput. Sci.* **33,** 548–559.

Gardner, S. P., and Thornton, J. M. (1991). *In* "Abstracts of Papers, 202nd American Chemical Society National Meeting," p. 32 (CINF). American Chemical Society, Washington, D.C.

Giacovazzo, C., Monaco, H. L., Viterbo, D., Scordari, F., Gilli, G., Zanotti, G., and Catti, M. (1992). "Fundamentals of Crystallography" (C. Giacovazzo, ed.). Oxford University Press, Oxford/New York.

Glusker, J. P., and Trueblood, K. N. (1985). "Crystal Structure Analysis: A Primer," 2nd Ed. Oxford University Press, New York/Oxford.

Goodford, P. J. (1985). *J. Med Chem.* **28,** 849–857.

Grindley, H. M., Artymiuk, P. J., Rice, D. W., and Willett, P. (1993). *J. Mol. Biol.* **229,** 707–721.

Güner, O. F., Henry, D. R., and Moock, T. E. (1993). *In* "Abstracts of Papers, 206th American

Chemical Society National Meeting," p. 86 (COMP). American Chemical Society, Washington, D.C.
Güner, O. F., Henry, D. R., Moock, T. E., and Pearlman, R. S. (1990, published 1992). *Tetrahedron Comput. Methodol.* 3(6C) 557–563.
Güner, O. F., Henry, D. R., and Pearlman, R. S. (1992). *J. Chem. Info. Comput. Sci.* **32,** 101–109.
Hall, S. R. (1991). *J. Chem. Info. Comput. Sci.* **31,** 326–333.
Hall, S. R., Allen, F. H., and Brown, I. D. (1991). *Acta Cryst.* **A47,** 655–685.
Hansch, C., Sammes, P. G., and Taylor, J. B. (eds.) (1990). "Comprehensive Medicinal Chemistry" (C. J. Drayton, vol. ed.), Vol. 6. Pergamon Press, Oxford/New York.
Haraki, K. S., Sheridan, R. P., Venkataraghavan, R., Dunn, D. A., and McCulloch, R. (1990, published 1992). *Tetrahedron Comput. Methodol.* 3(6C), 565–573.
Hendrickson, M. A., Nicklaus, M. C., Milne, G. W. A., and Zaharevitz, D. (1993). *J. Chem. Info. Comput. Sci.* **33,** 155–163.
Hendrickson, W. A., and Wüthrich, K. (eds.) (1993). "Macromolecular Structures 1993." Current Biology, Ltd., London.
Hicks, M. G., and Jochum, C. (1990). *J. Chem. Info. Comput. Sci.* **30,** 191–199.
Hodes, L., and Feldman, A. (1991). *J. Chem. Info. Comput. Sci.* **31,** 347–350.
Hoffmann, R., and Laszlo, P. (1991). *Angew. Chem.* (Intnl. Eng. Ed.) **30,** 1–112.
International Union of Crystallography (1987). "Crystallographic Databases: Information Content, Software Systems, Scientific Applications." Data Commission of the International Union of Crystallography, Bonn, Cambridge, and Chester.
IUPAC Commission on Nomenclature of Organic Chemistry (1976). *Pure Appl. Chem.* **45,** 11–30.
IUPAC-IUB Commission on Biochemical Nomenclature (1970). *J. Biol. Chem.* **245,** 6489–6497.
IUPAC-IUB Joint Commission on Biochemical Nomenclature (1983). *Eur. J. Biochem.* **131,** 9–15.
IUPAC-IUB Joint Commission on Biochemical Nomenclature (1984). *J. Biol. Chem.* **260,** 14–42.
Islam, S. A., and Sternberg, M. J. E. (1989). *Protein Eng.* **2,** 431–442.
Jones, T. A., and Thirup, S. (1986). *EMBO J.* **5,** 819–822.
Jones, T. A., Zou, J.-Y., Cowan, S. W., and Kjeldgaard, M. (1991). *Acta Cryst.* **A47,** 110–119.
Kabsch, W., and Sander, C. (1983). *Biopolymers* **22,** 2577–2637.
Kier, L. B. (1971). "Molecular Orbital Theory in Drug Research." Academic Press, New York.
Lam, P. Y. S., Jadhav, P. K., Eyermann, C. J., Hodge, C. N., Bacheler, L. T., Meek, J. L., Otto, M. J., Rayner, M. M., Wong, Y. N., Chang, C.-H., Weber, P. C., Jackson, D. A., Sharpe, T. R., and Erickson-Viitanen, S. (1994). *Science* **263,** 380–384.
Lawrence, M. C., and Davis, P. C. (1992). *Proteins: Struct. Funct. Genet.* **12,** 31–41.
Leach, A. R., Dolata, D. P., and Prout, K. (1990). *J. Chem. Info. Comput. Sci.* **30,** 316–324.
Lesk, A. M. (1979). *Commun. ACM* **22,** 219–224.
Lipkowitz, K. B., and Peterson, M. A. (1993). *J. Comp. Chem.* **14,** 121–125.
Marshall, G. R., Barry, C. D., Bosshard, H. E., Dammkoehler, R. A., and Dunn, D. A. (1979). *In* "Computer-Assisted Drug Design; ACS Symposium Series 112" (E. C. Olson and R. E. Christoffersen, eds.), pp. 205–226. American Chemical Society, Washington, D.C.
Martin, Y. C. (1992). *J. Med. Chem.* **35,** 2145–2154.
Martin, Y. C., Bures, M. G., and Willett, P. (1990). *In* "Reviews in Computational Chemistry" (K. B. Lipkowitz, and D. B. Boyd, eds.), pp. 213–263. VCH, New York, Weinheim, and Cambridge.
McPherson, A. (1989). "Preparation and Analysis of Protein Crystals." Robert E. Krieger Publishing Company, Malabar, Florida.
Milne, G. W. A., and Miller, J. A. (1986). *J. Chem. Info. Comput. Sci.* **26,** 154–159.

Mitchell, E. M., Artymiuk, P. J., Rice, D. W., and Willett, P. (1990). *J. Mol. Biol.* **212,** 151–166.
Moock, T. E., Henry, D. R., Ozkabak, A. G., and Alamgir, M. (1995). Submitted for publication.
Murrall, N. W., and Davies, E. K. (1990). *J. Chem. Info. Comput. Sci.* **30,** 312–316.
NBS Crystal Data Center (1982). "NBS Crystal Data File." NIST Office of Standard Reference Data (National Institute for Standards and Technology, formerly the National Bureau of Standards), Gaithersburg, MD.
Nicklaus, M. C., Milne, G. W. A., and Zaharevitz, D. (1993). *J. Chem. Info. Comput. Sci.* **33,** 639–646.
Nilakantan, R., Bauman, N., Dixon, J. S., and Venkataraghavan, R. (1987). *J. Chem. Info. Comput. Sci.* **27,** 82–85.
Nilakantan, R., Bauman, N., Haraki, K. S., and Venkataraghavan, R. (1990). *J. Chem. Info. Comput. Sci.* **30,** 65–68.
Pearlman, R. S. (1987). *Chem. Design Autom. News* **2,** 5–6.
Poirrette, A. R., Willett, P., and Allen, F. H. (1991). *J. Mol. Graph.* **9,** 203–217.
Poirrette, A. R., Willett, P., and Allen, F. H. (1993). *J. Mol. Graph.* **11,** 2–14.
Protein Data Bank (1989). "Atomic Coordinate and Bibliographic Entry Format Description." Protein Data Bank, Upton, NY.
Rhodes, G. (1993). "Crystallography Made Crystal Clear: A Guide for Users of Macromolecular Models." Academic Press, San Diego.
Richardson, D. C., and Richardson, J. S. (1992). *Protein Sci.* **1,** 3–9.
Ricketts, E. M., Bradshaw, J., Hann, M., Hayes, F., Tanna, N., and Ricketts, D. M. (1993). *J. Chem. Info. Comput. Sci.* **33,** 905–925.
Roberts, J. K. M., and Jardetzky, O. (1985). *In* "Modern Physical Methods in Biochemistry" (A. Neuberger, and L. L. M. van Deenen, eds.), Part A, pp. 1–67. Elsevier, Amsterdam/New York.
Rusinko, A., III, Skell, J. M., Balducci, R., McGarity, C. M., and Pearlman, R. S. (1993). "CONCORD, a Program for the Rapid Generation of High Quality Approximate 3-Dimensional Molecular Structures" (User's Manual, v.3.0.1). The University of Texas at Austin and Tripos Associates, St. Louis.
Schlunegger, M. P., and Grütter, M. G. (1992). *Nature* **358,** 430–434.
Shenkin, P. S., Yarmush, D. L., Fine, R. M., Wang, H., and Levinthal, C. (1987). *Biopolymers* **26,** 2053–2085.
Sheridan, R. P., Nilakantan, R., Dixon, J. S., and Venkataraghavan, R. (1986). *J. Med. Chem.* **29,** 899–906.
Sheridan, R. P., Rusinko, A., III, Nilakantan, R., and Venkataraghavan, R. (1989). *Proc. Natl. Acad. Sci. USA* **86,** 8165–8169.
Sprague, P. W. (1991). *In* "Proceedings of the Montreux 1991 International Chemical Information Conference" (Annecy, France, 23–25 September 1991), pp. 107–112. Infonortics Ltd., Calne.
Taylor, R., and Kennard, O. (1984). *Acc. Chem. Res.* **17,** 320–326.
Teig, S. L., Berezin, S., Greene, J. W., Kahan, S. D., and Ku, S.-L. (1993). *In* "Abstracts of Papers, 206th American Chemical Society National Meeting," p. 34 (CINF). American Chemical Society, Washington, D.C.
Tripos Associates (1992). "PROTEP User's Manual." Tripos Associates, Inc., St. Louis.
Ullmann, J. R. (1976). *J. Assoc. Comput. Mach.* **23,** 31–42.
Van Drie, J. H. (1993). *In* "Abstracts of Papers, 206th American Chemical Society National Meeting," p. 33 (CINF). American Chemical Society, Washington, D.C.
Van Drie, J. H., Weininger, D., and Martin, Y. C. (1989). *J. Comput.-Aided Mol. Des.* **3,** 225–251.

# 6

# Computer-Assisted Drug Discovery

**PETER GUND**
Molecular Simulations Incorporated
Burlington, Massachusetts

**GERALD MAGGIORA**
Upjohn Laboratories
Kalamazoo, Michigan

**JAMES P. SNYDER**
Emerson Center for Scientific Computation
Department of Chemistry
Emory University
Atlanta, Georgia

## I. THE DRUG DEVELOPMENT PROCESS

### A. Introduction

The discovery and development of a new chemical entity with demonstrated utility in ameliorating or curing disease is a long and arduous process. Industry statistics suggest that up to 10,000 compounds are synthesized and tested; up to 100 compounds are assessed for safety; and up to 10 compounds are tested clinically in humans for every drug that is approved for medical use. Even when a drug is approved for marketing, success is not assured; many marketed drugs fail to recoup the hundreds of millions of dollars required for their discovery and development because they are not sufficiently efficacious in practice or because undesirable side effects emerge from large-scale experience. On the other hand, a "blockbuster" new drug can earn billions of dollars for a company and its shareholders. So the quest for new therapies remains attractive, and tens of billions of dollars are spent industry-wide every year on drug discovery and development, resulting in the introduction of 30 to 60 new substances into medical practice annually. Yet, in some ways, we are losing ground in the battle against illness. Although major scourges like smallpox, bubonic plague, sepsis, polio, and tuberculosis have been largely eradicated, some are reemerging in resistant forms; other diseases like arthritis and most cancers are poorly treated; and new diseases such as AIDS and old diseases like the common cold remain unconquered.

Why is drug discovery and development so arduous? First, the odds of finding a new compound with just the right combination of activity, selectivity, stability, and safety are very unfavorable. Second, the clinical studies required to demonstrate statistically significant safety and efficacy are expensive and heavily regulated by government agencies. Completing the clinical studies and preparing a new drug application (NDA) for the U.S. Federal Drug Agency (FDA) typically takes 5 to 10 years and can cost over $200 million. This expensive development process is designed by government agencies to reduce risk to the public in two ways. First, the process requires pharmaceutical companies to prove safety of a new drug beyond reasonable doubt in order to prevent tragedies such as the unrecognized teratogenic effects of a tranquilizer that caused improper development of embryos in pregnant patients. Second, the studies must prove efficacy so that the public is protected from spending money on ineffective treatments.

## B. The Discovery and Development Process

The drug discovery and development process can be divided into the following phases: new lead discovery and lead optimization; preclinical and clinical lead development; and postmarketing surveillance (Table I). Actually the steps are more interconnected and simultaneous than indicated in the table; several leads are typically being worked on at all but the last stage so that if problems surface in a later stage, another lead compound may be brought forward without significant delay.

## C. New Lead Discovery Strategies

There are a number of ways of discovering new leads, as listed in Table II (review: Wermuth, 1992). Herbs and broths have traditionally been rich sources of active compounds, but it seems this strategy is becoming less productive, perhaps because the natural drugs available in abundance have mostly been discovered. Paul Ehrlich, in the early 1900s, appears to have been the first practitioner of screening, making a number of analogs and testing them to discover a syphilis cure. New scientific disciplines and new computer techniques have expanded the repertoire of new lead discovery, as indicated by the increasing use of genetic sequence and biomolecular structural information. Advances include the isolation and cloning of target enzymes and receptors, and the consequent ability to work with human rather than animal biomolecules; the classification of receptor subtypes, increasing the likely specificity of derived agonists and antagonists; the use

**TABLE I** The Drug Discovery and Development Process

1. New lead discovery
   - Find a repeatable, high-capacity assay related to the disease of interest
   - Continue to refine the assay, purify the biomaterial, get reproducible results
   - Isolate relevant enzymes/receptors; purify, characterize, and clone them
   - Get structural information on relevant enzymes and receptors
   - Choose one or more standard compounds for assay development/calibration
   - Screen known compounds randomly or by some selection criterion
   - Design (by some hypothesis), synthesize, and test new compounds
   - Use knowledge of related assays and structures to form hypotheses
   - Develop a feeling for how actives may be working mechanistically
2. Lead optimization
   - Develop secondary assays to check specificity and activity in animals
   - Synthesize and test congeneric structures
   - Develop structure–activity and/or mechanism models of action
   - Calculate/measure physical properties, correlate with activity
   - Check safety in lower animals and cells
3. Preclinical lead development
   - Develop large-scale synthesis
   - Animal safety studies including 3-year dog carcinogenicity tests
   - Drug delivery/elimination/metabolism studies
   - Drug formulations experiments
   - Dose-ranging studies in animals
4. Clinical development
   - Prepare investigational new drug application
   - Small-scale safety and dose-ranging tests in healthy humans
   - Develop clinical study protocols, obtain approval
   - Recruit clinical investigators and patients for study
   - Carry out the study
   - Analyze and report results
   - Present new drug application
5. Postmarketing surveillance
   - Collect usage and side effect reports

**TABLE II** Strategies for Discovering New Leads

Isolate active substances from natural products, e.g., penicillin, quinine
Modification of natural products, e.g., thienamycin
Broad screening of known synthetic substances, e.g., sulfa drugs
Derivatization and application of structure–activity data, e.g., cephalosporin
Structure-directed molecular design, e.g., carbonic anhydrase inhibitors
Mechanism-based molecular design, e.g., suicide inhibitors
Chemist's intuition, e.g., enalapril

of combinatorial synthesis and diversity screening to increase the amount of structure–activity information; and the use of leads in one project to jump-start discovery in another, possibly related project. The last-mentioned strategy is particularly exciting, and examples include use of renin inhibitors as leads for inhibiting another aspartate protease, HIV protease (Thaisrivongs *et al.*, 1991), and development of benzodiazepine and other G-protein-coupled receptor antagonist leads into specific antagonists of cholecystokinin and gastrin (Freidinger, 1992).

### D. Composition of Drug Discovery Teams

A strong characteristic of modern drug discovery and development teams is that they are usually large and diverse in terms of skills and disciplines (Table III). This is because so many aspects of the chemistry and biology of candidate drugs need to be explored, optimized, and documented. While productive research teams are able to surmount the differences of training and jargon, it remains challenging to form such effective teams and to keep them interactive and cooperative. We believe that computer methods can facilitate this interactivity and cooperation, and therefore materially enhance the productivity of these teams.

## II. THE PRACTICE OF COMPUTER-ASSISTED DRUG DISCOVERY (CADD)

### A. Current Practice of CADD in the Pharmaceutical Industry

CADD can contribute not only to the design of potent compounds, but to many of the steps of going "from concept to clinic." Some of the areas along this "path of discovery" are traditionally more well developed and/or more well adapted to computer assistance than others. For example, the general areas of molecular modeling and database methods have played important roles in many aspects of drug discovery, whereas tasks "downstream" from discovery have made very little use of these methods. For another example, the use of simulation techniques to model biological systems and subsystems (Lacey, 1992) is less well developed and is not traditionally considered to be part of the discovery/development process.

There are a number of trends that increase the need for serious computer support of these processes. In addition to the aforementioned need to share results across an interdisciplinary drug discovery team, there is a burgeoning amount of relevant data being generated both in-house and externally. For example, the approximately 4 billion base-pair sequences existing in the human genome are being systematically determined, and a research group that first teases structural or functional information out of a newly determined sequence is most likely to succeed in turning this sequence into a

**TABLE III** Disciplines and Skills of Drug Discovery and Development Teams

| |
|---|
| Physician (MDs) |
| Clinical research specialists |
| Statisticians |
| Microbiologists |
| Experimental biologists |
| Molecular biologists |
| Pharmacologists |
| Immunologists |
| Neuropharmacologists |
| Parasitologists |
| Psychologists |
| Toxicologists |
| Formulations chemists |
| Analytical chemists |
| Metabolic chemists |
| Chemical engineers |
| Process chemists |
| Medicinal chemists |
| Synthetic chemists |
| Biochemists |
| Enzymologists |
| Biophysicists |
| Crystallographers |
| NMR and other spectroscopists |
| Computational chemists |
| Molecular modelers |
| Scientific programmers |
| Chemical information specialists |
| Biological data specialists |
| Computational biologists |
| Computer systems, hardware and networking specialists |

novel therapeutic target (Venuti, 1990). Similarly, the trend toward chemical diversity synthesis and robotic synthesis threatens to overwhelm our traditional methods of registering and managing chemical structural data (Moos *et al.*, 1993). We believe that any company that does not seriously address the "data problem" in an effective manner will be seriously handicapped in the competition to discover new drugs. Moreover, the data problem will get worse, and at an increasingly rapid rate.

Furthermore, simply managing the burgeoning amounts of data will not be sufficient. Methods must be (and are being) developed to analyze these huge databases in novel ways, to tease out subtle structure–activity patterns, and to identify new leads. Such methods make best use of previously derived internal and published information, and can drastically improve the productivity and effectiveness of current research efforts.

Another related trend is the escalating level of competition among increasingly competent research groups. The group that moves most quickly to exploit a new research discovery and to turn that into an active new lead is most likely to own the key patents and to develop the first therapeutic agent in the new area. Effectively utilized computer methods can provide a major competitive advantage to a savvy drug discovery group.

## B. Management Structures of CADD Groups

CADD groups have existed and done well in various management structures in different pharmaceutical companies. A view of how CADD groups have become established, and why their contributions remain less than optimal, has already been published (Snyder, 1991). Table IV lists some of the reporting relationships of groups with which we are familiar.

We know of a number of cases where such groups have changed their position in the organization, with less than a revolutionary impact on their contributions to drug discovery. We offer the following opinions regarding the question of organizing such a group:

- A CADD group needs to work closely with a variety of disciplines, and any hierarchy tends to reinforce certain relationships at the expense of others; so all management structures represent compromises. Similarly, if groups which need to work together and are not doing so are combined organizationally, this may (or may not) enhance their collaboration but will probably reduce incentive for some other important collaborations.
- Any organizational structure will work if people make it work. On the other hand, some organizational structures may interfere with the smooth planning and prioritizing of, and collaborations on, research projects.
- Management structure de facto defines the boundaries for computational support within the discovery/development process. It also determines, to a very large extent, how and what resources will be distributed. Even the most highly integrated set of "cutting edge" laboratory and computer-based applications cannot overcome large and intractable "management barriers."

**TABLE IV** Some Reporting Relationships of CADD Groups in the Pharmaceutical Industry

| |
|---|
| Report to head of basic research |
| Report to head of synthetic or medicinal chemistry |
| Report to head of research administration/information services |
| Report to head of corporate computing services |
| Report to head of biophysics |
| Report to head of analytical chemistry |

• Most critical is management's expectations for CADD contributions to the projects and their efforts to reward the whole project team on a successful project—not only outstanding individual efforts.

## C. Contributions and Achievements of CADD Groups

A number of examples now exist that clearly show that CADD has made major contributions to a variety of drug design efforts; some published examples are listed in Table V.

Perhaps the most visible testament to those contributions is the recognition afforded to computational chemists on drug discovery teams by inclusion in new lead development awards and incentives, and by the inclusion of modelers on dozens of submitted patents from many pharmaceutical companies. But however satisfying these "home runs" are to the modeling practitioners, focusing on them detracts from the less dramatic contributions to discovery projects that modelers make every day (Table VI).

A CADD group most effectively contributes to a drug discovery project when it is contributing ideas frequently and interactively, with back-and-forth exchanges of ideas and experimental suggestions. In this type of collaborative environment, specific intellectual contributions of the CADD staff may become transformed into derivative ideas in which the initial CADD insight is diminished or forgotten. For example, relatively simple questions may be (fully or partially) answered or ideas and insights may be generated by merely manipulating "molecular images" on a computer screen, with subsequent exhaustive modeling studies confirming the original insight. While these "successes" alone certainly do not justify the amount of investment in hardware, software, and humanware needed to maintain a first-rate molecular graphics facility, it appears that most pharmaceutical companies now recognize the value and importance of their CADD groups. Although management likes to see the equivalent of "Page One" stories that clearly justify their investments, past acceptance of this criterion for success by CADD managers has been a major mistake. The CADD scientists are valuable members of the project team, but are not oracles delivering complete solutions from first principles. Their laboratories are their workstations and

**TABLE V** Some Examples of Successful Use of CADD for Drug Discovery[a]

| |
|---|
| Design of thymidylate synthase inhibitors as anticancer agents (Appelt *et al.*, 1991). |
| Design of HIV protease inhibitors as antiviral agents (Thaisrivongs, 1994). |
| Design of neutrophil elastase inhibitors as agents for emphysema (Hlasta and Pagani, 1994). |
| Design of carbonic anhydrase inhibitors as antiglaucoma agents (Baldwin *et al.*, 1989). |
| Discovery of novel sweeteners using a sweet-taste receptor model (Worthy, 1990). |

[a] Reviews: Boyd (1990), Kuntz (1992), and Navia and Murcko (1992).

**TABLE VI** Typical Contributions of CADD to Drug Discovery Projects

| |
|---|
| Suggestions of structures which, when retrieved from a compound collection or synthesized, were found upon testing to be active (or inactive as predicted) |
| Development of structure–activity relationships |
| Visualization of receptor models, pharmacophoric models, molecular alignments, or data models |
| Validating and/or prioritizing chemists' synthetic ideas |
| Reanalyzing available data to achieve new insights |
| Creative search of available structures to find new leads |
| Identification of preferred sites for structure elaboration |
| Development of models to improve drug transport, specificity, safety, or stability |
| Development of mechanistic insights |
| Use of leads in one area to derive new leads in a related assay |
| Establishment of useful databases of project structures and properties |
| Computation of physical or chemical properties to correlate with activities |

mainframes and software. Like other scientists on the project team, they generate hypotheses that do not hold up and perform (computational) experiments that do not work; like in other disciplines, the more creative scientists may have more failed experiments but also more successes. The "home run seeking" mentality of managers toward CADD scientists is undoubtedly changing, and the diverse roles that computations can and do play in the discovery process are beginning to be more fully appreciated. Consequently, expectations that CADD should produce the "big score" to justify its existence are waning, albeit slowly.

## D. The Case for Company Investments in CADD

A related question is whether CADD should be considered from a "return on investment" (ROI) or from a "resource" perspective. At Merck, for example, the growth of the CADD group was tied to the position that Stan Fidelman and other administrators took, that, like the libraries and animal facilities, CADD needs to be provided as basic support of the drug discovery process. Thus, while a charge-back system for use of the Cray supercomputer would have prohibited its use for all but the most critical of applications, open access to its power has enabled many useful studies to be done. Conversely, when the corporate area started charging for time-shared access in the 1970s, Merck's computer-assisted synthetic planning program became prohibitively expensive to run, further development was stopped, and this promising technique to assist the synthetic chemist became unavailable.

If one accepts these arguments concerning the value of incremental "small successes" in supporting the drug discovery team, then the "resource" perspective is sustained and reasonably large CADD investments are logical. From the ROI perspective, however, the payout from the CADD investment is less clear. However, pharmaceutical industry investment analysis is beleagered by uncertainties and risks—the first company to discover and patent an active new series may wipe out the investment of competing companies in this area—and an ROI analysis of CADD investments needs to factor in the unquantifiable probability that this methodology increases the chances for project success. Since this is by definition unquantifiable, the ROI analysis must be unreliable and the "resource" model for CADD support is preferred.

The bottom line for CADD is that, where companies have committed resources and support in the belief that this methodology can contribute to the drug discovery process, major contributions to projects have often been made. Where companies have not been convinced, such resources have not been made available and drug discovery teams operated (often productively) without the benefits of modeling support. Time will tell which set of companies are most successful. Certainly experience to date does not guarantee success to a company that invests in this area, but many companies that have invested have had good results and have increased their investment. One caveat, as pharmaceutical research becomes more structure and mechanism oriented and more competitive, even companies that have fared well in the past without a CADD component may not fare as well in the future.

## III. LIMITATIONS TO CADD SUPPORT

### A. Inherent Limitations of CADD Support

The first question here is, if we could be provided with substantial computer resources and the best available software, can we model the action of a drug molecule in an organism exactly? We cannot—for two reasons. First, we cannot calculate results totally accurately even for an *in vitro* system such as ligand binding to an enzyme. Second, we cannot hope to accurately model *in vivo* systems due to the incredible complexity and variability of organs and organisms. The difficulties involved in accurately modeling the enzyme inhibition process are set out in Fig. 1. Understanding how such an inhibitor exerts its clinical effect is more difficult still since the path such a molecule can take is long and convoluted (Fig. 2).

An organism by definition is dynamic and alive. Thermodynamics strictly applies to systems in equilibrium; that is not, however, sufficient to describe living systems since such systems have conspired to defeat the laws of thermodynamics, i.e., to defeat the tendency toward maximum entropy (maximum disorder). Kinetics applies to a reacting system, but organisms exhibit extremely complex, time-dependent and cyclical kinetics. A healthy organism is in a state of stasis, which is dynamic but regulated to remain

Enzyme in Solution (free energy, solvation, counter-ions, multiplicity of conformations, dynamics, cofactors, regulants, membrane proximity)

+

Inhibitor in Solution (free energy, solvation, counter-ions, concentration effects, multiplicity of conformations, dynamics, propensity for crystallization or coagulation)

↓

Enzyme-Inhibitor Complex in Solution (free energy of complex, counter-ions, solvation, concentration effects, multiplicity of conformations, unproductive vs. productive binding modes, "induced fit" conformations, polarization and charge-transfer effects, dynamic effects, binding reversibility, competitive inhibitors, co-factors, enzyme regulants, quaternary structure, supramolecular structure, effect of inhibitor on enzyme action and recycling)

**FIGURE 1** Factors involved in computing enzyme–inhibitor interactions.

within a range of acceptable, i.e., living, states. At another level, an organism appears to be one step away from chaos (Cambel, 1993) since a rather mild perturbation to the system may result in drastic effects, e.g., heart defibrillation, stroke, clinical depression, cancer, and even death. These are profoundly difficult systems, states, and behaviors to attempt to completely model by computer!

If we cannot model medically relevant processes completely, then why attempt to model them at all? The answer is because simulation can provide insights not available by any other scientific technique. Indeed, we can argue that simulation represents a third paradigm of science: illuminating theory and stimulating experiment. If we cannot model processes exactly, should we then be content to just run low-level, empirical simulations? Perhaps, but if we have the resources to compute at higher levels of theory, our results are more likely to be an accurate and reliable predictor of experiment, and the results have increased credibility when we present them. There are trade-offs between performing simulations at as high a level of rigor as possible and performing them rapidly and approximately and in close collaboration with experimentalists. One may draw the analogy with significant figures in mathematics; the simulation need be no more accurate than the model will sustain. Thus, detailed molecular dynamics calculations may be neither required nor appropriate if we are trying to understand hepatic circulation in a human.

Drug administration/ingestion → nonselective binding → transport through body compartments → metabolism/synergistic effects → action in undesired locations and receptors → arrival at site of action → full or partial inhibition/antagonism/agonism of receptor → consequent up- or down-regulation of the receptor → immunological and allergic effects → drug interactions → molecule inactivation/modification/excretion

**FIGURE 2** Path of a molecule in a clinical setting.

It is our observation that the degree of rigor of modeling results is often an issue between CADD staff and other members of the drug discovery team. Sometimes the CADD scientist feels the need to write a fairly involved computer program, to perform validation studies with some relatively newly obtained software, or to perform an exhaustive study in response to a rather carelessly phrased question. It is generally wise for the modeler to clarify the aims of the question before investing weeks or months in performing the wrong study! In other cases, the modeler may be able to perform a crude study quickly and to discuss the results with chemists or biologists, but the results may be dismissed because they are not sufficiently rigorous. In such cases, the project may suffer more than the modeler; even a crude model of a process, if considered with an open mind, can cause one to think deeply about what is really known about the process and what, if any, additional information could be computed or measured. Note that this process can spiral downward, exacerbating the situation: the rejected modeler, unable to get a chemist interested in testing the model, may spend many more weeks at the uncomplaining computer terminal, trying to increase the rigor and impressiveness of the model, whereas the chemist, hearing nothing further from the modeler, assumes nothing more is being done and feels neglected in turn. It takes a strong-minded computational modeler (or biochemist or biophysicist) to continue to carry proposals to an unappreciative chemist (or biologist); but such persistence is often rewarded!

## B. State of Current Computational Models

There are limitations to the applicability and reliability of current methodology for drug discovery problems, as discussed in other chapters of this book and as summarized here (Table VII). There is currently a profusion of computational models, of varying accuracy, in this fast-moving field.

Although all of these methods can be improved, and undoubtedly will continue to be improved, it is our feeling that the current state of most of these methods is sufficient to ensure that they can be applied effectively for drug discovery. It is important to realize that developing "perfect" methods may not be the rate-limiting step for project success. Rather, the most important function of CADD methods may be to enhance the speed and effectiveness of the discovery and development process, and this can in many cases be accomplished by what may appear to be rather primitive methods.

## C. Software and Hardware Constraints

There are hardware limitations to the size and accuracy of problems that can be addressed, but some of these constraints are vanishing as new, more powerful and cost-effective computer technologies are unveiled each year. The best way to participate in this "hardware revolution" is to use

**TABLE VII** Accuracy of Current Computational Methods

| |
|---|
| Molecular mechanics: accurate geometries and energies of drug-sized molecules, provided appropriate force-field parameters have been determined. Macromolecular structures (proteins, nucleic acid) are computed less accurately, and solution behavior is approximately reproduced. Protein–ligand geometries and energies can be approximated with work. Difficulties include proper description of solvent and counter-ion effects, reproducing dynamical behavior, and having appropriate parameters for all molecules of interest. |
| Quantum mechanics: accurate geometries and *in vacuo* energies (including relative conformational energies) for small drug-sized molecules with large basis set ab initio methods including electron correlation. Some loss of accuracy for more approximate methods, including local density functional and semiempirical methods. Advances in the field have produced excellent values of solvation free energies of small molecules (e.g., AMSOL; Cramer and Truhlar, 1991). |
| Monte Carlo and molecular dynamics methods: increased likelihood of reproducing the multiple conformations that exist in solution, but still not rigorous. Limitations include the sampling problem, i.e., ensuring that sufficient conformations have been sampled and that all important conformations have been included, and proper inclusion of solvent effects. |
| More approximate methods: drug–enzyme docking; *de novo* procedures for constructing molecules *in vacuo* and *in situ*; 2-D and 3-D similarity/dissimilarity methods, statistical methods (QSAR, QSPR), etc. These are covered in other chapters. |

present hardware resources effectively and to build the case that additional hardware resources will result in more accurate models and more timely contributions to ongoing drug discovery projects. For timely solutions are the sine qua non of an effective CADD effort.

Another limitation is due to the inability of many methods to be used in concert because of software incompatibility. Although the idea of software and hardware "open architecture" has been bandied about in the industry for years, this is only beginning to be addressed by the software industry, especially by molecular modeling software vendors. We believe that to be truly effective in advancing drug discovery, various available methods (including methods developed in-house) need to be "packaged" in a consistent, easy-to-use manner. Software vendors can then concentrate on improving the reliability and utility of packages they uniquely offer instead of spending massive amounts of time and resources reproducing functionality already offered by other vendors.

### D. Organizational Issues: The CADD Conundrum

Another limitation to the effectiveness of CADD support is related to the organization and composition of project teams. The issue here is that most CADD groups are expected by their companies to contribute to the success of their supported drug discovery projects, but project success requires the derivation of active molecules in the project assays. Since CADD

**TABLE VIII** Factors Associated with Successful CADD Support of Drug Discovery Projects

| |
|---|
| Early assignment of CADD staff to project/program teams |
| Small, focused teams with clear, realistic expectations of CADD contributions |
| Positive, enthusiastic, respected, and knowledgable project leader(s) |
| Management orientation to project "teams," rewarding the whole team |

scientists do not synthesize molecules, they can only influence project success indirectly. However, chemists in most cases have their own ideas of molecules they could profitably make, and in the absence of compelling incentives, they prefer to work on their own ideas rather than on someone else's, since any resulting success is theirs alone. Thus, CADD scientists find themselves being held accountable for project success without the authority to influence project directions.

This is a large and difficult problem; however, there appears to be two approaches to partially solving the CADD conundrum. First, the CADD scientist should be as interactive and supportive of project goals and chemist's goals as possible, and acknowledge chemist and biologist contributions to turning the model and suggested tests into practical experiments; in other words, be a collaborative and supportive team member. Second, research management should ensure that the entire team is rewarded for incremental contributions which result in a successful drug candidate instead of rewarding only the most obvious contributors and perhaps stimulating counterproductive competition among project team members.

While organizational considerations may not guarantee the success of CADD contributions to drug discovery projects, they can minimize impediments to such contributions. Gund *et al.* (1992) suggest a number of factors associated with project success (Table VIII).

## IV. PROPOSALS FOR MAXIMIZING CADD TECHNOLOGY

An important consideration is how to deal with the artificial boundaries between various disciplines along the path of discovery. One way is through the development of an integrated CADD environment based, for example, on the popular "Windows" metaphor, where scientists can gain access to an extensive collection of relevant data, analyze this data with a variety of applications programs, and generate new data from, for example, molecular modeling. This type of seamless integration of computer functionality can be thought of as a type of "electronic notebook" and will be of growing importance in the future. It will enhance the efficiency of the discovery/development process and will provide a suitable means for handling and unifying the complex information obtained from the systems being studied (cf. Lacey, 1992).

As mentioned earlier, the exploding size of databases will require multidimensional data analysis methods (including fuzzy sets, neural nets, and nonparametric statistical methods). System analysis tools can also be of help but mostly as tools for conceptualizing biological systems rather than for computing their properties. Cultivation of a "systems mindset" could enable all scientists in the pharmaceutical industry to deal with the realities of complex biological systems in a more effective and holistic fashion.

The composition of a CADD group is quite important. If CADD is viewed as having the broad scope suggested here, it is essential that such a group be multidisciplinary and this does not mean just computational chemists and programmers. It should also include highly computer-literate scientists from several other disciplines such as biostatistics, biochemistry, biophysics, bioengineering, molecular biology, and mathematics as well. In addition, it would appear to be essential that all of the computational sciences be housed in the same unit and in a common area if possible. This provides an environment that fosters the type of synergy that will allow computational methods to attain their maximum effectiveness and have maximum impact on the discovery/development process. Another critical point is that members of a CADD unit should not only have good programming skills and be knowledgeable in their primary discipline, they should also be intellectually flexible so that they can be of assistance to a variety of projects. The ability to interact with their scientific colleagues is a very desirable trait. Moreover, for CADD to be effective in a company setting, its practioners must be good salesmen.

## V. CONCLUSIONS

We have seen that drug discovery is a complex and difficult process and that it usually requires a great deal of effort (and luck) on the part of a large and multidisciplinary research team to succeed. We have considered the techniques and contributions available to the computational scientist in support of the discovery team, and we have seen that computational contributions may be substantial and diverse. We believe the sum of contributions to projects in various pharmaceutical companies justifies the significant resources that the most enlightened companies are entrusting to a discipline that only indirectly affects a drug discovery project. We have, however, also seen that molecular modelers are far from being able to design one exact compound as a new lead or as the optimal member of a series. We have seen some of the limitations of current methods in this field, and have offered hope for future advances in accuracy and utility.

The appropriate attitude for a computational scientist, then, is humility in the face of the variability and perversity of nature, not arrogance in providing "better" (i.e., more theoretically defensible) answers than the lowly experimentalists. If there is an image for the modeler on the drug discovery team, we suggest it is the worker ant, working with its colleagues to build a tunnel by removing a grain of sand at a time.

## REFERENCES

Appelt, K., Bacquet, R. J., Bartlett, C. A., Booth, C. L. J., Freer, S. T., Fuhry, M. A. M., Gehring, M. R., Herrmann, S. M., Howland, E. F., Janson, C. A., Jones, T. R., Kan, C.-C., Kathardekar, V., Lewis, K. K., Marzoni, G. P., Matthews, D. A., Mohr, C., Moomaw, E. W., Morse, C. A., Oatley, S. J., Ogden, R. C., Reddy, M. R., Reich, S. H., Schoettlin, W. S., Smith, W. W., Varney, M. D., Villafranca, J. E., Ward, R. W., Webber, S., Webber, S. E., Welsh, K. M., and White, J. (1991). Design of enzyme inhibitors using iterative protein crystallographic analysis. *J. Med. Chem.* **34,** 1834–1925.

Baldwin, J. J., Ponticello, G. S., Anderson, P. S., Christy, M. E., Murcko, M. A., Randall, W. C., Schwam, H., Sugrue, M. F., Springer, J. P., Gautheron, P., *et al.* (1989). Novel topically active carbonic anhydrase inhibitors for the treatment of glaucoma. *J. Med. Chem.* **32,** 2510–2513.

Boyd, D. B. (1990). Successes of Computer-Assisted Molecular Design. *In* "Reviews in Computational Chemistry," Vol. 1, p. 355–371. VCH Publishers.

Cambel, A. B. (1993). "Applied Chaos Theory." Academic Press, New York.

Cramer, C. J., and Truhlar, D. G. (1991). General parameterized SCF model for free energies of solvation in aqueous solution. *J. Amer. Chem. Soc.* **113,** 8305–8311.

Freidinger, R. M. (1992). Non-peptide ligands for peptide receptors. *In* "Medicinal Chemistry for the 21st Century" (C. G. Wermuth, ed.), p. 233. Oxford Press, London.

Gund, P., Maggiora, G. M., and Snyder, J. P. (1992). "Chemical Design and Automation News," pp. 30–33. Waltham, MA.

Hlasta, D. J., and Pagani, E. D. (1994). Human leukocyte elastase inhibitors. *Annu. Rep. Med. Chem.* **29,** 195–204.

Kuntz, I. D. (1992). Structure-based strategies for drug design and discovery. *Science* **257,** 1078–1082.

Lacey, M. E. (1992). The systems approach as a paradigm for pharmaceutical research: A proposal. *Drug Info. J.* **26,** 243–249.

Moos, W. H., Grant, D. G., and Pavia, M. R. (1993). Recent advances in the generation of molecular diversity. *Annu. Rep. Med. Chem.* **28,** 315–324.

Navia, M. A., and Murcko, M. A. (1992). The use of structural information in drug design. *Current Opinions in Structural Biology* **2,** 202–210.

Snyder, J. P. (1991). Computer-assisted drug design. I. Conditions in the 1980s. *Med. Res. Rev.* **11,** 641–662.

Thaisrivongs, S. (1994). HIV protease inhibitors. *Annu. Rep. Med. Chem.* **29,** 133–144.

Venuti, M. C. (1990). The impact of biotechnology on drug discovery. *Annu. Rep. Med. Chem.* **25,** 289–298.

Wermuth, C. G. (ed.) (1992). "Medicinal Chemistry for the 21st Century." Oxford Press, London.

Worthy, W. (1990). Synthetic sweeteners developed by design. *Chem. Eng. News* April 30, p. 8.

# 7

# Modeling Drug–Receptor Interactions

**KONRAD F. KOEHLER**
Department of Medicinal Chemistry
Istituto di Ricerche di Biologia Molecolare
00040 Pomezia
Rome, Italy

**SHASHIDHAR N. RAO**
Drug Design Section
Searle Research and Development
Skokie, Illinois 60077

**JAMES P. SNYDER**
Emerson Center for Scientific Computation
Department of Chemistry
Emory University
Atlanta, Georgia 30322

## I. RECEPTORS—INTRODUCTION AND DEFINITION

### A. Macromolecular Targets

The potential targets of drugs are as varied as the molecular components that make up the organism or pathogen (Kenakin, 1990). These sites of action include proteins, nucleic acids, lipids, and saccharides. By far, the most common targets are proteins such as enzymes, membrane-bound receptors, ion channels, antibodies, and structural proteins. Another common drug interaction site is the nucleic acids including nuclear and mitochondrial DNA, as well as messenger RNA.

Since proteins are the most prevalent organic molecules contained within organisms, it is not surprising that they are the most frequent drug targets. Enzymatic proteins are critical for basic metabolic housekeeping as well as

*Guidebook on Molecular Modeling in Drug Design*

for the generation and degradation of other molecular components of the cell. Drugs whose sites of action are enzymes most frequently function as enzyme inhibitors (Page, 1990b), but allosteric activation is another possible mechanism of action (Corton *et al.*, 1995).

Integral membrane receptors are a common means by which cells receive information from their environment (Limbird, 1986). This information or signal is transmitted through the membrane by the receptor to an "effector" located on the cytosolic side of the cell membrane. These effectors may include enzymatic active sites (Fantl *et al.*, 1993), ion channels (Hille, 1992), or binding sites for intracellular proteins (Gilman, 1987). A common mechanism for receptor signal transmission is through binding of the receptor to a G-protein which in turn activates other enzymes (Gilman, 1987). These enzymes most often catalyze the conversion of triphosphate nucleotides to cyclic variants which function as second messengers (Schacter *et al.*, 1988; Chinkers and Garbers, 1991). These messengers in turn pass on the signal through a variety of mechanisms including activation of other enzymes or ion channels.

Drugs can act either as an agonist which mimics the effect of an endogenous chemical signal (e.g., a hormone) or alternatively as an antagonist which blocks the signal. Signals that receptors receive are often greatly amplified through ion channel (Hille, 1992), second messenger (Berridge, 1987), or other mechanisms. Hence drugs which act at receptors are capable of profound effects on the cell.

Ion channels themselves may be targets of drugs, either through an allosteric receptor site located on the same protein or through direct interaction with the ion transport pore (Hille, 1992). Direct binding to the pore usually implies blockade of the ion channel. Local anesthetics, such as lidocaine, interact directly with sodium channels, probably by occluding the ion pore. It has been proposed that general anesthetics interact indirectly with ion channels through the intermediacy of cell membranes. However, accumulating evidence strongly suggests that the site of action is directly with ion channels themselves (Franks and Lieb, 1994; Moody *et al.*, 1994).

Nucleic acids are also frequent targets of drugs. The interaction may be either direct (e.g., intercalation, covalent binding, or cleavage) or indirect through proteins which regulate the replication and expression of genes (Laduron, 1992). Since DNA is central to the function and replication of cells, selective DNA cleavage agents have found application in killing cancerous or foreign pathogenic cells. For selectivity, these agents in general rely on the increased metabolic rate and replication of the targeted vs normal host cells. Consequently, side effects caused by killing rapidly dividing host cells, such as those comprising the endothelial, are often unavoidable. Nucleotide analogs may be phosphorylated by host kinases and in turn are recognized as substrates by viral polymerases. These phosphorylated derivatives may suppress viral replication either through inhibition of the polymerase or through incorporation into viral DNA or RNA which halts further extension of the viral nucleic acid. To minimize toxicity, these agents must be selective for viral vs host polymerases. DNA and RNA direct the production of proteins, and hence agents that can selectively bind to genes or messenger RNA and therefore suppress the production of targeted proteins have therapeutic application (Cohen and Hogan, 1994). These strategies include antisense (Cook, 1991; Mol and van der Krol, 1991) and triplex agents (Maher *et al.*, 1991).

## B. Sources of Structural Information

Modern drug discovery has increasingly relied on detailed three-dimensional (3D) structural information of the macromolecular target (Greer *et al.*, 1994). The atomic coordinates of the site of drug action are most frequently derived from X-ray crystallographic (Glusker, 1994) or nuclear magnetic resonance (NMR) (Zuiderweg *et al.*, 1993) structural determinations. Electron diffraction may provide useful lower resolution information (Glaeser and Downing, 1993), especially for membrane-bound macromolecules for which the required crystallization for X-ray diffraction is difficult.

In the absence of a detailed 3D experimental structure, models of the macromolecular target may be used as surrogates. For example, if the primary sequence of protein of interest is known and it shares a certain degree of sequence homology with one or more proteins for which detailed structural information is available, it is possible to construct a homology model of the target protein (Greer, 1990; Bajorath *et al.*, 1993; Fetrow and Bryant, 1993; Thornton and Swindells, 1993; Johnson *et al.*, 1994; Rost and Sander, 1994). Approximately 28% of the entries in sequence databases share at least 25% sequence homology with one or more of the entries in the Brookhaven database (Chothia, 1992). Therefore, approximately one-third of the sequences stored in the current databases are amenable to homology modeling. Furthermore, it has been estimated that the current Brookhaven data bank contains representatives of 120 different families and a conservative estimate of the total number of different protein folds at approximately 1000 (Chothia, 1992). Therefore, at least 10% of new sequences that are discovered may be homology modeled.

The first step in such a homology modeling building exercise is to determine the registration or alignment of the target sequence onto the experimental structure allowing for insertions and deletions. This may be accomplished with various dynamic programming algorithms pioneered by Needleman and Wunsch (1970). The performance of these methods to locate distantly related sequences is strongly dependent on the quality of the substitution scoring matrices and gap penalties used (Pearson, 1995). More sophisticated methods which emphasize alignments within structurally conserved regions of the experimental structures and evolutionary conserved regions of sequences of related proteins have been developed (Thompson *et al.*, 1994). Also, "threading programs" that determine which structure is compatible with a given sequence (the inverse protein-folding problem) may be useful in testing the alignment or in carrying out the alignment itself (Fetrow and Bryant, 1993; Wodak and Rooman, 1993; Bryant and Altschul, 1995). These methods are based on a statistical analysis of high resolution crystallographic structures to determine the frequencies with which specific amino acid residue pairs (e.g., Glu/Arg, Phe/Trp, etc.) are in contact. Since alternative alignments of a given sequence onto a tertiary protein structure will change the types of residues that are in contact, these statistically determined contact frequencies may then be used to judge the relative probabilities of alternative threadings.

The second step of homology modeling is to mutate the amino acid residues of the experimental structure to match those of the target protein. If more than one experimental structure is available, it is often useful to create an average composite structure of the conserved regions of the protein and to use this hybrid in the model building (Taylor *et al.*, 1994). Insertions and deletions usually occur in the loop regions of proteins and since the conformations of these loops are highly variable, it is usually necessary to either build these *de novo* (Shenkin *et al.*, 1987) or to search the structural databases for loops from other proteins which possess the same sequence length and have a similar relative geometry of the anchor residues (Topham *et al.*, 1993). Next, the assignment of side chain conformation is made (Dunbrack and Karplus, 1993; Eisenmenger *et al.*, 1993). The final step is refinement and examination of the structure to make sure it is reasonable.

Numerous sequence databanks are available for use in multiple sequence alignments. The principle ones are the GenBank (Benson *et al.*, 1994), Protein Identification Resource (PIR) (George *et al.*, 1994), and the EMBL/SWISS-PROT (Bairoch and Boeckmann, 1994) databases. Many of these sequence depositories are accessible over the internet (see Table I). Finally, a useful compilation of sequence and alignment databases is contained in the review by Johnson *et al.* (1994). A widely used collection of software tools for searching and analyzing these sequence databanks is the GCG (Genetics Computer Group) package (Devereux *et al.*, 1984).

The source of the experimental structure(s) used in homology modeling is most commonly the Brookhaven Crystallographic Database [(Bernstein *et al.*, 1977; Abola *et al.*, 1987), available via anonymous ftp:pdb.pdb.bnl.gov (130.199.144.1)]. Crystallographic structures stored in the small molecule Cambridge database (Allen *et al.*, 1979) may also be of use for the modeling of protein ligands.

Membrane-bound proteins, especially receptors, are an important class of drug targets. However, the lack of detailed structural information has hindered structure-based design efforts for these receptors. A few low resolution structures of membrane-bound proteins such as bacteriorhodopsin have

**TABLE I** Depositories of Sequence Databases Available over the Internet[a]

| Organization | Name | Address | Databases held |
|---|---|---|---|
| National Center for Biotechnology Information (NCBI) | NCBI repository | ncbi.nlm.nih.gov (130.14.20.1) | GenBank |
| European Biology Molecular Laboratory | EMBL Anonymous Ftp Server | ftp.embl-heidelberg.de (192.54.41.33) | EMBL |
| Weizmann Institute of Science (EMBnet Israel national node) | DNA and protein sequence analysis (DAPSAS) ftp server | sunbcd.weizmann.ac.il (132.76.64.79) | EMBL, PIR |
| Basel Biozentrum Biocomputing server (EMBnet SWISS national node) | Basel EMBNet ftp server | bioftp.unibas.ch (131.152.8.1) | EMBL, PIR |
| National Institute of Genetics (Japan) | National Institute of Genetics ftp server | ftp.nig.ac.jp (133.39.16.66) | EMBL, GenBank, PIR |

[a] Courtesy of Mark Dalton, Cray Research.

become available and reveal that the membrane-spanning portions of these proteins are composed of a juxtaposition of parallel and antiparallel $\alpha$-helices. Since the receptor sites for these proteins appear to be located within membrane-spanning units and since the secondary structure of these units is certain, homology modeling becomes feasible (Reithmeier, 1995).

One class of membrane-bound receptors appears to be evolutionary related and is composed of seven membrane-spanning units. Furthermore, this family of receptors shares a common signal transduction mechanism, namely a binding site for G-proteins, and hence this family of proteins is known as the G-protein-coupled receptors (GPCRs). Numerous models have been developed for GPCRs (Hibert *et al.*, 1993; Kontoyianni and Lybrand, 1993; Hoflack *et al.*, 1994; Strader *et al.*, 1994), including a general GPCR model (Roper *et al.*, 1994) and models for the dopamine D2 (Teeter *et al.*, 1994) and angiotensin-II receptors (Underwood *et al.*, 1994).

An important issue to be considered in any structure-based design or homology building exercise is the quality of the experimental structures on which the modeling is based. A 3.0-Å resolution crystallographic structure is in general the minimum necessary for detailed structure-based design since it is not possible to resolve the positions of side chain atoms at lower resolution. For the location of ordered water molecules, generally a structure with at least a 2.5-Å resolution is required. For 2-Å or better resolution data, refinement should in theory lead to a mean error in atomic positions of 0.25 Å or less (Janin, 1990). An estimate of the typical error that can be expected from a crystallographic structure analysis may be obtained from comparison of 32 pairs of duplicate independent protein structure determinations which reveal a 0.5-Å RMS deviation of C-$^{\alpha}$ atoms. For a more comprehensive discussion of the accuracies of X-ray crystallographic structures and of other experimental methods of structure determination (neutron and electron diffraction, NMR, and microwave) as well as the accuracies of theoretical methods (molecular and quantum mechanics), the reader is referred to the excellent monograph edited by Domenicano and Hargittai (1992).

Several software packages have been developed to check the quality of experimental protein structural models including PROCHECK (Morris *et al.*, 1992), PROSA (Sippl, 1993; Ortner *et al.*, 1994), and QPACK (Gregoret and Cohen, 1990). These methods are now routinely applied by crystallographers during structure refinement. Furthermore, these methods may be used to identify problem areas in protein homology models.

## C. Ligand–Receptor Interaction

As mentioned in Section IA, the macromolecular sites of action through which drugs mediate their effects are usually proteins, although other bipolymers (DNA, membranes, etc.) have been targeted. An understanding of what forces are responsible for the binding of drugs to proteins may be obtained by first considering what forces drive protein folding since these two processes share many characteristics in common.

The observed structure of proteins is in large part a consequence of the hydrophobic effect (Dill, 1990). Secondary amides form much stronger hydrogen bonds to water than to other secondary amides (Eberhardt and Raines, 1994) and therefore protein folding must be primarily driven by hydrophobic collapse of the unfolded protein. Furthermore, a statistical analysis of the frequency with which amino acid residues of various types associate with each other in protein structures has demonstrated that the hydrophobic effect provides a dominant contribution to attractive amino acid residue "contact potential" (Bryant and Lawrence, 1993). As a consequence, proteins generally bury hydrophobic residues in the core while exposing hydrophilic residues to the exterior. Hydrophilic groups, if found in the interior, almost invariably are paired with complementary hydrophilic residues to form hydrogen-bonding interactions which compensates for the desolvation of these polar functionalities.

In contrast to the overall organization of the protein structure described earlier, ligand-binding clefts in proteins often expose hydrophobic residues to solvent and may contain partially desolvated hydrophilic groups that are not paired with complementary hydrogen-bonding residues. Furthermore, solvation of these polar functional groups may not be optimal due to steric constraints imposed by the protein. The desolvation penalty of these hydrophilic groups is paid for by favorable interaction (mainly hydrophobic) elsewhere in the protein structure. Hence these "sticky" binding pockets represent nonoptimally folded portions of the protein, and the protein as a whole may be thought of as incompletely folded. The interaction of a binding cleft with a complementary ligand is favored energetically since this interaction buries exposed hydrophobic patches and forms favorable electrostatic contacts with partially desolvated hydrophilic groups in the receptor. Hence the association of the ligand with the protein may be thought of as the last step in the protein-folding process. Consistent with this viewpoint, the binding of ligands to protein receptors is known to increase the stability of the protein toward thermal denaturation (Kleanthous *et al.*, 1991; Morton *et al.*, 1995). In other words, the stability of the protein is sacrificed in order to create a binding site and this loss of stability may be "recaptured" by binding an appropriate ligand.

As with protein folding, the principle forces driving drug binding are thought to be the hydrophobic effect (Blokzijl and Engberts, 1993) and electrostatic interactions (including hydrogen bonding). It has been proposed that the hydrophobic component (which is largely nondirectional) contributes primarily to affinity whereas the hydrogen bond, because of its highly directional nature, contributes principally to specificity (Fersht, 1984; Miyamoto and Kollman, 1993).

There is a great deal of controversy concerning the strength of the hydrogen bond and its contribution to drug–receptor binding. Much of this controversy stems from the difficulty of partitioning the contributions to the binding event between the hydrophobic effect and hydrogen bonding. It is nearly impossible to devise a model system in which hydrogen bonding is the sole contributor to binding. Furthermore, analysis of binding events must take into account the unfavorable entropic contribution resulting from immobilizing the ligand. Some studies have assumed that the ligand is completely immobilized upon binding whereas this is rarely the case. Overestimates of this entropic penalty lead to corresponding overestimates of the intrinsic strength of a hydrogen bond.

From site-directed mutagenesis experiments, the free energy of uncharged hydrogen bond formation has been estimated to be in the range of −0.5 to −1.8 kcal/mol (Fersht, 1987; Pace, 1992; Thorson *et al.*, 1995), while a thermodynamic study of amide–amide bond formation in water concluded that uncharged hydrogen bonds are worth −3 to −6 kcal/mol (Doig and Williams, 1992). However, a reinterpretation of the latter data (correcting for residual motions in the complex and accounting for hydrophobic effects) suggests that the original estimate was too high and that a more realistic estimate of hydrogen bond strength is in the range of −1 to −3 kcal/mol (Williams, 1992). A statistical analysis of the contribution of ligand functional groups to protein binding suggests that the average hydrogen bond is worth −1 to −3 kcal/mol (Andrews *et al.*, 1984). Reanalysis using an adjustable entropy term to again account for residual motion in the complex suggests that this estimate (−1 to −3 kcal/mol) is too high (Andrews, 1993). Another statistical analysis, this time of the association free energy of protein–protein complexes, indicates that the average uncharged hydrogen bond contributes −0.24 kcal/mol to complex formation (Horton and Lewis, 1992).

Hydrogen bonds formed between ligand and receptor are at the expense of bonds broken between ligand and solvent and between receptor and solvent. Hence the net enthalpic contribution of hydrogen bonding to affinity is often close to zero or even can be unfavorable. For example, a microcalorimetry analysis of the binding of rapamycin to both the wild-type FK506 binding protein (FKBP-12) and a Tyr-82 → Phe mutant (Y82F) reveals that the hydrogen bond between Tyr-82 and rapamycin is enthalpically disfavored (Connelly *et al.*, 1994). In contrast, a thermodynamic analysis of the same system reveals that this unfavorable enthalpic contribution is more than compensated by a favorable entropic contribution. The net result is that binding to the wild-type protein is slightly favored over the mutant. X-ray crystallographic analysis reveals that two water molecules are released from Tyr-82 upon association of rapamycin with FKBP-12, thus accounting for the favorable entropic term (Connelly *et al.*, 1994). An exceptionally careful and detailed free energy perturbation simulation of the Y82F FKBP-12/FK506 thermodynamic cycle (Pearlman and Connelly, 1995) supports the view that the more favorable binding of FK506 to the native binding protein relative to the Phe-82 mutant is primarily a result of entropy effects.

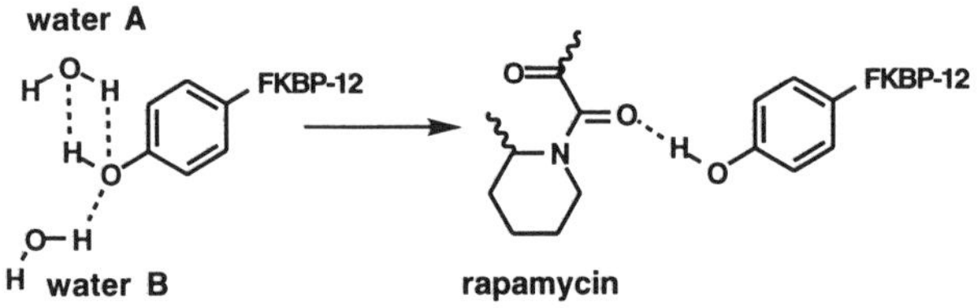

Charge-assisted hydrogen bonds and salt bridges appear to be significantly stronger than neutral hydrogen bonds. The strength of hydrogen bonds in which one or both partners is charged has been estimated from protein mutagenesis data to be in the range of −3 and −6 kcal/mol (Fersht, 1987; Pace, 1992). However, protein engineering and site-directed mutagenesis experiments show that these salt bridges and hydrogen bonds are significantly weakened if solvent exposed (Dao-Pin *et al.*, 1991; Serrano *et al.*, 1992). The thermodynamically measured propensities of amino acid residues to form $\beta$-sheets correlate with the hydrogen exchange rates of the backbone amide hydrogen atom; this correlation has been attributed to strengthening of the hydrogen bonding interaction by shielding it from solvent (Bai and Englander, 1994). However, solvent shielding also involves a substantial desolvation penalty for both $\beta$-sheet (Yang and Honig, 1995) and salt bridge (Hendsch and Tidor, 1994) formation. It is therefore not surprising that relatively few salt bridges exist in the core of proteins (Barlow and Thornton, 1983). At ligand–receptor interfaces, some of this penalty may be paid for in the protein-folding process where hydrogen bonding and salt bridge forming groups located in receptor sites may be partially desolvated. Hence ligand interactions with these desolvated hydrogen bonding centers with complementary functional groups may be especially favorable.

Hydrogen bonding and salt bridge interactions are normally observed between pairs of heteroatoms with an intervening hydrogen atom. However, the $\pi$ electron clouds of aromatic rings frequently appear to act as hydrogen bond acceptors in crystal structures (Levitt and Perutz, 1988). In addition, electron-rich aromatic rings can stabilize cations, e.g., in the complex of acetylcholine with acetylcholinesterase (Sussman *et al.*, 1991). What is remarkable about the crystallographic structure of this complex is that the quaternary ammonium group of acetylcholine is found buried deep within the protein with no carboxylic acid functionality in the immediate vicinity with which to electrostatically stabilize the charge. Instead, the quaternary ammonium group is surrounded by a cluster of electron-rich aromatic groups. Studies of interactions between cations and $\pi$ systems in aqueous media have been reported by Kearney *et al.* (1993) employing NMR, circular dichroism, and *ab initio* quantum mechanical methods. They conclude that in addition to tetraalkylammonium, sulfonium and guanidinium cations also exhibit substantial binding to $\pi$ electron cloud aromatic systems.

Other pairs of interacting functional groups such as $\pi$ clouds of aromatic rings with aromatic (Burley and Petsko, 1985) or aliphatic (Chakrabarti and Samanta, 1995) C–H moieties, while not possessing high intrinsic interaction energy, may still contribute significantly to drug binding. As a result of the relatively low desolvation energy of aliphatic and aromatic groups relative to heteroatom-based hydrogen bond donors and acceptors, the net enthalpies of C–H/$\pi$ interactions may approach those of hydrogen bonds (Chakrabarti and Samanta, 1995).

Halogens bonded to carbon are frequently contained within drug molecules. Despite the high polarization of carbon–halogen bonds, halocarbons, with the possible exception of fluorine, rarely function as hydrogen bond acceptors. To the contrary, a statistical analysis of the Cambridge crystallographic database reveals that oxygen and nitrogen atoms are frequently located within sub van der Waals contact distance with halogen atoms indicative of a stabilizing interaction (Ramasubbu *et al.*, 1986). This is a somewhat surprising result since simple electrostatic arguments would predict that these interactions would be destabilizing. These halogen–heteroatom interactions must therefore be partially covalent in character.

Proteins that interact with drugs are typically enzymes or membrane-associated receptors. Hence drugs may be classified as either substrates or inhibitors if they interact with the former or agonists or antagonists if they interact with the latter.

**Classification of Drugs**

| Protein (host) | Enzyme | Receptor |
|---|---|---|
| Ligand (guest) | Substrates<br>Inhibitors | Agonists<br>Antagonists |

Ligands for receptors normally bind via a noncovalent reversible mechanism, although exceptions certainly exist (e.g., photoaffinity labels). Inhibitors of enzymes generally exhibit a larger range of modes of action which include noncovalent reversible, covalent reversible, and covalent irreversible or suicide inhibition (Bartlett *et al.,* 1990). Enzymes preferentially bind transition states (or more properly reaction intermediates) and hence may not optimally bind substrates since part of the binding energy is used for catalysis (Page, 1990a). Depending on the nature of the particular enzyme, the binding energy may be applied to accelerate the reaction through a variety of mechanisms (Knowles, 1991), including substrate desolvation, locking the substrate in a reactive conformation (Lightstone and Bruice, 1994), or by inducing strain or distortion into the bound substrate (the "rack" mechanism). In contrast, inhibitors have evolved or have been designed for high affinity association with enzymes. Hence inhibitor affinities for enzymes often exceed the corresponding substrate affinities by several orders of magnitude (Page, 1990b). Compelling experimental confirmation of the rack mechanism of the enzyme catalysis in action is provided by the crystallographic structure of dehydro-trypsin complexed with a canonical proteinaceous inhibitor (Marquart *et al.,* 1983). In this crystal structure, Ser-195 is dehydrated so that nucleophilic attack on the scissile amide carbonyl is impossible. Yet the enzyme is still able to significantly distort the amide carbonyl group from planarity. Consequently, transition state analogs in which this amide carbonyl group is replaced by a tetrahedral center (e.g., phosphonate ester) have much higher affinities for trypsin compared to peptide substrates.

The type of binding (substrate-like or transition state-like) of a series of inhibitors and their corresponding substrates for a given enzyme may be inferred from kinetic parameters. A linear correlation between $K_i$ and $K_m$ is indicative of substrate-like binding whereas a correlation between $K_i$ for the inhibitors and $K_{cat}/K_m$ for the corresponding substrates strongly suggests transition state-like binding (Bartlett and Marlowe, 1983; Brady and Abeles, 1990). Consequently, kinetic analysis may be used to determine whether a series of putative transition-state analog inhibitors is binding to the enzyme as intended. If not, a high priority should be given to optimizing the functional group designed to act as a transition state mimic since this optimization has significant potential for increasing potency.

Agonists are somewhat analogous to enzyme substrates in that part of the binding energy is co-opted for another purpose, namely to transmit a signal by inducing a confirmation or aggregation shift in their receptors that may detract from their binding affinity (Limbird, 1986). However, agonist interactions with receptors may not be an equilibrium phenomenon since binding may be coupled to an energy-releasing event such as ATP or GTP hydrolysis. The receptors may thus cycle between several states (e.g., resting, activated, and inactivated), each with differing affinities for the same ligand. Hence receptors often exhibit high and low affinity states for agonists not but for antagonists. Furthermore, the ratio of affinity of agonists for the high vs low affinity states has been demonstrated to be proportional to the efficacy of the ligand (Lefkowitz *et al.*, 1993).

Binding to both types of drug targets (i.e., enzymes and receptors) appears to be governed by the curious phenomenon of entropy/enthalpy compensation, wherein increases in the enthalpy of binding are compensated by decreases in entropy and vice versa over a wide series of ligand–receptor interactions (Gilli *et al.*, 1994). This compensation may in part be a consequence of the trade-off between maximizing enthalpically favorable van der Waals and electrostatic interactions to produce a closely packed interface at the expense of the entropically unfavorable reduction in conformational degrees of freedom. An early report which suggested that receptor–agonist interactions are enthalpy driven whereas receptor–antagonist interactions are entropy driven (Weiland *et al.*, 1979) appears not to be general (Borea *et al.*, 1992). At least for the dopamine D2 receptor, the thermodynamic nature of the binding seems to be controlled more by structural parameters (the binding of hydrophobic ligands tends to be entropy driven) than by the pharmacological effect (agonist or antagonist) of the ligand (Testa *et al.*, 1987). Furthermore, the relative contributions of entropy vs enthalpy to agonist binding may differ between the high and low affinity states (Fraeyman *et al.*, 1994).

Enthalpic contributions to binding are thought mainly to derive from favorable electrostatic interactions including hydrogen bonding and contributions from the solvent (Chervenak and Toone, 1994). Entropic contributions are due to the release of ordered water from surfaces of complementary shapes on the ligand and receptor (Fersht, 1984). Cases in which antagonist binding is primarily entropy driven (i.e., the hydrophobic effect) are fully consistent with Ariëns (1987) observation that many agonists can be converted into antagonists by the addition of hydrophobic groups to the ligand. These are hypothesized to interact with "accessory antagonistic hydrophobic interaction sites" on the receptor (Ariëns, 1987). Conversely, cases are known in which the addition of a hydrophobic group results in a switch from antagonism to agonism, e.g., with the $AT_1$ receptor antagonist L-159,282 and the agonist L-163,101 (Underwood *et al.*, 1994) which differ by an isobutyl moiety.

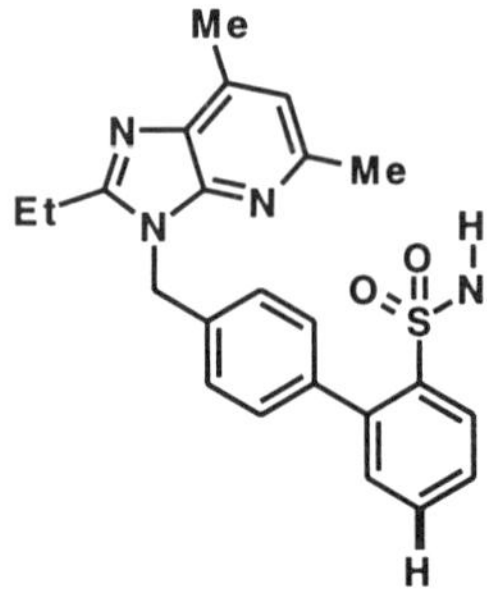

L-159,282 (Antagonist)

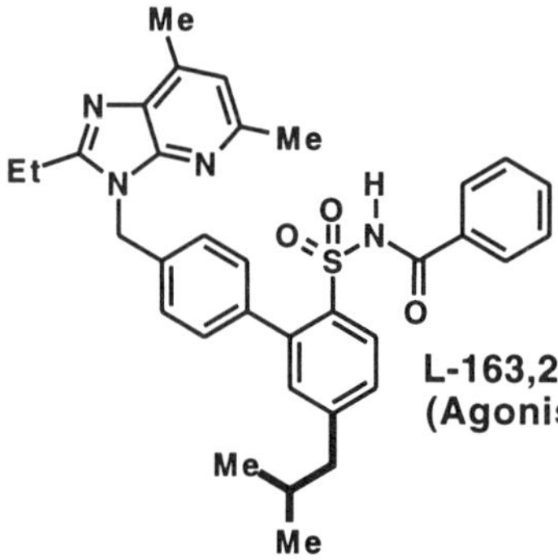

L-163,202 (Agonist)

## II. RECEPTOR FITTING—X RAY AND OTHER EXPLICIT STRUCTURES

### A. Utility

Depending on the type of information available, two distinct modeling strategies may be employed in drug design. If an explicit structure of the drug receptor target is not available, but several competitive ligands that bind to the receptor are known, the receptor structure may be inferred based on what binds to it. This strategy is referred to as receptor mapping and is discussed in more detail in Section III. If, on the other hand, an explicit 3D structure is available from crystallographic or NMR determination or from homology modeling (Section I,B), receptor fitting may be used.

There are two basic situations in which receptor-fitting procedures may be applied: (1) *de novo* design of a structurally novel ligand and (2) the refinement of an existing lead. As explained in more detail in Section II.C.2, 3D database searches and automated docking programs may be of assistance in *de novo* design. As for refinement, if an experimental structure of the lead complexed with the protein target is available, then careful examination of this structure may suggest ways in which the ligand can be modified to enhance affinity. For example, an experimental structure of the complex may indicate hydrophobic pockets or hydrogen-bonding opportunities in the vicinity of the bound lead which are not being taken advantage of. Conversely, ligand functional groups that possess high temperature factors in the X-ray crystallographic structure of the complex are loosely bound and therefore probably do not contribute significantly to affinity. These disordered functional groups are good candidates for optimization.

Receptor fitting, used either for refinement or for *de novo* design, may yield many synthetic ideas or database hits, and these ideas need to be prioritized for synthesis or screening, respectively. Hence there is obvious utility in developing and employing accurate methods for estimating the relative binding affinities of candidate ligands. For this purpose, very approximate but computationally efficient empirical scoring functions (Section

II.C.3) and more accurate but computational demanding free energy perturbation procedures (Section II.D) have been developed.

## B. Binding-Site Properties

### 1. Stereoelectronic Factors

Enzymes and receptors are highly discriminating, and hydrogen bonding is believed to contribute in large part to this specificity (Fersht, 1984). Therefore it is not surprising that ligand–receptor complexes possess numerous intermolecular hydrogen-bonding contacts (Klebe, 1994a). Hydrogen bonds are also frequently observed to bridge protease–inhibitor and antibody–antigen protein interfacial regions (Janin and Chothia, 1990). Backbone amide groups in proteins account for fully one-third of the hydrogen bonds seen between ligands and proteins (Klebe, 1994a), whereas by definition the remaining two-thirds of ligand hydrogen bonds are to amino acid side chains. The most commonly observed amino acid residues at receptor sites in terms of absolute numbers (in decreasing order) are glycine, serine, arginine, and tyrosine (Villar and Kauvar, 1994). The role of glycine may be to increase the flexibility of the backbone and hence the plasticity of the receptor site. When compared to the frequencies that are present in proteins generally, arginine, histidine, tryptophan, and tyrosine are overrepresented at binding sites whereas proline is underrepresented. The side chain hydroxyl groups of serine and tyrosine residues are frequently found to hydrogen bond with ligands. The arginine side chain is also a common hydrogen bond partner (involved in 16.8% of protein–ligand hydrogen bonds) and is most often associated with carboxylic acid and phosphate groups on the ligand (Klebe, 1994a). In addition to arginine–acid salt bridges, other types of charge-assisted hydrogen-bonding interactions are frequently seen.

Strong hydrogen-bonding directional preferences are revealed by an examination of protein (Thanki *et al.*, 1988; Boobbyer *et al.*, 1989; Tintelnot and Andrews, 1989; Ippolito *et al.*, 1990; Wade *et al.*, 1993; Böhm, 1994a) and especially small molecule X-ray crystallographic structures (Murray-Rust and Glusker, 1984; Vedani and Dunitz, 1985; Boobbyer *et al.*, 1989; Vedani and Huhta, 1990; Wade *et al.*, 1993; Klebe, 1994b). Furthermore, a qualitative agreement exists between the geometry of hydrogen-bonding interactions seen in small molecule crystal structures compared to those in protein–ligand complexes (Klebe, 1994b; Pascard, 1995).

Other types of functional group interactions also display directional preferences. For example, oxygen and nitrogen atoms tend to approach halogen–carbon bonds head on (Het $\cdots$ X–C angle $\sim 165°$; Het = N,O; X = Cl, Br, I) (Ramasubbu *et al.*, 1986). In contrast, aromatic rings tend to enage in edge-to-face interactions (Burley and Petsko, 1985).

Nonpolar functional groups, not capable of hydrogen bonding, can also contribute to specificity. A prime example is limonene, a chiral hydrocarbon terpine devoid of H-bonding capability. The mirror images of limonene clearly interact very differently with olfactor receptors as the R-(+)-enantiomer has a citrus smell whereas the S-(−)-enantiomer is harsh and turpentine like. Another example is that of carvone in which the (+)-isomer has the fragrance of caraway whereas the (−)-isomer possesses the odor of spearmint (Friedman and Miller, 1971).

**Limonene Enantiomers**

$H_3C$ $CH_3$ $H_3C$ $CH_3$

(+) (R)-form
citrus fragrance

(–) (S)-form
harsh turpentine smell

**Carvone Enantiomers**

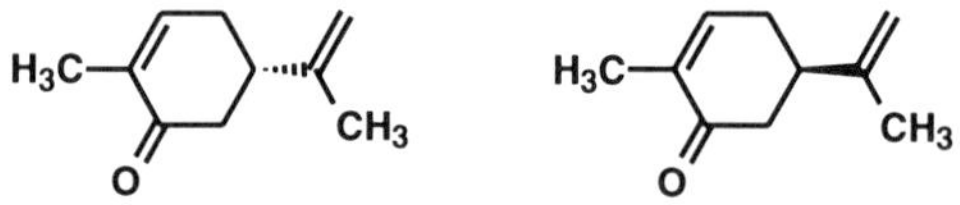

$H_3C$ $CH_3$ O $H_3C$ $CH_3$ O

(+) (S)-form
caraway

(–) (R)-form
spearmint

Enantiomers of halocarbon anesthetics also display modest but statistically significant differences in pharmacology (Moody *et al.*, 1994). These differences in potency provide strong evidence that proteins rather than membranes are the primary site of action of general anesthetics (Franks and Lieb, 1994). Furthermore, these anesthetics are very hydrophobic with exceptionally limited hydrogen-bonding potential, again demonstrating the ability of nonpolar groups to influence selectivity.

Receptor- and enzyme-binding sites are highly chiral both in terms of the disposition of hydrogen-bonding functionality and hydrophobic pockets. Therefore, they often display large differences in affinity for enantiomeric pairs of ligands (Casy, 1993). Conversely, differences in the affinity of enantiomeric ligands have been used as evidence for specific receptor involvement and to gain further insight into drug–receptor interactions (Gualtieri, 1993). Consequently, it is natural to consider chiral ligands as drug candidates. The principle advantage of chirality is that it affords a means of enhancing potency and selectivity. The principle disadvantage is the increased complexity of chiral drug synthesis (Stinson, 1994). Increases in potency are frequently accompanied by increases in the ratio of activity between enantiomers; this correlation is known as Pfeiffer's rule (Pfeiffer, 1956). Observations of "anti-Pfeiffer" behavior in some series of drugs (Van de Waterbeemd *et al.*, 1987) were later reinterpreted as "non-Pfeiffer" behavior after normalizing the data for the degree of physicochemical difference (i.e., steric and electrostatic potential) between the superimposed pairs of enantiomeric ligands (Seri-Levy *et al.*, 1994).

## 2. Receptor Flexibility

The classical definition of substrate–enzyme interaction was provided by Emil Fischer (1894) who viewed it as one between a "lock and key." This concept of rigid guest and host was later extended to selective drug–receptor interactions by Paul Erhlich (1913). However, accumulating evidence from X-ray crystallographic (Jorgensen, 1991) and molecular dynamics studies (Brooks *et al.*, 1988) demonstrates that considerable differences in conformation between occupied and unoccupied receptors frequently occur. The "hand and glove" analogy is therefore a more appropriate description for drug–receptor interactions. Typically, amino acid side chains in the binding clefts undergo the most motion (Janin and Chothia, 1990), but main-chain atoms may also experience sizeable displacements.

One extreme example is that of HIV protease in which X-ray crystallographic studies have shown that the tips of the "flaps" of this enzyme are capable of undergoing a 7-Å movement (Fitzgerald and Springer, 1991). Another example is provided by the changes in structure observed on inhibitor binding to the enzyme purine nucleoside phosphorylase (PNP). An initial drug design effort was based on the uncomplexed protein structure, but this modeling study was not successful in predicting new inhibitors (Montgomery *et al.*, 1993). However, crystallographic analysis of PNP/inhibitor complexes revealed significant movement in the loop made up of residues 241–260 relative to the uncomplexed protein (Erion *et al.*, 1993). Structure-based design starting from the PNP/inhibitor complex was far more fruitful (Montgomery *et al.*, 1993) and led to the successful prediction of several potent PNP inhibitors prior to chemical synthesis (Erion *et al.*, 1993). A third example is the conformation of antibody-binding sites on binding antigens that appear to be unusually susceptible to change (Wilson and Stanfield, 1993). This receptor flexibility in part accounts for the remarkable diversity of antigens that antibodies are able to recognize.

A qualitative estimate of the flexibility of various portions of a protein may be obtained from X-ray crystallographic temperature factors or from molecular dynamics simulations. Temperature factors may be displayed using color-coded atoms whereas MD flexibility may be projected onto color-shaded surfaces (Zachmann *et al.*, 1995). Flexibility estimates may be of direct use in ligand design for they indicate which parts of the receptor may tolerate variations in the size of ligand substituents (Morton and Matthews, 1995). Furthermore, protein loops with high temperature factors that line the receptor site may indicate portions of the receptor that are susceptible to large conformational changes on ligand binding. Indeed, X-ray crystallographic analyses of HIV-1 protease reveal that the backbone atoms that comprise the "flaps" that surround the binding pockets possess relatively large temperature factors, consistent with the large movement that these loops are capable of undergoing (Fitzgerald and Springer, 1991).

In predicting changes in receptor conformation due to side chain movements, rotamer libraries may be consulted (Dunbrack and Karplus, 1993; Vriend *et al.*, 1994). However, Schrauber *et al.* (1993) have found that from 5 to 30% of the side chain conformations observed in high resolution crystallographic structures differ substantially from the common rotameric states. As an alternative, systematic conformer searches followed by molecular mechanics energy evaluation are surprisingly effective in predicting experimentally observed protein side chain conformations (Eisenmenger *et al.*, 1993). A variant of this conformational search procedure has been successfully applied to predicting the stability of various mutant enzyme–inhibitor complexes (Wilson *et al.*, 1991) and has led to the development of a mutant enzyme with altered substrate specificity (see Section II,D,1).

### 3. Tight Binding

From the preceding discussion, it should be obvious that a prerequisite for tight binding is a high degree of shape (Sufrin *et al.*, 1981; Lawrence and Colman, 1993) and electrostatic (Chau and Dean, 1994) complementarity between ligand and receptor. One of the tightest binding ligand–receptor pairs known is biotin and strepavidin ($K_a \approx 100$ f$M$). A high resolution X-ray crystallographic analysis (Weber *et al.*, 1992b) reveals that in addition to shape complementarity between this ligand and receptor resulting in the release of five ordered water molecules from the binding pocket of apostrepavidin, a tetrahedral oxyanion hole and two hydrogen bond-accepting groups in the receptor cause a polarization of the ureido moiety of biotin (see Fig. 1). This in turn results in an exceptionally strong electrostatic stabilization of this complex. However, a molecular dynamics simulation reveals that a similar hydrogen-bonding interaction exists between biotin and solvent (Miyamoto and Kollman, 1993). Furthermore, a free energy

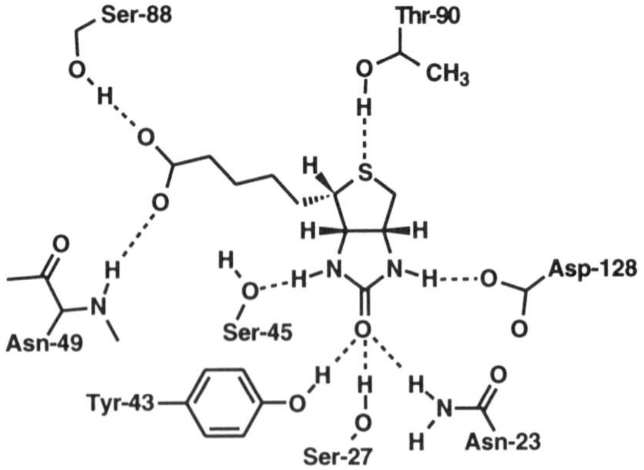

**FIGURE 1** Schematic diagram of the complex between biotin and streptavidin (Weber *et al.*, 1992b). The oxyanion hole is formed by Tyr-43, Ser-27, and Asn-23.

perturbation analysis suggests that van der Waals attraction contributes more to binding than does electrostatics. This conclusion was obtained by running separate perturbations of the two components of the nonbonded interactions (Miyamoto and Kollman, 1993).

Conformational flexibility and strain energy of the ligand also play critical roles. A multiple regession analysis of the contributions of various functional groups to binding demonstrates that, on average, each freely rotating bond in a ligand reduces binding free energy by 0.7 kcal/mol (Andrews *et al.*, 1984). Many examples are known where rigidification of a flexible ligand causes a substantial boost in affinity. For example, freezing of two rotational degrees of freedom in a flexible $H^+/K^+$-ATPase inhibitor resulted in a rigid analog which displayed a 150-fold boost in potency (Kaminski *et al.*, 1991).

**$H^+/K^+$ — ATPase Inhibitors ($IC_{50}$ in μM)**

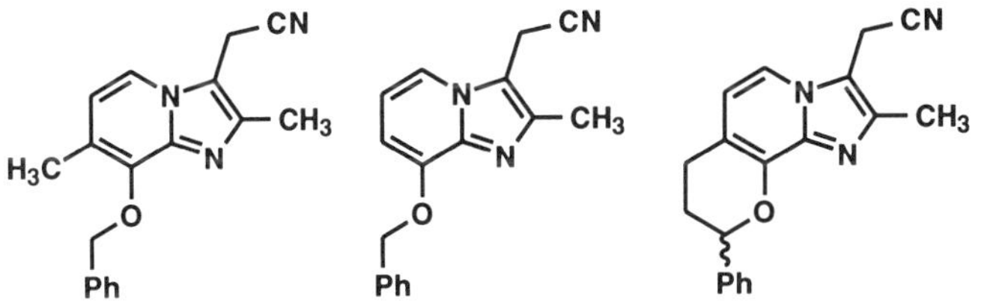

| | | |
|---|---|---|
| 13.7 ± 0.7 | 1.6 ± 0.4 | 0.09 ± 0.2 |

Nature has also exploited this principle in the evolution of serine protease inhibitors in which the binding loops of canonical inhibitors undergo little change in conformation from the free to the bound state (Hubbard *et al.*, 1991). In part, inhibitor rigidification accounts for the exceptionally high affinity displayed by these proteinaceous blockers. However, rigidifying a flexible ligand into the wrong conformation will of course produce an inactive compound.

Counter examples of flexible ligands which nevertheless display high affinity binding are known. One of these is the octapeptide angiotensin-II (A-II) which displays high affinity for both its membrane-bound receptor $AT_1$ with a $K_a$ of ~1 n*M* ($K_a$ varies slightly as a function of species and tissue) and for antibody MAb131 ($K_a$ = 7.4 n*M*). Titration calorimetry shows that binding is favored both enthalpically and entropically, but that enthalpy makes the major contribution (Murphy *et al.*, 1993). The large loss of configurational entropy on binding of the very flexible octapeptide is apparently more than compensated by the release of solvent molecules during the binding process. Consistent with this hypothesis, crystallographic analysis of the A-II/MAb complex reveals that a large amount of surface area is buried during complex formation (725 $Å^2$ for MAb and 620 $Å^2$ for A-II) and that ~60% of this buried surface area is nonpolar (Garcia *et al.*, 1992).

Strain energy of the bound conformation of the ligand also detracts from the affinity. In the design of inhibitors of PNP, consideration of the conformational strain energy was shown to be valuable (Secrist *et al.*, 1993). In all cases, subsequent X-ray crystallographic analysis demonstrated that the observed bound conformation matched one of the lowest energy conformers obtained through an exhaustive molecular mechanics conformational search. In contrast, a study that compared the relative energies of the protein bound vs global energy minimum structures of 33 ligands contained in the Brookhaven database using the CHARMm molecular mechanics force field and AM1 and PM3 semiempirical hamiltonians concluded that the bound conformations are often highly strained, with strain energies frequently exceeding 10 kcal/mol (Nicklaus *et al.*, 1995). For the purpose of the energy comparisons, the X-ray crystallographic coordinates of the bound conformations were first partially minimized holding torsional angles fixed and fully optimizing bond lengths and angles. However, a preliminary study of the conformational energetics of these same ligands demonstrates that the use of position constraints in place of torsion constraints results in conformations with significantly lower conformational energy *and* simultaneously a closer fit to the crystallographic data (unpublished results). In addition, if the relative energies of the bound vs global energy minimum are evaluated with the AMBER* force field, including an aqueous solvent continuum correction, an upper limit of 5 kcal/mol of strain energy is obtained (unpublished results).

## C. Ligand-Binding Predictions

### 1. Binding Modes and the Creation of "Nonnatural" Active Sites

Attempts to design ligands and to predict their affinities are made considerably more complex by the existence of alternative binding modes (Mattos and Ringe, 1993). For example, NMR experiments have demonstrated that both the coenzyme and the substrate of dihydrofolate reductase (DHFR), $NADP^+$ and folate, respectively, have more than one binding orientation (Roberts, 1991). The pteridine ring of folate can bind to DHFR in both a productive and a nonproductive conformation depending on the protonation state of Asp-26, and the substrate binds in each mode with roughly equal affinity. The nicotinamide ring of coenzyme $NADP^+$ may bind to the enzyme or hang free in solution while the rest of the coenzyme is bound. Once again, DHFR binds the coenzyme in each mode with approximately equal affinity.

A series of closely related trifluoroacetamide dipeptide elastase inhibitors have been shown by X-ray crystallographic analysis to bind very differently (Mattos *et al.*, 1994). While some strong inhibitor functional group preferences are seen (e.g., if present, lysine always binds to the $S_2$ subsite), other functional groups appear to have no clear preference (e.g., the trifluoroacetyl group can occupy either the oxyanion hole or the $S_1$ pocket). In another example, relatively small changes in the structure of antiviral "WIN" compounds cause a reversal in the binding mode to rhinovirus coat protein (Badger *et al.*, 1988).

Furthermore, ligands may induce conformational changes in protein structure to open up binding pockets that do not exist in the absence of ligand. Examples of this phenomena are seen in the X-ray crystallographic structures of free hemoglobin and hemoglobin complexed with various ligands (Abraham *et al.*, 1983; Perutz *et al.*, 1986) and in free HIV-1 reverse transcriptase vs complexes with nonnucleoside inhibitors (Ding *et al.*, 1995). These binding pockets would be extremely difficult to predict without a detailed 3D experimental structure of the ligand–receptor complex. Ultimately the existence of alternative binding modes and receptor flexibility (Section II,B,2) may require that X-ray crystallography (Appelt *et al.*, 1991) or NMR spectroscopy be brought into the iterative design, synthesis, and testing cycle (Section VI,A). Experimental structure determination is at present the only sure way to verify that the conformation of the protein has not changed and that the binding mode of the inhibitor is as expected.

### 2. Docking and Database Searching

Before an X-ray crystallographic or NMR structure may be used for the design of ligands, it must be examined to determine where the ligands are likely to bind. In the absence of an experimental structure of the protein of interest complexed with a ligand, it is not a trivial exercise to locate these sites. Most high affinity receptors are located on a concave surface of the protein. However, there may be several concave surfaces and it is difficult to determine which of these is responsible for binding. Fortunately, once a protein X-ray crystallographic structure is available, it is relatively easy to obtain a structure of the ligand complex through Fourier difference maps, provided that the ligand can be "soaked" into the protein crystal. Likewise, interpretation of an NMR stucture of a ligand–protein complex is usually straightforward once the structure of the protein itself has been solved. However, even if an experimental structure of the protein/ligand complex is available, alternative binding modes (Section II,C,1) may make extrapolation of the structure of one complex to another risky, especially if the two ligands belong to different chemical classes.

Once the binding site has been identified, it is often necessary to refine or augment the experimental structure. This is especially true of protein X-ray crystallographic structures that lack hydrogen atom coordinates. Most molecular modeling software packages have facilities to add these hydrogen atoms, but the user should carefully check the positions of the added atoms. The assignment of protonation states and the location of hydrogen atoms on tautomeric centers require special attention. Solvent-exposed aspartate, glutamate, arginine, and lysine side chains as well as C- and N-terminal groups that have p$K_a$'s far from physiological pH are almost always in their charged state. Histidine with a p$K_a$ near physiological pH is often protonated whereas the more basic NE2 site is more often protonated in the neutral state (Tanokura, 1983). When buried, the acidic or basic amino acid residues mentioned earlier are usually paired to form salt bridges. While it is difficult or impossible to determine from the experimental structure whether these salt bridges are in their charged or neutral tautomeric state, it is usually desirable to neutralize charged residues wherever possible since these charges create difficulties in molecular mechanics refinement of docked ligands due to inadequate treatment of electrostatic effects by the force field. For this reason, charged surface residues may also need to be neutralized. The interiors of proteins often contain extended hydrogen bonding networks and the tautomeric states of heteroatoms involved in these networks frequently need adjustment to maximize the number of favorable hydrogen bonds.

Water molecules are frequently observed within the binding pockets of protein X-ray crystallographic structures. Receptor site water molecules that are highly disordered (as judged by their temperature factors) are weakly bound and therefore are likely to be displaced by appropriately positioned ligands. These water molecules can safely be deleted from the structure. In contrast, tightly bound water molecules are often observed to form bridging hydrogen bonds between protein–protein interfacial regions (Janin and Chothia, 1990; Bhat *et al.*, 1994; Buckle *et al.*, 1994). Bridging water molecules are seen as well in the structures of many HIV protease inhibitor complexes (Fitzerald and Springer, 1991). Therefore, retention of these water molecules in the receptor cleft used for docking may be important. However, even tightly bound water molecules may be displaced by the right ligand, and displacement of these water molecules contributes favorably to the entropy of binding. In fact, the displacement of specific, highly ordered water molecule was used explicitly in the successful design of a highly potent HIV protease inhibitor (Lam *et al.*, 1994).

If the design starts with an experimental structure that is a covalent protein/ligand complex and if the design of a noncovalent ligand is desired, it may be necessary to geometry optimize the residue involved in the covalent linkage in the absence of ligand. The covalent bond will distort the structure of the receptor relative to the free protein, making the binding cavity smaller than if it were bound to a noncovalent inhibitor.

Docking of ligands to proteins is a difficult problem since it involves optimization of six degrees of freedom (three rotational and three translational) between the ligand and the protein as well as optimization of internal torsional degrees of freedom in the ligand. Therefore, it is often easier to start with a crystallographically determined protein/ligand complex and to "mutate" the structure of the ligand into the desired structure. The modeled structure may be geometry optimized using molecular mechanics in the presence of the rigidly held protein. Again, it must be emphasized that the possibility of alternative binding modes (Section II,C,1) makes this strategy risky, especially if the ligand mutation represents a large structural change. If a crystallographic structure of the protein complexed with a relatively close analog of the ligand is available, "ligand-based docking" may be performed (McMartin and Bohacek, 1995). In this procedure, one or more conformations of the candidate ligand are fitted to the crystallographic structure of the known ligand by optimizing the similarity in electrostatic and steric potentials using a program such as SEAL (Kearsley and Smith, 1990). The experimental structure of the "template" ligand is then deleted, leaving the candidate ligand docked to the protein. In addition, the conformation of the fitted ligand may be simultaneously optimized during the fitting (McMartin and Bohacek, 1995).

Interactive docking may be performed using high resolution graphics with the six degrees of freedom controlled by the user with a mouse or a dial box. The docking may be assisted through the display of the solvent-accessible surface (van der Waals plus aqueous solvent probe radii) of the receptor and ligand (Lee and Richards, 1971; Connolly, 1983). Docking of the ligand with the display of hydrogen atoms turned off to the solvent-accessible surface of the receptor is an especially useful tactic since this depiction is relatively uncluttered. Furthermore, optimal van der Waals contact between the two binding partners may then be approximated by placing the heavy (i.e., nonhydrogen) surface atoms of the ligand on the solvent-accessible surface of the receptor. To assist the user in maximizing chemical complementarity between ligand and receptor, the surface may be color coded to distinguish between hydrogen bond accepting and donating sites as well as hydrophobic regions (Bohacek and McMartin, 1992). Manual docking may also be facilitated through the identification of receptor "hot spots" by calculating GRID-based interaction energies between the receptor and various molecular mechanics probes representing typical functional groups present in ligands (Goodford, 1985). More recently, virtual reality approaches to docking have been developed wherein the user "becomes" the ligand and experiences the forces felt by the ligand during the docking process through tactile feedback (Illman, 1994).

GRID was applied to analyzing the binding site of the influenza virus sialidase enzyme and was instrumental in the development of more potent inhibitors (von Itzstein *et al.*, 1993). The analysis started with the X-ray crystallographic structure, the enzyme/sialic acid substrate complex. The GRID program was used to predict energetically favorable substitution of a known unsaturated sialic acid inhibitor Neu5Ac2en ($K_i = 1\ \mu M$). More specifically, a protonated primary amine GRID probe located a hot spot near the 4-hydroxy moiety of Neu5Ac2en. Hence it was predicted that substitution of the 4-hydroxy group by a basic functional group such as amino or guanidino should significantly enhance affinity. Furthermore, the proximity of the 4-position Neu5Ac2en when complexed with sialidase enzyme to Glu-199 suggested the formation of a salt bridge with this protein residue. Gratifyingly, the measured $K_i$'s for 4-amino- and 4-guanidinyl-Neu5Ac2en were 50 and 0.2 n*M*, respectively, which represents an increase in potency of 5000-fold for the guanidinyl derivative vs the unsubstituted Neu5Ac2en lead. The 4-guanidinyl compound showed promising *in vivo* antiviral activity in a ferret model when applied intranasally (von Itzstein *et al.*, 1993) and now is in human clinical trials (Hayden *et al.*, 1994).

While subsequent X-ray crystallographic analysis confirmed the overall conservation in the binding mode for the 4-guanidinyl derivative (Varghese *et al.*, 1995), the 4-guanidinyl group did not form a salt bridge with Glu-119 as predicted in the original design (von Itzstein *et al.*, 1993). Instead the guanidinyl side chain is stabilized by hydrogen bonding to Glu-227, Asp-151, and a backbone carbonyl group. Interestingly, a full molecular mechanics minimization of the 4-guanidinyl-Neu5Ac2en/sialidase complex including crystallographically located water molecules and a cap of explicit water molecules about the binding site suggested that salt bridge formation between the 4-guanidinyl group and Glu-119 is favorable (Varghese *et al.*, 1995). However, full optimization of the complex was also accompanied by relatively large movements in the protein structure. In contrast, when the heavy atoms of the protein were constrained to their crystallographic coordinates, a close correspondence between the predicted and crystallographically determined binding mode of the 4-guanidinyl-Neu5Ac2en was obtained.

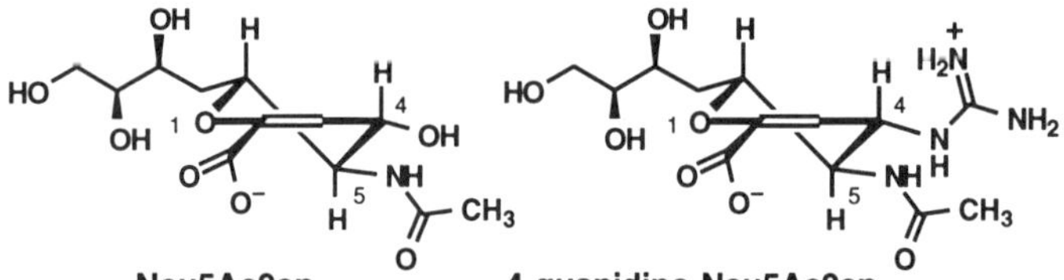

**Neu5Ac2en**
**$K_i$ = 1 μM**

**4-guanidino-Neu5Ac2en**
**$K_i$ = 0.2 nM**

Automated docking procedures have also been developed (Blaney and Dixon, 1993; Humblet and Dunbar, 1993; Lybrand, 1995) which allow a much more thorough exploration of alternative binding modes as compared to interactive docking. One of the principle goals of these automated methods is to make them sufficiently fast such that searches of large databases of candidate ligands become feasible. Early methods, such as the DOCK program (Kuntz *et al.*, 1982), fill the receptor site with a minimum number of spheres of adjustable size. Volume matching between this negative image of the receptor and candidate ligands is subsequently performed. These early methods focused on shape complementary and ignored electrostatic or chemical complementary. Later methods have taken into account electrostatics using a GRID-like potential for the receptor site (Böhm, 1994b; Kuntz *et al.*, 1994; Miller *et al.*, 1994).

Sources of machine-readable collections of small molecule structures which have already been synthesized and which are suitable for Dock-type or pharmacophore searches include the Cambridge Structural Database (CSD) (Allen *et al.*, 1979), the National Cancer Institute drug database (Milne *et al.*, 1994), and the Fine Chemical Directory (FCD) (Henry *et al.*, 1990). The FCD is especially useful for this purpose since it is composed of compounds that can be purchased on the open market. However, it should be pointed out that the percentage of "drug-like" molecules in the FCD is relatively low. It therefore may be desirable to select a subset of such databases for pharmacophoric searches. For example, restricting the search to compounds with molecular weights between 250 and 600, which contain only "normal organic" atoms (i.e., no heavy metals, boranes, etc.), at least one ring, at least two heteroatoms each attached to carbon, and which contain no reactive functionality (i.e., diimides, anhydrides, acid halides, etc.), will ensure a much higher percentage of drug-like molecules in the hit list. In addition, many pharmaceutical firms have created large collections of proprietary compounds that have been synthesized in-house.

Construction of pharmacophore-searchable databases of proprietary or commercially available compounds is facilitated by automated procedures for the conversion of 2D to 3D structures (Pearlman, 1993). Alternatively automated methods for generating collections of hypothetical structures which conform to user specification may be used to create 3D structural databases (Lauri and Bartlett, 1994; Ho and Marshall, 1995). More recently, conformationally flexible 3D database searching techniques have been developed which include on-the-fly generation of conformations (Hurst, 1994) and storing of representative conformations (Kearsley *et al.*, 1994). Finally, *de novo* design software packages have been developed in which computer-generated molecular repertoires "evolve" within the receptor site of interest, and ligands from these repertoires are "selected" for based on their predicted binding affinities (Humblet and Dunbar, 1993; Bohacek and McMartin, 1994; Lewis and Leach, 1994).

Examples of successful application of 3D database searching techniques include DOCK searches of the FCD which resulted in the identification of several selective micromolar inhibitors of two parasitic proteases (Ring *et al.*, 1993). An additional interesting feature of this work was that it was based entirely on protein homology models since X-ray crystallographic structures were unavailable for either enzyme.

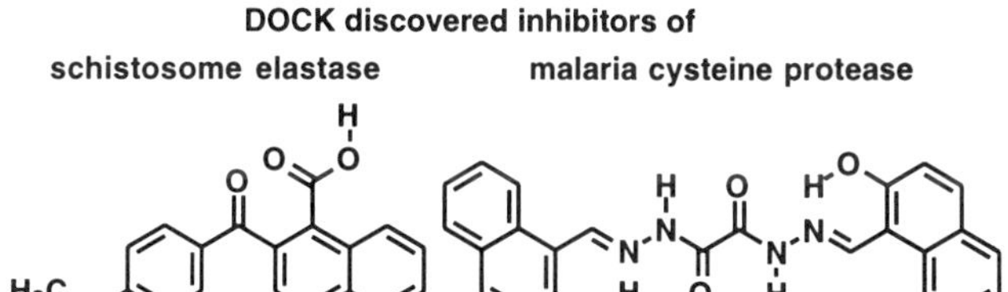

**DOCK discovered inhibitors of**

**schistosome elastase** — $K_i = 3\ \mu M$

**malaria cysteine protease** — $K_i = 6\ \mu M$

A 3D pharmacophore search (Section III) of the CSD identified a scaffold which, after extensive modification, led to the development of a highly potent inhibitor of HIV-1 protease (Lam *et al.*, 1994). The initial pharmacophore was based on a modeled $C_2$ symmetric inhibitor (Section VI,C,1,b) that bound directly to the catalytic aspartate residues and thus displaced a highly ordered water molecule. The DGEOM program was used to dock the modeled inhibitor into the HIV active site.

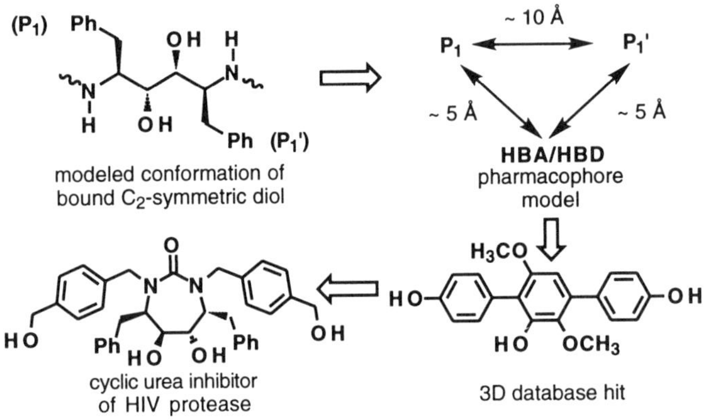

modeled conformation of bound $C_2$-symmetric diol

~ 10 Å; ~ 5 Å; ~ 5 Å

HBA/HBD pharmacophore model

cyclic urea inhibitor of HIV protease

3D database hit

An instructive cautionary tale of complications that may arise in docking is provided by DOCK predictions (Rutenber *et al.*, 1993) of the binding of a thioketal analog of haloperidol (UCSF8). X-ray crystallographic analysis of the complex of HIV-1 protease with UCSF8 revealed that UCSF8 bound 5 Å distant from the DOCK predicted orientation. However, this prediction was made on the assumption of a rigid enzyme and inhibitor. The X-ray complex further revealed that the conformation of the protease was halfway between that of the uncomplexed protein and protein complexed with peptide-like inhibitors. Furthermore, the conformation of UCSF8 was considerably different from that used for the prediction. If the conformation of the protein and inhibitor is taken from the X-ray structure of the complex, DOCK predicts the correct binding orientation. This example underscores the importance of accounting for flexibility in both the receptor (Section II,B,2) and the ligand (Section V,B).

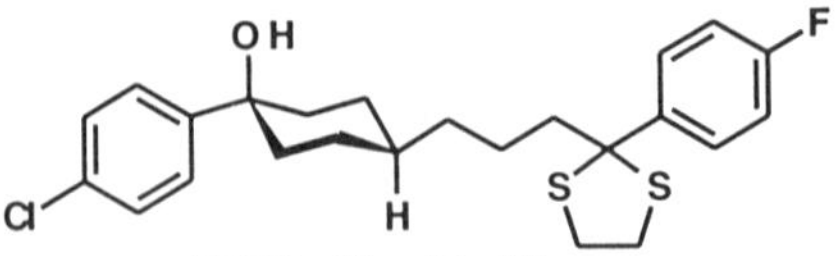

UCSF8 ($K_i = 15\ \mu M$)

### 3. Empirical Scoring Functions

When searching large databases of small molecules for prospective ligands, it is desirable to sort the candidate ligands on the basis of their predicted binding affinity. Free energy perturbation (Section II,D) at present is clearly inappropriate for this purpose because of its computational expense. A number of more approximate methods therefore have been developed. These include scoring based on a molecular mechanics interaction enthalpy between the ligand and the receptor (Blaney *et al.*, 1984; Menziani *et al.*, 1989; Meng *et al.*, 1992). In an early study of binding of thyroid hormone analogs to prealbumin, it was necessary to add a correction term for the differential solvation energy of the analogs (Blaney *et al.*, 1984) in order to achieve a reasonable agreement with the experimental binding affinities. To speed the calculation, only a shell of protein atoms in the immediate vicinity of the bound ligand is included in the calculation. Alternatively, the molecular mechanics calculation can be recast using the fast Fourier transform which allows all atoms of the ligand and the protein to be included in the calculation at a reasonable computational cost (Kurinov and Harrison, 1994). This latter enhancement increases the accuracy of the predictions. Agreement with the experiment appears promising, but it should be pointed out that the method was parameterized against only one protein (trypsin) and a relatively limited number of ligands and was subsequently tested against a small number of ligands not included in the training set.

An intermolecular molecular mechanics energy term was also used to predict the inhibitory concentrations ($pIC_{50}$'s) of a closely related series of HIV protease inhibitors in which the $P_1'$ and $P_2'$ positions were varied (Holloway *et al.*, 1995). In these calculations, the X-ray crystallographically determined structure of the protease was held fixed while the inhibitor was subjected to full geometry optimization in the presence of the enzyme. A training set of 33 inhibitors was used to develop a linear regression equation between $pIC_{50}$ and the molecular mechanics intermolecular interaction energies while the intramolecular energies of the ligands were excluded from the correlation. An $R^2$ of 0.78 was obtained for the training set. This correlation is remarkable given that it ignored entropy and desolvation effects. The presumption is that these other effects are more or less constant over the range of inhibitor studied.

Based on this regression equation, $pIC_{50}$ predictions were made for a set of 16 inhibitors not included in the training set prior to chemical synthesis. The mean error between the predicted and measured $pIC_{50}$'s was one log unit and the measured $pIC_{50}$'s spanned five orders of magnitude. If the one significant outlier was excluded from the test, the mean error dropped to 0.8 log units. An examination of this outlier is particularly instructive. Energy minimization of this inhibitor in the absence of enzyme to the nearest local minimum resulted in a 14-kcal/mol drop in energy. This value is much larger when compared to the corresponding values in the other 15 inhibitors. Hence a likely reason for the overprediction of the inhibitory potency of this analog is that the calculations did not take into account the internal strain energy of the bound inhibitor.

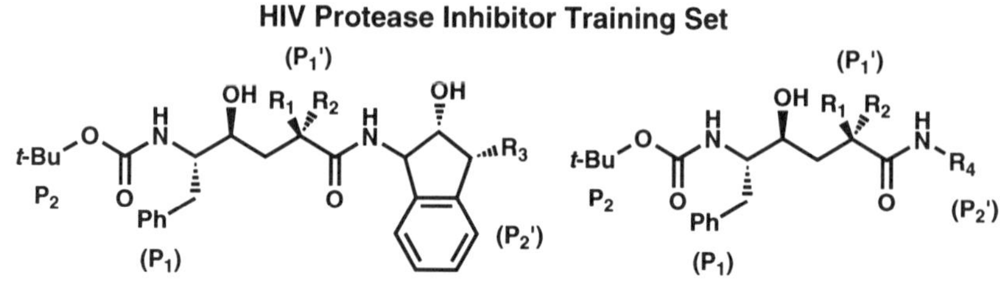

**HIV Protease Inhibitor Training Set**

Even more approximate methods for estimating binding affinities have been developed. These estimates are based, in part, on the proposition that the hydrophobic contribution to binding is proportional to the interfacial hydrophobic solvent-accessible surface area (ASA) (Nicholls *et al.*, 1991; Sharp *et al.*, 1991). However, others have argued that either solute volume (Ben-Naim and Mazo, 1993) or molecular surface area (MSA) (Jackson and Sternberg, 1995) are more appropriate correlates. The use of MSA in place of ASA eliminates the discrepancy between the microscopic estimate of the hydrophobic effect using ASA of hydrocarbons transferred from pure liquid to water and the macroscopic estimate based on surface tension at the hydrocarbon–water interface. Furthermore, MSA may be superior to ASA for estimating binding affinities since the former is a more sensitive measure of surface complementarity.

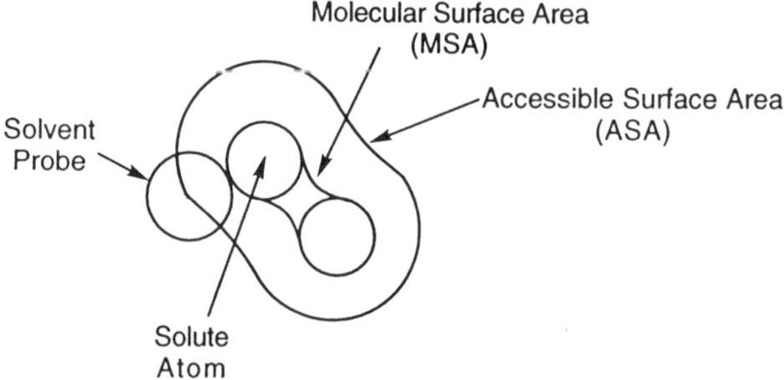


Binding affinity prediction methods include a parameterized procedure based on counting the number of hydrogen-bonding interactions and estimating the hydrophobic contact surface area (Bohacek and McMartin, 1992). This method has been extended by distinguishing between ionic and neutral hydrogen bonds, summing the number of rotatable bonds to account for conformational entropy loss of the ligand, and by parameterizing against a wider training set of ligand/protein interactions (Böhm, 1994a). This extended method also contains a penalty function to account for hydrogen bonds with less than ideal geometry, but neither does it account for internal strain energy of the ligand nor does it account for bad steric contacts between receptor and ligand. The latter is precluded by performing a bump check in the docking process. Interestingly, a rough qualitative agreement exists between the coefficients obtained from the scoring function's parameterization and experimentally determined estimates of the strengths of hydrogen bonding (Fersht, 1987; Shirley *et al.*, 1992) and lipophilic contacts (Sharp *et al.*, 1991). This is a good indication that the scoring function's coefficients are reasonable and are not an artifact of the parameterization. Therefore the method should be fairly robust for predictions of binding affinities of ligands for proteins not included in the training set. This is verified by cross validation statistics as well as by predictions of binding affinities of several test ligands segregated from the training set. However, the predictions made by these empirical methods are at best only semiquantitative (standard errors of approximately 2 kcal/mol), but their computational efficiency makes them an attractive scoring technique for database screening.

Related methods for predicting the association constants between proteins have also been developed (Horton and Lewis, 1992; Freire, 1993; Krystek *et al.*, 1993). One of these methods (Horton and Lewis, 1992) is a parameterized procedure based on the change in the solvent-accessible surface area for individual atoms between the unassociated proteins and the protein complex. Polar and nonpolar atoms are treated separately, and the "polar" atoms that engage in interprotein hydrogen bonding and salt bridges are incorporated in the polar interaction term. All other atoms, including polar atoms not involved in these ionic interactions, are included in the nonpolar interaction term. The individual interactions in the polar term all contribute favorably to protein association. In contrast, individual interactions in the nonpolar term either detract from the affinity if they involve polar atoms or contribute to the affinity if the atoms are nonpolar. Finally, a term was added for the loss of rotational and translation entropy upon association of the two proteins. For 15 protein complexes for which experimental association constants are available and for which the rate of association is diffusion controlled, the least squares correlation ($R^2$) between experimental and predicted binding affinities was 0.92. For cases in which kinetic evidence indicated intermediates in the association, the difference between the calculated and the experimental affinity was used to estimate the reorganization free energy required for the association. Finally, an examination of the parameter coefficients revealed that the average neutral hydrogen bond contributed $-0.24$ kcal/mol to association whereas the average charge-assisted hydrogen bond and salt bridge contributed $-1.9$ and $-2.5$ kcal/mol, respectively, to affinity.

A somewhat more sophisticated method for estimating free energies of binding which takes into account solvation, conformational entropy, and ligand strain energy contributions to the binding process has been developed (Vajda *et al.*, 1994). In this method, a reasonable assumption is made that the attractive van der Waals forces between solvent–ligand and solvent–receptor are approximately equal to that experienced between ligand–receptor and solvent–solvent and hence largely cancel during the binding process. Modified atomic solvation parameters (Eisenberg and McLachlan, 1986) were used to estimate the energy differences between the free and the associated state. An estimate of the reduction in conformational entropy to the free energy of binding was made by counting the number of freely rotatable bonds that are frozen. This method was applied to the predictions of free energies of association of (1) canonical proteinaceous inhibitors for serine proteases, (2) biotin and biotin analogs for streptavidin, and (3) peptides for the MHC receptor. The correspondence between experiment and calculated values was very encouraging. In the second test case, the results compared very favorably with considerably more expensive free energy perturbation calculations (Miyamoto and Kollman, 1993).

If the attractive van der Waals force between solvent–ligand and solvent–receptor are approximately equal to those between ligand–receptor and solvent–solvent, why should there be a correlation between binding affinity and intermolecular molecular mechanics interaction energies which include a van der Waals term? It may be that this intermolecular van der Waals term is an indirect measure of buried interfacial surface area. Hence molecular mechanics-based approaches may obtain the right answer for the wrong reason.

### 4. Packing

Folded proteins frequently contain cavities. From X-ray crystallographic analysis, cavities that are lined with polar functional groups often contain water molecules whereas those that are surrounded by hydrophobic groups appear to be empty (Hubbard *et al.*, 1994; Williams *et al.*, 1994). However, many of these "empty" cavities may actually contain positionally disordered water molecules not observable crystallographically but detectable by NMR (Ernst *et al.*, 1995). Creating a methylene-sized hole in the interior of a protein not filled by water has been estimated by thermodynamic measurements of the stability of mutant proteins to cost about 1 kcal/mol (Jackson *et al.*, 1993). Recognition of this phenomena has led to the development of software to judge the quality of protein-folding models based on the packing density within the protein interior (Gregoret and Cohen, 1990). Empty cavities, especially prevalent along the "collar" region located at the periphery of the ligand–protein interface, cause a reduction in affinity (Nicholls *et al.*, 1991). However, attempts to apply this "cavity penalty" to improve the predictions of ligand-binding affinity have so far been unsuccessful (Krystek *et al.*, 1993). Bridging water molecules are frequently seen at the interface between the two partners of protein complexes (Janin and Chothia, 1990). Hence the cavity penalty's failure to improve binding predictions probably arises from the fact that it is difficult to anticipate the number and location of water molecules at receptor–ligand interfaces and hence the total volume of these cavities. A method for predicting the location of water molecules in the first solution sphere of proteins has been described (Vedani and Huhta, 1990) but it has not been tested on protein–ligand interfacial regions.

Good shape complementarity exists at the interface region between many protein–protein complexes (Lawrence and Colman, 1993) and these binding events are normally entropy driven (Fioretti *et al.*, 1993; Tello *et al.*, 1993) consistent with the release of ordered solvent from the surfaces of the binding partners. Other protein–protein complexes, especially those involving antibodies, show less shape complementarity (Lawrence and Colman, 1993), and the interfaces between these proteins are characterized by large numbers of bridging water molecules to fill the voids (Bhat *et al.*, 1994; Buckle *et al.*, 1994). These protein–protein interactions are normally enthalpy driven (Kelley and O'Connell, 1993). While analogous systematic comparisons of the shape complementarity and thermodynamics of association of small molecule ligand with protein receptors have not yet been undertaken (Hitzemann, 1988), these interfaces undoubtedly share many of the same structural features and thermodynamic properties characteristic of protein–protein interfaces.

## D. Free Energy and the Elusive Entropy Factor

### 1. Static Enthalpy Calculations

Since the mid-1980s, a number of investigations have attempted to establish theoretical methods with which to calculate free energies rather than enthalpic energies of physicochemical processes. In the traditional molecular mechanics and quantum mechanics computations, inter- and intramolecular interactions are evaluated for a given frozen configuration of the system, typically located at a local energy minimum. Often, these computations are done in the absence of explicit solvation, which is mimicked through an appropriate dielectric. Such evaluations obviously do not account for entropic factors such as those arising due to conformational changes in the interacting molecules, to displacement of water molecules bound to the ligand, and to the macromolecular target followed by rearrangement of the solvent structure.

Nevertheless, the enthalpic interaction terms evaluated as molecular mechanics energies may provide useful qualitative insights and are helpful in understanding the relative binding affinities for related systems. For example, X-ray crystal structure conformations of peptide and nucleotide backbones have been rationalized through numerous molecular mechanics and quantum mechanics calculations on model peptide (Krimm and Bandekar, 1986) and nucleotide systems (Saenger, 1984), respectively. Base stacking interactions in polynucleotides and the neighbor exclusion principle among DNA intercalators are two additional examples where experimental observations have been rationalized on the basis of molecular mechanics calculations (Rao and Kollman, 1987). In an approach combining molecular mechanics with quantum mechanics, Weiner *et al.*, (1985) studied the reaction pathway simulating formamide hydrolysis by hydroxide ion in aqueous and gas phases. In this approach, the solutes were treated quantum mechanically while the surrounding solvent system was treated with a molecular mechanics force field.

In an interesting study aimed at altering the substrate specificity of $\alpha$-lytic protease, Wilson *et al.* (1991) employed molecular mechanics to estimate the relative binding affinities of mutant enzymes for three tetrapeptide boronic acid transition state analogs. Substrates with a large hydrophobic $P_1$ group (e.g., leucine) are slowly cleaved by the wild-type enzyme compared to substrates with small hydrophobic groups (e.g., alanine). The goal of the study was to design a mutant enzyme with enhanced cleavage efficiency for substrates with leucine at the $P_1$ position and, as an additional requirement, to design a mutant with good selectivity for leucine over a closely related amino acid residue type, isoleucine. Hence boronic acid tetrapeptides with either leucine or isoleucine and, as a benchmark, an alanine $P_1$ substituent were docked to the X-ray crystallographic structure of $\alpha$-lytic protease. The binding affinity estimates were based on a molecular mechanics enthalpy term (using AMBER force field parameters) plus a solvation term (Wesson and Eisenberg, 1992). All possible single point mutations of the three residues that line the $P_1$ specificity pocket were evaluated ($20^3 = 8000$). For each of these mutations, all combinations and permutations of side chain conformations [selected from a rotamer library (Ponder and Richards, 1987; Dunbrack and Karplus, 1993)] of the residues that line the specificity pocket were energy evaluated ($\sim 10^6$ rotamer combinations). A Met-192 $\rightarrow$ Val mutation was predicted to give an enzyme with high leucine $P_1$ substrate cleavage efficiency and leucine vs isoleucine specificity. The mutant protein was cloned and had a rate acceleration of three orders of magnitude vs the wild-type enzyme for substrates containing a $P_1$ leucine residue, very close to what was predicted. Furthermore, the mutant enzyme had a 200-fold preference for leucine over isoleucine. However, this preference is considerably less than what was predicted. An examination of the model isoleucine boronic acid/mutant protease complex revealed a number of bad contacts between the substrate isoleucine side chain and protein side chains that line the $P_1$ specificity pocket. These bad contacts could be relieved by torsional rotation of these side chains. This points out one of the primary shortcomings of the study: the use of rigid rotamers which results in the underpredictions of affinities for some substrate/enzyme pairs. Nevertheless, this study clearly demonstrates that binding affinity estimates from molecular mechanics enthalpy including a solvation correction term can give qualitatively correct results that are useful for the design of proteins and, by analogy, ligands.

Another noteworthy feature of this study is that it successfully accounted for receptor flexibility through explicit examination of side chain rotamers.

### 2. Molecular Dynamics

In contrast to the "static" molecular mechanics approaches, molecular dynamics and Monte Carlo simulation provide an opportunity to calculate average properties (enthalpies and internal energies) and conformations of molecular systems as a function of time (Brooks, 1995). The average properties and the differences among them for various systems are more representative of the real-time dynamics of molecules compared to the corresponding static model values. Several examples employing these approaches are documented in the literature. For example, Pranata and Jorgensen (1991) have studied the conformational profiles of FK506 in water through molecular dynamics simulations. They concluded that the solution structure of FK506 maintains a close similarity to the X-ray structure that was employed as the starting point for molecular dynamics simulations. However, their simulated structure is at large variance with the models derived from the NMR data on FK506 (Karuso *et al.*, 1990; Rosen *et al.*, 1990). The reason for this discrepancy may lie in the fact that the dynamics simulations were carried out with only one starting structure instead of multiple low energy conformers that this molecule is capable of adopting (Pranata and Jorgensen, 1991).

However, the fundamental difficulty in extracting useful molecular properties resides in large statistical fluctuations found in calculating internal energies and enthalpies. The significance of this difficulty cannot be overemphasized, particularly in light of the fact that one must evaluate small differences between large numbers, each of which is subject to large statistical variations during the simulations. Hence the correlation of calculated relative enthalpy differences using the dynamics approaches with the corresponding experimental observations is not entirely satisfactory.

Nonetheless, molecular dynamics may provide useful qualitative guidelines for modification to the ligand structure which are likely to enhance affinity. In one study, the optimization of a series of $\beta$-carbolinone-based elastase inhibitors (Fig. 2) was guided by molecular modeling (Veale *et al.*, 1995). Whereas vacuum phase molecular mechanics optimizations were unable to distinguish between favorable and unfavorable structural modifications to this series of inhibitors, molecular dynamics calculations which included explicit water molecules were able to provide qualitatively correct predictions of relative binding affinities. For example, gas phase calculations indicated a favorable interaction between a 8-carboxamide substituent and a backbone amide hydrogen atom of Gly-218 of the enzyme whereas aqueous MD simulations revealed that this group showed no tendency to form hydrogen bonds with the enzyme. The latter simulation is consistent with the experimental result that showed the parent system and the 8-carboxamide

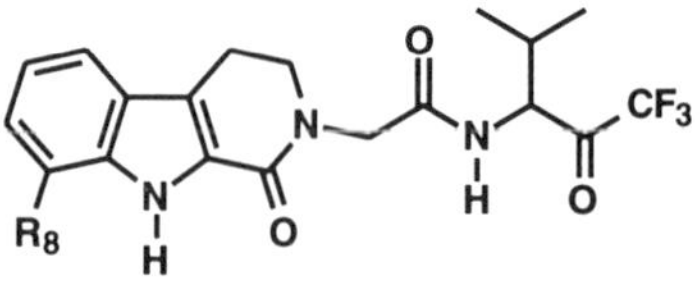

| $R_8$ | $K_i$ (nM) |
|---|---|
| H | 290 ± 40 |
| $CO_2H$ | 22 ± 9 |
| $CONH_2$ | 730 ± 160 |

**FIGURE 2** Structure–activity relationships of substitution at the 8 position of $\beta$-carbolinone trifluoromethylketone elastase inhibitors (Veale *et al.*, 1995).

analog to be essentially equipotent. In contrast, the 8-carboxylate substituent was shown to form a long-lived hydrogen-bonding interaction with the enzyme in an aqueous molecular dynamics simulation. Consistent with this modeling result, the 8-carboxylate analog showed an order of magnitude increase in potency relative to the parent system.

A number of factors may explain the greater affinity of the 8-carboxylate inhibitor relative to the 8-carboxamide analog. These include (1) the more favorable charge-assisted (carboxylate) vs neutral (carboxamide) hydrogen bond and (2) a "spectator" electrostatic repulsion (Pranata *et al.*, 1919) between the inhibitor 8-carboxamide NH and the enzyme amide hydrogen atom of Gly-218. Furthermore, the aqueous MD simulation with the 8-carboxylate inhibitor revealed that, in addition to the interaction with Gly-218, a long-lived interaction between the carboxylate of the inhibitor with the guanidinium group of Arg-177 is observed. The 8-carboxamide analog showed no propensity to interact with Arg-177 during an aqueous MD simulation.

### 3. Free Energy Perturbation Theory

The advent of supercomputers and high speed workstations has made the art of simulations leading to the calculations of relative free energies of solvation and binding a reality, allowing for more meaningful and satisfying comparisons with experimental data. The so-called free energy perturbation (FEP) approach has been employed in conjunction with either molecular dynamics or Monte Carlo to study the relative solvation and binding energies in a number of closely related systems. A few examples of such applications will be discussed following a qualitative outline of the methodology.

The basis for the free energy perturbation approach is in the statistical mechanics perturbation theory first developed by Zwanzig (1954). Since 1985, this approach has been applied to a number of problems in biomolecular interactions with an emphasis on validating the reproducibility of the force field to predict experimental observations such as solvation and free energies of binding. Conceptually, two closely related systems S and S′ are treated as a "perturbation" of one another. For example, benzene and toluene are regarded as interconvertible through the *mutation* of the methyl group of toluene into a hydrogen atom (or *vice versa*, Fig. 3). It must be emphasized that this mutation is a physically unrealistic phenomenon and that it is carried out strictly as a computational "experiment." During the course of the free energy perturbation calculation, system S is "transformed" into system S′ through a series of intermediate states, each one closely related to its neighbor. Thus, at any given stage of the simulation, the observed state has characteristics of both the starting (S) and the final system (S′). A coupling parameter, λ, is typically used to denote the state of transformation, and the Hamiltonian of the system at any given value of λ is defined by

$$H = \lambda H_s + (1 - \lambda)H_s' \qquad 0 < \lambda < 1.$$

This type of coupling ensures a smooth conversion of the system S to S′ through several stages which allow for adjustment in the surrounding system (e.g., the aqueous milieu). Simulations (either molecular dynamics or Monte Carlo) are performed at each intermediate state, and the ensemble average free energy is evaluated using the expression

$$G(\lambda) = -RT \ln < \exp(-H(\lambda)/RT>.$$

The coupling parameter λ may in practice be varied by two approaches. The first one, known as the windowing technique, allows λ to assume discrete values by dividing its entire range (0 to 1) into *N* windows (not necessarily of equal size). Let us consider the example of mutating a methyl group (S) into a hydrogen atom (S′) (as in perturbing toluene into benzene, Fig. 3). In a typical simulation using the window technique, this mutation

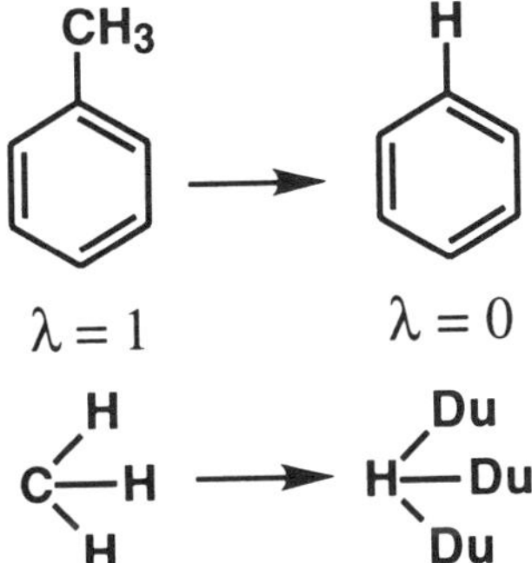


**FIGURE 3** Extrema in the perturbation of toluene (λ = 1) into benzene (λ = 0) in a hypothetical free energy perturbation experiment. "Du" represents a dummy atom possessing a zero van der Waals radius and charge.

is performed over 20 steps, with λ varying by 0.05 over each step. In the forward direction, λ varies from 1 to 0 whereas in the reverse direction, it varies from 0 to 1. Thus at a given value of λ, for example 0.45, the van der Waals and charge characteristics of the system are a mixture of 45% of toluene and 55% of benzene. In the course of the simulation, the next value of λ is either 0.40 or 0.50, depending on the direction of the simulation (forward or reverse). At each value of λ, molecular dynamics equilibration is carried out and data are collected for free energy evaluation.

In the second approach, known as time-dependent perturbation or slow growth, λ is defined as a continuous function of time. Thus if the total simulation time is specified to be 100 psec for the mutation in the example just described, the values of λ at the end of 25, 35, and 52 psec of simulations would be 0.25, 0.35, and 0.52, respectively. With either of the two approaches, simulations are typically carried out in both the "forward" ($S \rightarrow S'$) and "backward" ($S' \rightarrow S$) directions. The perturbations in either direction should give the same free energy difference. Lack of agreement between the two experiments is an indication that the calculations have not sufficiently converged and that the experiments should be repeated with longer equilibration times and/or smaller increments of λ.

The free energy perturbation approach has been implemented in a number of computational chemistry software packages available in the public domain. Of these, AMBER-UCSF (version 4.0) (Pearlman *et al.*, 1991), BOSS (Jorgensen, 1992), GROMOS (Hermans *et al.*, 1984), and CHARMm (Brooks *et al.*, 1983) have been most commonly employed in evaluating differential free energy changes in a number of cases of protein–substrate interactions, nucleic acid polymorphism, and of small molecule solvation.

Since the simulations are carried out as a series of hypothetical transformations in which each configuration of the entire molecular system is adjusted, the entropy component of the free energy is included in the free energy evaluation. More recent versions of some of these molecular dynamics programs (e.g., AMBER) include algorithms to calculate the enthalpy and entropy contributions to the free energy. The mathematical details leading to the calculation of the total free energy term are described in the literature (van Gunsteren and Mark, 1992; Kollman, 1993).

While FEP is the only method developed to date that has been able to consistently and accurately predict relative free energies of binding for a wide range of proteins and ligands, its high computational expense limits its application in drug design. A large part of the calculation is devoted to making sure that the conformational space is adequately sampled; this lengthy sampling time makes FEP impractical for use as a drug design "screening" tool. A more efficient algorithm for conformational sampling has been described (Guarnieri and Still, 1994) which employs a mixed Monte Carlo/stochastic dynamics (MC/SD) approach. This hybrid procedure allows for much faster convergence than either MC or SD used alone. This method has been extended to preferentially sample low energy portions of the potential energy surface identified by a standard grid or stochastic conformational search (Senderowitz *et al.,* 1995). This later enhancement, called jumping between wells (JBW), has been estimated to converge at a rate of a million fold faster compared to standard MD or SD approaches.

Another reason for the substantial computational costs of FEP is the need to slowly mutate one ligand into another and hence most of the time is spent on simulating uninteresting intermediate configurations. A promising approach called the linear interaction energy (LIE) approximation which removes the necessity of this mutation has been reported (Åqvist *et al.,* 1994; Åqvist and Mowbray, 1995). In this method, two MD simulations are performed: on the ligand in aqueous solution and on the solvated protein–ligand complex. The binding affinity is then estimated from a weighted difference in the electrostatic and van der Waals interaction terms between the free and the bound state. The critical empirical parameter which converts the van der Waals energy difference into a "hydrophobic" contribution to binding is called $\alpha$ which was estimated to be 0.16 based on a training set of four inhibitors of endothiapepsin (Åqvist *et al.,* 1994). It is not clear whether the value of this parameter will be transferable to other systems, but one very encouraging result is that this same parameter was used to successfully predict both the absolute and relative binding affinities of two sugars for glucose-binding protein (Åqvist and Mowbray, 1995). This is a remarkable outcome given the training set of ligands which possess considerable hydrophobic character in marked contrast to the sugars in the test set which are largely hydrophilic.

### 4. Applications of Free Energy Perturbation

The free energy perturbation theory has been applied to a number of problems in the areas of protein–ligand interactions, site-directed mutagenesis in proteins, nucleic acid structures, and solvation of small molecules. The principal emphasis in these applications has been on validating the procedure in terms of the reproducibility of experimental observations within reasonable error ranges. Very few applications using this theory in a predictive fashion are documented. In Sections IV and V, we will briefly discuss a few cases of such applications and their implications for the process of computer-aided drug design.

The FEP approach has been employed in a number of investigations of protein–substrate interactions, more successfully when a high resolution X-ray crystal structure of the complex is readily available (van Gunsteren and Mark, 1992; Kollman, 1993). Calculations have been done both to evaluate differential binding of ligands to a common enzyme and to understand the impact of site-directed mutations in enzymes on binding and catalysis of substrates. A few examples will be discussed to illustrate the power and limitations of these computational experiments.

Bash and co-workers (1987) have calculated the relative binding energies of two thermolysin inhibitors, one of them a phosphonamidate and the other a phosphonate ester. In this study, the mutations between the two inhibitors were carried out using the windowing technique and the crystal structure of the complex with the phosphonamidate inhibitor (Tronrud *et al.*, 1987). The conformation of the phosphonamidate inhibitor in the X-ray crystal structure (Tronrud *et al.*, 1987) was used in the calculations to determine the relative solvation energy differences (in the presence of TIP3P waters) with the phosphonate ester. The calculated relative binding of the inhibitors to the enzyme ($4.21 \pm 0.54$ kcal/mol) was in good agreement with the experimental observations (4.1 kcal/mol) (Bartlett and Marlowe, 1987).

**Thermolysin Inhibitor**
**FEP Experiment**

Cbz–NH–CH₂–P(=O)(O⁻)–NH–Leu–Leu ⇌ Cbz–NH–CH₂–P(=O)(O⁻)–O(Du)–Leu–Leu

**phosphonamidate** **phosphonate**

An earlier study also carried out calculations starting from a model of the complex built using computer graphics techniques starting from the X-ray coordinates of thermolysin complexed with an irreversible inhibitor (Holmes *et al.*, 1983). The corresponding result ($4.38 \pm 0.06$ kcal/mol) was very close to that obtained using the X-ray structure of the complex with the noncovalently bound inhibitor. This remarkable quantitative correlation between theory and experiment is attributed to the high resolution of the crystal structure data and to the practically identical conformations of the two inhibitors in the active site of the enzyme. Also noteworthy is the small magnitude of the error bars, which make the calculated values practically identical to the experimental binding constants. Similar quantitative agreement was also obtained for the relative binding affinity of benzamidine and *p*-fluorobenzamidine to trypsin (Wong and McCammon, 1986). This result is not surprising since the structural change occurring during the mutation of one substrate into the other is smaller than that used in the studies of Bash and co-workers (1987).

An application of the free energy perturbation approach to the study of binding of diastereomeric hydroxyethylene amide isostere inhibitors (see Fig. 4) to HIV-1 protease (Rao *et al.*, 1992) was not as successful as the two examples described earlier. Experimentally, the S-diastereomer was found to bind better than the R-diastereomer by a factor of 80 (corresponding to a free energy difference of 2.6 kcal/mol) (Rich *et al.*, 1991). In marked contrast to this experimental value, the theoretical simulations, using the slow growth approach, yielded a free energy difference of $- 2.16 \pm 0.65$ kcal/mol. However, a value of $3.37 \pm 0.64$ kcal/mol was obtained by the imposition of *constraints on the catalytically significant aspartate residues.*

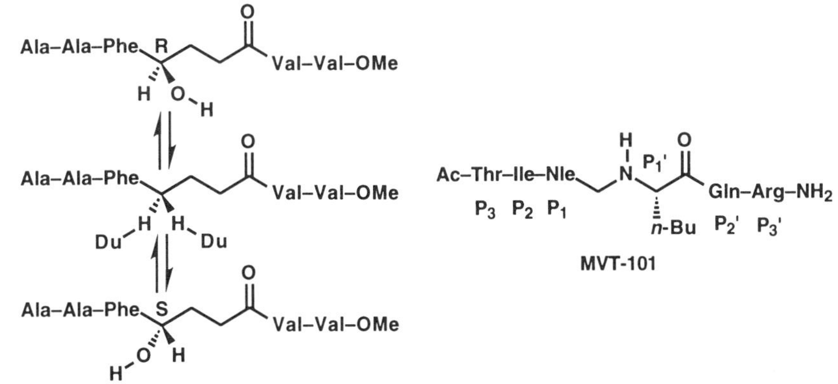

**FIGURE 4** Stepwise free energy perturbation of the R to S configurations of diastereomeric hydroxyethylene HIV-1 protease inhibitors (Rao *et al.*, 1992). The simulation was based on the crystal structure of MVT-101 complexed with HIV-1 protease.

These discrepancies are particularly surprising in light of the high resolution of the crystal structure data (of HIV-1 protease complexed with MVT-101) used in modeling the hydroxyethylene amide isostere inhibitor in the active site of the enzyme. A crystal structure analysis of the complex with this inhibitor (Murthy *et al.*, 1992) demonstrated that the conformation of the latter is practically identical to that used in the modeled structure. Hence, the deviations between calculated and experimental values may be interpreted as arising from the conformational dynamics of key active site side chains. Since similar discrepancies were also observed in the studies of pepstatin binding to rhizopus pepsin (Rao and Singh, 1991b) (an aspartyl protease) and of Ro 31-8959 binding to HIV-1 protease (Rao and Murcko, 1994), these observations taken together suggest that conformational mobility in the enzyme active site is of prime importance in calculating relative binding energies using FEP techniques. In contrast, good correlations between experimental and calculated free energy differences were found for a series of HIV-1 protease inhibitors where the mutation was distant from the catalytic aspartates (Reddy *et al.*, 1994).

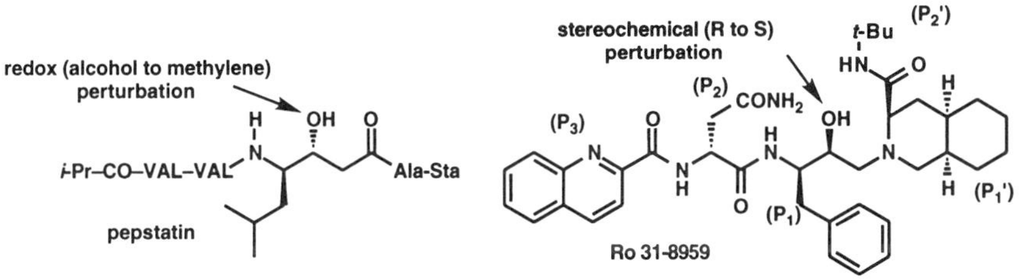

The 4.0 version of the AMBER program (Weiner *et al.*, 1984) calculates free energies, which takes into account the energy changes associated with

dihedral angle variations as the mutations proceed during the course of the dynamics simulations. Employing such a strategy may better account for the variations in free energies due to conformational mobility, giving rise to potentially lower discrepancies. Differential conformational free energies can also be evaluated using the Monte Carlo simulations in the framework of the BOSS program.

Many of these HIV-1 protease inhibitor FEP experiments assumed a single protonation state for the catalytic aspartates (one protonated and the second deprotonated). However, another study (Harte and Beveridge, 1993) which examined all four possible aspartate protonation states (0/0, −/0, 0/−, −/−) showed closest agreement between the molecular dynamics structure when either the diprotonated 0/0 state (for U85548E, an hydroxyethylene inhibitor) or the dianonic −/− state (for MVT-101, a reduced amide inhibitor) was used. Hence, discrepancies between experimental and FEP-calculated affinities may reflect incorrect assumptions about the protonation state of the catalytic aspartate residues. The latter study (Harte and Beveridge, 1993) in turn assumed that the nitrogen atom of the reduced amide of MVT-101 is protonated; however, this is a questionable assumption since the $pK_a$ of a reduced amide flanked by two peptide groups as in MVT-101 is considerably reduced. While an amine normally has a $pK_a$ in the range of 10–11, a reasonable estimate of the $pK_a$ of the basic center in MVT-101 is 6.1 (Perrin *et al.*, 1981).

Free energy perturbation simulations of the effect of single residue mutations on binding and catalysis have been studied in the case of subtilisin, a bacterial serine protease (Rao *et al.*, 1987; Warshel *et al.*, 1988; Mizushima *et al.*, 1991). The main goal of these investigations was to calculate variations in the $K_m/K_{cat}$ ratios as a function of mutations in key residues of subtilisin. This serine protease is a particularly interesting target for site-directed mutagenesis (Bryan *et al.*, 1986; Braxton and Wells, 1991) in light of the unusual stabilization of the transition state oxyanion by a backbone N-H and by the side chain of Asn-155 instead of two backbone N-H groups, as is normally found in mammalian serine proteases (e.g., trypsin and chymotrypsin). Enzyme kinetic parameters have been obtained for a number of Asn-155 mutants (Braxton and Wells, 1991). The calculated free energy changes of binding and catalysis of a peptide substrate to an Ala-155 mutant were calculated to be within 0.4 kcal/mol of the experimental values (Rao *et al.*, 1987; Mizushima *et al.*, 1991). Most of the calculated changes were seen to effect the rate of catalysis rather than binding, qualitatively consistent with the reduction in experimental $k_{cat}$ values. The success of these investigations should be viewed in light of the fact that while a high resolution crystal structure of subtilisin was employed for the enzyme component, no X-ray information was available for the complex itself. The complex was model built using computer graphics packages, keeping in mind the similarities in specificities of subtilisin and chymotrypsin for peptide substrates with a large aromatic side chain at position $P_1$. The molecular dynamics simulations conducted during one of these studies (Rao *et al.*, 1987) unexpectedly suggested the possible role of Thr-220 in affecting the binding and catalysis. A later investigation (Warshel *et al.*, 1988) reported that mutation of this residue to alanine mainly affected the rate of catalysis.

## III. RECEPTOR MAPPING—STRUCTURALLY ILL-DEFINED BIOLOGICAL RECEPTORS

### A. The Pharmacophore Concept

While the previous section dealt with molecular modeling methods applied in the context of availability of explicit receptor structure, this section deals with molecular modeling challenges in situations most often encountered in drug design, where the three-dimensional structure of the therapeutically relevant receptor is not known. Typically in such cases, the structures of the ligands are well characterized and sometimes their biologically relevant conformations can be deduced unambiguously. The receptor structure may then be inferred or "mapped" from such ligands and employed to (a) rationalize the existing SAR and to (b) design novel drug candidates. The relative orientation of essential functional groups necessary for biological activity of the ligands define a pharmacophore. A pharmacophore should ideally account for the binding of a structurally diverse set of ligands to a common receptor.

The most fundamental assumption in the derivation of a pharmacophore model for a collection of ligands is that they all bind to the same receptor site instead of at topologically distinct but allosterically linked sites. This implies that functional groups common to all the ligands are oriented similarly within the receptor active site and thus interact with complementary receptor components analogously. Another assumption is the occupancy of comparable volumes in the receptor active site by the key recognition elements on the ligands. These assumptions are inspired by the necessity to correlate the derived pharmacophore model to the biological activities in the relevant receptor-binding assays.

The general approach to pharmacophore development is initiated by the identification of bioisosteric functional groups (Lipinski, 1986; Burger, 1991), common to all highly potent ligands, considered essential for biological activity. From an examination of the structure activity relationship (SAR) data of the training set, it is often possible to distinguish parts of the molecules that are essential for receptor binding from those that are optional or deleterious. These essential functional groups are then matched in the 3D space of the energetically accessible conformations of the ligands. Features commonly employed in identifying a match include atoms, centroids of collections of atoms, electron lone pair positions, surfaces, volumes, and steric and electrostatic potentials. In addition, features located in the receptor may also be mapped. For example, hydrogen bond vectors extending from the ligand represent potential hydrogen-bonding sites located on the receptor. Use of these receptor pharmacophoric features removes the arbitrary requirement that ligand atoms must overlap. A more important requirement of a pharmacophore is that it allows ligand receptor groups to interact with common receptor points. Examples of the use of hydrogen bond extension vectors in the formulation of a pharmacophore model include a general central nervous system (CNS) pharmacophore (Lloyd and Andrews, 1986) and a benzodiazepine receptor inverse agonist model (Diaz-Arauzo *et al.*, 1991).

In an ideal situation, it is preferable to have one or more active rigid molecules to use as a starting template. In their absence, semirigid and fully flexible molecules are utilized in the derivation of a pharmacophore. All possible combinations of conformations and pharmacophoric groups are then examined to identify overlays which allow good alignment of essential pharmacophoric elements. If the ligands are rigid or if the active conformation is known through some other means, programs such as SEAL, which align molecules to maximize overlap of steric and electrostatic fields they present, may be used to develop a pharmacophore model (Kearsley and Smith, 1990). The advantage of this technique is that it does not rely on preconceived notions of bioisosteric groups and novel alignments that other programs may miss are often obtained. However, methods such as SEAL are too computationally expensive to handle large numbers of conformations.

Since pharmacophoric searches require examining large numbers of conformations and alignments, these searches must be automated to be practical. To address this problem, a variety of pharmacophore identification programs have been developed and are described in the literature (see Table II). Since the magnitude of the pharmacophore search problem is often very large, each one of these software programs has employed various approximations and trade-offs to make the problem tractable. The following critical issues must be addressed in automated pharmacophore identification with reference to how these software programs deal with these issues.

*Underdetermined problem.* Unless fairly rigid active analogs and/or a large number of structural diverse active ligands exist, pharmacophore mapping is usually an underdetermined problem. Therefore, one is often forced to make simplifying assumptions such as a rigid receptor. Even with these assumptions, multiple solutions capable of explaining the pharmacological data are often obtained, and therefore one can only speak of the relative probabilities of the various models. If the problem is reduced to a relatively few solutions, 3D QSAR methods such as CoMFA, which take advantage of the quantitative affinity data (Cramer *et al.*, 1988), may help to distinguish between these models (Allen *et al.*, 1992; Nicklaus *et al.*,

**TABLE II** Automated Pharmacophore Generation Methods

| | | Ability to handle | | | | | |
|---|---|---|---|---|---|---|---|
| Program name | Reference | Alternate mappings | Inactives | Chirality | Conformations | Energy | Volume |
| APEX | Golender and Vorpagel (1993) | Yes | Yes | Yes | Low | No | Yes |
| APOLLO | Koehler *et al.* (1988); Snyder *et al.* (1993) | Limited | No | Yes | Moderate | Yes | Yes |
| AUTOFIT/RECEPS | Kato *et al.* (1992) | Yes | No | Yes | High | Yes | No |
| Catalyst | N/A | Yes | Yes | Yes | Moderate | Yes | Yes |
| Constrained Simulated Annealing | Hodgkin *et al.* (1993) | No | No | Yes | High | Yes | No |
| DHYDM | Ghose *et al.* (1995) | No | No | Yes | Moderate | Yes | No |
| DISCO | Martin *et al.* (1993) | Yes | No | Yes | Low | No | No |
| Energy Filtered DG | Linschoten *et al.* (1986, 1990) | No | No | No | High | Yes | No |
| Ensemble DG | Sheridan *et al.* (1986) | No | No | No | High | No | No |
| Genetic Algorithm | Payne and Glen (1993) | Yes | No | Yes | Moderate | Yes | Yes |
| GERM | Walters and Hinds (1994) | Yes | Yes | Yes | No | No | Yes |
| Receptor/CSEARCH | Dammkoehler *et al.* (1989) | Limited | No | Yes | High | No | No |

1992). Furthermore, a 3D QSAR neural network algorithm called Compass has been developed in which the alignment and selection of conformation of flexible molecules are made an integral part of the model development (Jain *et al.*, 1994). This latter approach uses an iterative procedure of training a neural network to predict activities, then using this neural net to select conformers and alignments for each molecule which are predicted to have the highest activity, and then iterating to convergence. The method is somewhat limited, however, in the number of conformations it can consider.

*Alternative mappings and conformational flexibility.* It is difficult to explore both simultaneously. Programs such as APEX (Golender and Vorpagel, 1993) and DISCO (Martin *et al.*, 1993) which employ a clique search algorithm (Brint and Willett, 1987) concentrate on alternative mappings of pharmacophoric points and possess a somewhat limited ability to process large numbers of conformations. In contrast, approraches such as ensemble DG (Sheridan *et al.*, 1986), Receptor (Dammkoehler *et al.*, 1989), and APOLLO (Koehler *et al.*, 1988; Snyder *et al.*, 1993) can process large numbers of conformations but with limited ability to test alternative mappings.

*Inactives.* Methods which allow the user to specify which functional groups to include in the pharmacophoric search implicitly take into account inactives since a comparison between actives and inactives is used to determine the essential pharmacophoric groups. However, it is not always possible to determine the pharmacophoric groups by inspection of the SAR alone. Most methods focus on active compounds and ignore closely related inactive structures during the search. However, the inactives may provide important supplementary information that is critical for obtaining solutions to undetermined problems. A few methods, such as APEX, do take advantage of the inactive data. However, inactive ligand data should be used very cautiously since inactivity may have little to do with the intrinsic activity of a ligand but rather with properties such as chemical or metabolic instability, or insolubility.

*Volume.* All other factors being equal, pharmacophore models that allow for the highest degree of common volume overlap of superimposed ligands have the highest probability of being correct since these overlays reduce the chance that any one ligand will be located in a sterically forbidden region of the receptor. Most methods, however, ignore volume analyses because they are relatively expensive calculations. Volume analyses are often performed after the pharmacophore model has been developed (Sufrin *et al.*, 1981). A few methods such as Catalyst employ excluded volume in the analysis. Other methods such as APOLLO first reduce the number of candidate models using an energy filter and a fast least-squares fitting routine. The remaining models are then subjected to a relatively expensive volume comparison. Finally, the models are ranked based on a composite score which is a function of the sum of the ligand conformational energies, the RMS deviation of the fitted points, and a percent common volume overlap.

*Conformational energy.* Only the Hodgkin Monte Carlo pharmacophore search and APOLLO take explicit account of conformational energy during the active conformation search. The philosophy behind APEX, Catalyst, DISCO, and DHYDM (Ghose *et al.*, 1995) is to restrict the active conformation search to low energy bioaccessible conformations, but these four programs do not explicitly use energy to rank the candidate pharmacophore models. While conformational energy is not explicitly mentioned in the Itai work (Kato *et al.*, 1992), the presumption is that low energy conformations are used to generate the pharmacophore. RECEPTOR and Ensemble distance geometry do not use conformational energy in the search for active conformation, except crudely through bump checking. The modified distance geometry approach of Linschoten *et al.* (1986, 1990) at least screens out high energy conformers in the creation of the distance geometry matrix, but since the matrix only contains distance ranges, it is likely that at least some high energy conformers are embedded within these distance ranges. If conformational energy is not considered in the pharmacophoric search, then the resultant models should be examined to make sure that they contain energetically reasonable conformations.

Because of the various trade-offs these programs are forced to make, each program has its own strengths and weaknesses. Programs such as AUTOFIT (Kato *et al.*, 1992), Constrained Simulated Annealing (Hodgkin *et al.*, 1993), ensemble DG, DHYDM, and Receptor excel at problems in which the identity and correspondence of pharmacophoric groups are certain, but the active conformation of highly flexible ligands is not. Programs such as APEX and DISCO which exhaustively explore alternative mappings of pharmacophoric groups have difficulty treating large numbers of conformations. Other methods such as APOLLO, Catalyst, and Genetic Algorithm (Payne and Glen, 1993) represent a compromise between these two extremes and are thus able to examine moderate numbers of alternative mappings and conformations simultaneously.

Since the pharmacophore identification problem is often underdetermined and hence a single unambiguous solution is not possible, an exhaustive examination of alternative mappings and/or conformations may not always be necessary. In these situations, a stochastic or optimizing searching procedure (e.g., Genetic Algorithm and GERM) may be appropriate. While these methods are not guaranteed to find the optimal solution, they will usually find a very good solution.

Another strategy that might be applied in situations in which there are simultaneously large numbers of conformation and alternative mappings is to first reduce the number of conformations with an energy filter. The number of conformations may further be reduced through a cluster analysis to select representative low energy conformers (Kearsley *et al.,* 1994). However, one must exercise great caution in using an energy screen. Because of the large numbers of conformers that must be evaluated, molecular mechanics is the only practical method for evaluating these energies. Geometry optimization that holds torsional angles fixed ensures a uniform sampling of conformational space. In addition, one must make certain that the force field used contains high quality parameters for the ligand in question. Even with good parameterization, molecular mechanics gas phase calculations usually favor compact, folded conformations in which the number of intramolecular hydrogen bonds is maximized. However, in an aqueous or receptor environment, these intramolecular hydrogen bonds are replaced with hydrogen bonds to solvent or protein. A screen based on the internal strain energy of the ligand would therefore be more relevant. Aqueous solvent continuum calculations (Still *et al.*, 1990; Wesson and Eisenberg, 1992) are relatively inexpensive and place extended and compact conformations on a more equal footing. Hence screens based on solvent continuum energy-optimized structures are probably more relevant than gas phase energies.

In summary, the matching procedure ideally identifies a unique set of conformations of the ligands, capable of qualitatively explaining their biological activity at a common receptor site. While a pharmacophore hypothesis provides a very valuable qualitative tool for computational and medicinal chemists in the design of novel drug candidates, it must be emphasized that it lacks the ability to quantitate the performance of one ligand against another in the context of receptor binding. For example, a typical pharmacophore cannot provide insights into relative free energies of binding of two closely related ligands. Unlike the case of structure-based drug design, pharmacophore models normally do not indicate which regions of the pharmacophore are solvent exposed, although the SAR on which the pharmacophore model is based may provide hints. For example, portions of the ligand that freely tolerate modification without significantly affecting potency are likely to be solvent exposed. Conversely, regions of the ligand that are very sensitive to substitution probably directly interact with the receptor and therefore, by definition, are shielded from solvent. However, pharmacophores provide a basis for the building of a pseudoreceptor model, as discussed in Section IV of this chapter. Furthermore, pharmacophores may provide alignment rules that are a prerequisite for 3D QSAR studies (Kubinyi, 1993).

## B. Practical Utility

An important yardstick for determining "the goodness" of a pharmacophore model is its ability to design active compounds (novel or otherwise) as potential drug candidates. Since the early to mid-1980s, medicinal chemists have typically utilized the key features of pharmacophore models in the following ways, while being appreciative of their strengths and limitations. A few of these schemes parallel the approaches taken in the design of drug candidates in the presence of an explicit three-dimensional structure of the receptor target.

1. The three-dimensional characteristics of the pharmacophore may reveal regions of space where additional chemical functionalities could be accommodated. Such modifications may lead to either the enhancement or the retention of potency. At the very least, these new analogs will augment the SAR and can be used in an iterative fashion to improve the pharmacophore model. Sterically tolerated regions of the receptor may be obtained by superimposing the active analogs consistent with the pharmacophore model and calculating the union volume of these active analogs (Sufrin *et al.*, 1981). Conversely, forbidden regions of the receptor may be obtained through the subtraction of the inactive from the active volume, leaving behind regions of space unique to the inactive analogs.

2. The *de novo* design of new drug candidates in which the core structure is replaced with a novel structure which preserves the relative orientation of key pharmacophoric groups. The programs NEWLEAD (Tschinke and Cohen, 1993) and HOOK (Eisen *et al.*, 1994) have been developed explicitly for this purpose. Furthermore, database-searching programs such as CAVEAT (Lauri and Bartlett, 1994) may be used to identify novel scaffolds which permit proper orientation of these pharmacophoric functional groups.

3. Discovery of novel drug candidates through the searching of 2- and 3D databases of compounds which match the pharmacophore hypothesis (Martin, 1992; Humblet and Dunbar, 1993; Bures *et al.*, 1994). Three-dimensional searches in particular are capable of identifying structurally novel and hence patentable ligands.

## C. A Case Study in New Lead Design

While many after the fact rationalizations of pharmacological activity have been reported using receptor mapping techniques, relatively few examples exist where pharmacophores have been employed in the design and a priori prediction of activity of new structural classes of ligands. One particularly successful published design using receptor mapping led to the discovery of Merrell Dow's selective 5-$HT_{1A}$ receptor antagonist MDL-72832 (Hilbert *et al.*, 1988). Four ligands (methiothepin, spiperone, propranolol, and buspirone) were used in the development of the pharmacophore (see Fig. 5). While all four ligands showed reasonable affinity for the 5-$HT_{1A}$ receptor, only one, buspirone, was reasonably selective for this receptor subtype.

Low energy conformations of each of these ligands were generated, and intersections of distance and angle maps among the three pharmacophoric centers yielded a single solution which allowed good alignment of the phar-

methiothepin spiperone

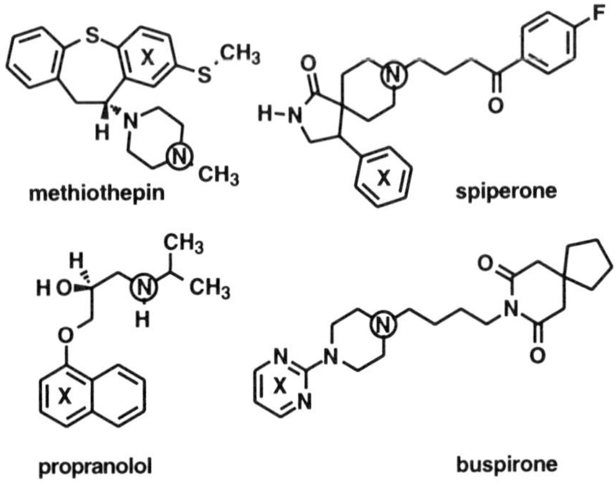

propranolol buspirone

**FIGURE 5** Ligands used in the development of a predictive 5-$HT_{1A}$ pharmacophore (Hibert *et al.*, 1988). Three features were included in the model and are represented by a "circled N" (a basic nitrogen atom), an "X" (an aromatic ring centroid), and a vector normal to the ring and passing through the centroid.

macophoric features. This alignment was further refined using a "multifit" procedure in which the ensemble of superimposed ligands was geometry optimized in the presence of distance constraints between the corresponding pharmacophoric centers. The pharmacophore obtained had a distance of 5.6 Å between the ring centroid and the basic nitrogen atom, and the nitrogen atom was located 1.6 Å above the plane of the aromatic ring. It should be pointed out that this model is reminiscent of an earlier proposed general CNS pharmacophore (Lloyd and Andrews, 1986). Finally, the union volume or a "receptor included volume" (Sufrin *et al.*, 1981) was determined for the superimposed ensemble of ligands in their active conformation. This volume maps out the region of space accessible to 5-$HT_{1A}$ antagonist ligands.

Based on this 5-$HT_{1A}$ pharmacophore model and volume map, a 2-substituted 1,4-benzodioxan deriviative (MDL-72832) was proposed as a 5-$HT_{1A}$ receptor ligand. Furthermore, based on the volume map, it was predicted that the S-enantiomer should display a higher affinity for the 5-$HT_{1A}$ site compared to the R-enantiomer. Gratifyingly, MDL-72832 displayed high affinity for the 5-$HT_{1A}$ receptor ($pIC_{50}$ = 9.1) which is significantly more potent than the most potent ligand in the training set (buspirone, $pIC_{50}$ = 7.7). Furthermore, consistent with the prediction, the S-enantiomer was at least 36 times as potent as the R-enantiomer (most of the activity in the less active enantiomer was in fact due to contamination by the more active isomer). The success of this study is attributable, in part, to the availability of a highly structurally diverse training set of ligands. This diversity allowed for the construction of a robust and predictive pharmacophore model.

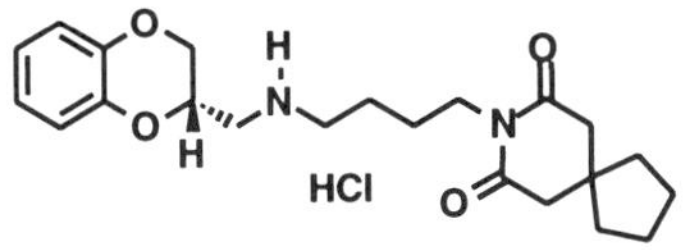

HCl

MDL-72832

## IV. PSEUDORECEPTORS—A BRIDGE BETWEEN FITTING AND MAPPING

### A. The Philosophy

The pharmacophore concept (described in the previous section), while qualitatively very useful, lacks the quantitative component of the interaction between the receptor and the ligand since the precise location of the receptor functionalities is not known. A few computational chemistry programs (e.g., APEX, Catalyst, and DISCO) have introduced the ability to generate a pharmacophore hypothesis based on biological activity. An extension of this approach that attempts to bring a quantitative element to receptor mapping has been through the construction of pseudo- and minireceptors which mimic the physiological receptors in their binding of ligand.

Here, the *bioactive* conformation of the most potent ligand is used as a basis for building a molecular framework consisting of functionalities complementary to those of the ligands. This framework could be constructed either from well-known organic functional groups or from amino acids that are representative of residues likely to be found in the receptor. The fundamental assumption in this construction scheme is that the ligands manifest their biological activity either directly or indirectly through their interaction with a small but critical subset of the entire receptor. Thus, despite the differences in the structures of the pseudoreceptor and the physiological receptor, the former provides access to similar primary interactions as in the latter. Consequently, studying binding phenomena (through calculations of interaction and solvation free energies) between the ligands and the receptor subset (pseudo- or minireceptor) enables one to correlate with experimental observations. Upon suitable validation of a pseudo- or minireceptor, it could be employed either in the *de novo* design of new leads or in the expansion of an existing SAR.

### B. Construction of Reduced Protein Models

The field of pseudoreceptor modeling has not yet been explored as extensively as structure-based approaches (Section II) or conventional pharmacophore modeling (Section III). In our laboratory, a pseudoreceptor was built around a collection of conformationally constrained cyclopropyl glutamates as NMDA agonists (Snyder *et al.*, 1992). The pseudoreceptor consisted of charged functional groups that complemented the charged character of the ammonium and carboxylate moieties of the ligands as well as provided hydrogen-bonding interactions with them. We explored numerous computational protocols encompassing molecular mechanics, quantum mechanics, molecular dynamics, and free energy perturbation methods, with and without the inclusion of explicit solvent (water), to derive models that were consistent with experimentally observed $K_i$'s. Our studies emphasized the critical role of electrostatics which need to be accounted for in the development of a pseudoreceptor by demonstrating closer agreement with experimental studies in a distance-dependent dielectric medium as opposed to a medium with explicit water molecules as solvent. The role of the dielectric constant in influencing the correlation between calculated and experimental relative-binding energies has also been highlighted by Gussio *et al.* (1992) in their pseudoreceptor docking study of 4,5-$\alpha$-epoxymorphinans.

In contrast to our NMDA pseudoreceptor where the essential chemical functionalities are connected to each other through chemical bridges, an unconnected version has been used to study the relative free energies of binding associated with the adenosine deaminase inhibitor, nebularine and its analog (Hansen and Kollman, 1990). Two different pseudoreceptor sets were investigated. In the first set, the pseudoreceptor was represented by acetic acid and in the second, it was taken from a collection of amino acids surrounding a hydrophobic pocket in a known enzyme. Both of these receptor models were treated with and without solvent, and the relative free energies of binding were calculated using molecular dynamics and free energy perturbation methods. The calculated free energy differences are argued to be qualitatively more consistent with the single moiety pseudoreceptor than with the disjoint amino acid minireceptor.

In contrast to an "organic" pseudoreceptor, Momany *et al.* (1989) constructed a protein pseudoreceptor common for CCK7 and a benzodiazepine. They built a polyalanine protein of 100 residues with helical structures for the first and the last 30 residues; the rest were involved in $\beta$-structure, bends, and loops. Subsequently, one of the many low energy conformations of CCK7 was docked into the protein followed by "mutations" in the latter to maximize interactions with the ligand using protein contact rules. Their studies did not report any correlation of binding energies between the pseudoreceptor and CCK7 on the one hand and experimental data on the other.

Walters and Hinds (1994) have outlined a computational approach to constructing receptor models using genetic algorithms. In their approach, the receptor is represented not as a collection of joint or disjoint chemical functions, but as a collection of atom types around a group of ligands with varying activities. The receptor is then optimized to maximize the correlation between ligand–pseudoreceptor interactions (calculated using molecular mechanics) and the observed biological potencies. This method has been applied to the development of a receptor model for peptidic sweeteners and their analogs, and the calculated potencies were found to be within one log unit of the experimental data.

## C. Structural Correlation with Experiment

In general, a pseudoreceptor differs from a physiological receptor in that the former may not have either the composition or the three-dimensional structure of the latter. However, applications of the receptor mapping program YAK (Vedani *et al.*, 1993) to construct peptidic pseudoreceptors for the active sites of the enzymes human carbonic anhydrase I and thermolysin demonstrate a reasonable correspondence of the theoretical models with the X-ray data. In this study, amino acid residues are placed in space around a collection of ligands that are used to define a pharmacophore model and on each of which vectors associated with directional interactions (lone pair vectors, hydrogen-extension vectors and hydrophobicity vectors) are generated. The vectors are employed to guide the placement of amino acid residues (e.g., the terminal N$\zeta$ hydrogen in the side chain of lysine is located at or about the tip of an average hydrogen extension vector from the ligands). This is followed by energy optimization of the intermolecular interactions. The amino acids selected were those in the active sites of the two enzymes. Four potent sulfonamide and three highly active phosphoramidate inhibitors were chosen as ligands for carbonic anhydrase I and thermolysin, respectively. In the case of carbonic anhydrase I, it was shown that the catalytic zinc and associated metal bound histidines were reproduced best with average root mean square deviations of less than 0.3 Å from the experimental positions. The corresponding deviations for other active residues were higher (between 0.8 and 3.6 Å). The atomic positions in the pseudoreceptor for the thermolysin active site residues were within 1.9 Å of the corresponding experimental positions. These results indicate that pseudoreceptor building of enzyme active sites offers reasonable promise at least for key residues, when the critical pharmacophoric centers on the ligands are identified. The YAK approach was also employed to build peptidic pseudoreceptors of dopaminergic and $\beta$-adrenergic agonists (Vedani *et al.*, 1995) qualitatively and semiquantitatively consistent with experimental data. It has been shown that the relative binding energies of various ligands calculated using these pseudoreceptors are within 1 kcal/mol of the corresponding experimental values.

The pseudoreceptor approach to constructing receptor models appears promising, but is faced with a few problems that need to be addressed in the course of its evolution. One of these problems concerns the issue of conformational flexibility, both in the ligands (Section V,B) and in the receptor model (Section II,B,2). The ligand conformational flexibility may be constrained to conform to its presumed bioactive conformation obtained through receptor mapping strategies (Section III). There is a greater degree of uncertainty in the conformational flexibility of the pseudoreceptor which, in mimicking a receptor binding/active site, is required to maintain its characteristic rigid framework. In the case of enzymes and membrane-bound receptors, the rigid framework is provided by regions of the macromolecule outside of the binding site. In the pseudoreceptor, this external region is typically a solvent or a vacuum environment. The resultant conformational uncertainty can contribute to deviations in the calculated relative binding energies compared to the experimental data.

Another significant issue is the role of electrostatic interactions (represented by Coulombic potential) between the ligands and the pseudoreceptors, particularly when the former are charged. This problem is no less significant in a well-defined binding site of an enzyme or a protein of known structure. However, the rigid framework of the latter sets a defined electrostatic environment in the active site in which variations in ligand interactions can be more consistently related to the experimental data. In the present schemes of pseudoreceptor construction, the spatial disposition of the various fragments or amino acids is governed primarily by the requirement of optimal interactions with the ligands. In the absence of guidelines for optimization of intramolecular interactions, it is not clear how one can construct an "ideal" pseudoreceptor without iterating through many cycles. Thus, complexities arising out of conformational flexibility and electrostatics clearly point to the need for developing strategies to build more "enzyme-like" or "receptor-like" pseudoreceptors.

## V. ROLE OF SOLVENT—MODELS AND LIMITS

### A. Aqueous and Nonaqueous Solvent Models

Water profoundly influences most biochemical phenomena and the binding of drugs to their macromolecular targets is no exception. The aqueous solvent plays a pivotal role in determining the conformational preferences of drugs and receptors as well as modulating the forces experienced between binding partners. The hydrophobic effect, which drives many receptor–ligand associations, is a direct consequence of the unfavorable interaction of nonpolar solutes with water. Hence, accounting for solvation effects is essential for a qualitative understanding of the interaction of drugs with their receptors and for the quantitative prediction of binding affinities. Accordingly, numerous methods have been developed to model the effects of solvents on both small and macromolecular solutes ranging from solvent continuum models to inclusion of discrete solvent molecules to hybrid methods (Smith and Pettitt, 1994; Tomasi and Persico, 1994). These techniques have been developed in both the molecular mechanical (MM) and quantum mechanical (QM) frameworks.

Solvation energy is the net difference between the energetic cost of creating a cavity in the solvent and the energy released due to attractive dispersion and electrostatic forces between solvent and solute. Electrostatic forces are reinforced by polarization in both solvent and solute. Entropy also plays a role since the structure solvent surrounding solutes may be more ordered compared to that of bulk solvent. The solvent also serves to reduce the intramolecular interaction of solute charges through dielectric screening and therefore can alter the conformational preferences of solutes.

Polarization in the solvent with discrete solvent models is accounted for by the reorientation of solvent dipoles. Because of computational costs, discrete solvent simulations are normally carried out only with MM methods. Solvent polarization effects can be mimicked through the use of continuum solvation models by inclusion of a reaction field, and both MM and QM variants of this strategy have been developed. Hybrid methods in which the first solvation shell is modeled with explicit solvent molecules whereas the bulk solvent is treated as a continuum have also been experimented with. Finally, adequate methods for mimicking the polarization of the solute may only be accomplished with QM methods. The following solvation methods that have attracted the most widespread attention are listed in an approximate order of increasing sophistication.

### 1. Distance-Dependent Dielectric

Force fields generally account for electrostatic effects using Coulomb's law

$$V_{ij} = \frac{q_i q_j}{4\pi \varepsilon r_{ij}},$$

where $V_{ij}$ is the force felt between two charges $q_i$ and $q_j$, $r_{ij}$ is the distance separating the two charges, and $\varepsilon$ is the dielectric constant of the medium. In order to simulate the electrostatic screening effect of the solvent, the dielectric constant used in the molecular mechanics simulation may be increased from the value for vacuum ($\varepsilon = 1$) to that of water ($\varepsilon = 80$). An intermediate dielectric constant might be approriate for the environment of a binding site. However, this simple treatment may overestimate the forces felt between charges at long distance as these charges interact through bulk solvent and therefore are more heavily screened. A slightly more sophisticated correction is to employ a distance-dependent dielectric where the dielectric constant in Coulomb's law is replaced by $dr_{ij}$ so that the force decays as $1/r_{ij}^2$ (Pickersgill, 1988). The proportionality constant $d$ is normally set to a value of 1 (protein) or 4 (water). The distance-dependent dielectric does reduce the unrealistically large forces experienced at large distances; however, it does not completely eliminate it. Furthermore, charges may interact through the solute rather than the solvent so it is often unclear what value of $d$ is most appropriate.

### 2. Atomic Solvation Parameters

Empirical solvation-free energies based on accessible surface areas and atomic solvation parameters (ASPs) have been developed and applied to molecular mechanics and dynamics simulations of peptide and protein structures (Schiffer *et al.*, 1992; Wesson and Eisenberg, 1992; Williams *et al.*, 1992; von Freyberg and Braun, 1993; von Freyberg *et al.*, 1993; Smith and Honig, 1994). With these methods, a solvation correction term $\Delta G_{\text{solv}}$ is added to the molecular mechanics energy and is calculated using the formula

$$\Delta G_{\text{solv}} = \sum_i \sigma_i \text{SA}_i,$$

where $\text{SA}_i$ is the solvent-accessible surface area of the $i$th atom and $\sigma_i$ is an empirical solvation parameter which is a function of the atom type. The $\sigma_i$'s are determined by parameterization most commonly using the octanol/water partition coefficient. These parameterized methods account for the hydrophobic effect but only account for the electrostatic polarization component of the solvation energy in an averaged way.

### 3. Reaction Fields

The solvent polarization contribution to solvation energy may be estimated using the generalized Born equation

$$\Delta G_{\text{pol}} = -\frac{1}{2}\frac{q^2}{r_b}\left(1 - \frac{1}{\varepsilon}\right),$$

where $q$ is the charge, $r_b$ is the effective radius of the ion, and $\varepsilon$ is the dielectric constant of the solvent. The polarization contribution to solvation may be combined with an empirical atomic solvation term to account for the hydrophobic effect (Still *et al.*, 1990). The combined generalized Born–surface area (GB/SA) method is then used as a correction to the molecular mechanics energy. It is important to note that the solvation energies calculated with the GB/SA method only work reliably when the atomic charges are reasonable. Typically the atom-centered point charges are best determined by fitting them to reproduce the electrostatic potentials (ESPs) obtained from a quantum mechanical wave function. The AMBER force field (Weiner *et al.*, 1984; Cornell *et al.*, 1995), which is based on *ab initio* ESP-fitted charges, contains realistic default charges and works reasonably well with the GB/SA technique (Still *et al.*, 1990).

One of the shortcomings of the simple Born equation is that it assumes a spherical or elliptical cavity whereas drugs and their receptors are normally irregularly shaped. The GB/SA method handles this irregularity by treating each charged atom separately and summing the polarization energy over these atoms. Another method to treat irregularly shaped solute–solvent boundaries is to place the solute on a 3D grid and estimate the charges and dielectrics on the grid points by solving the Poisson–Boltzmann equation (Gilson and Honig, 1988, 1991). This method determines the electrostatic component to the solvation-free energy. A parameterized method for determining the total solvation energy based on the finite difference Poisson–Boltzmann method has been reported (Sitkoff *et al.,* 1994) and is called PARSE (parameters for solvation energy). A third approach that has been employed is to surround the solute with dipoles fixed on a regularly spaced grid. The polarization of the solvent in response to solute charges is then simulated by adjusting the length and direction of the dipoles by solving a Langevin type formula. This procedure has been parameterized for proteins and is referred to as the protein dipoles Langevin dipoles (PDLP) method (Lee *et al.,* 1993).

#### 4. Explicit Solvent

The most economical way of incorporating discrete solvent in molecular simulations is to include only a shell of solvent molecules surrounding the solute (Arnold and Ornstein, 1994). With this method, the solute–vacuum interface is replaced by a solvent–vacuum interface. A more realistic but computationally demanding method involves placing the solute in a solvent box with periodic boundary conditions (Levitt and Sharon, 1988) where solvent molecules leaving one side of the box reappear at the opposite side so that the density of the system is maintained at a constant value.

Relative solvation-free energies may be estimated by employing free energy perturbation calculations in a discrete solvent bath. In a classic example of this strategy, the relative solvation-free energies of ethane and methanol were calculated. In the computer simulations, one of the methyl groups of ethane was *mutated* to a hydroxyl while the surrounding TIP3P water molecules (Jorgensen *et al.,* 1983) underwent continuous adjustment during the mutation. Both the OPLS (Jorgensen and Tirado-Rives, 1988; Pranata *et al.,* 1991) and the AMBER force fields (Singh *et al.,* 1987) were applied to this problem, and the calculated relative solvation energies were quantitatively consistent with experimental observations. This success story is of particular significance in light of the fact that the mutation involved a change from a hydrophobic methyl group to a hydrogen bond-donating and -accepting hydroxyl group.

Similar experiments were carried out to calculate relative solvation-free energies of methane, ethane, and propane (Flesichman and Brooks, 1987; Rao and Singh, 1990, 1991a). These studies predicted the energy differences to be within 1 kcal/mol of the experimental values. However, even smaller discrepancies would have been expected given that the mutations in these calculations were reasonably simple, i.e., between two hydrophobic groups ($CH_3$ to H and vice versa). Relative solvation energies of cationic systems such as $NH_4^+$, $(CH_3)_4N^+$, $(CH_3CH_2)_4N^+$, and closed shell ions like $Na^+$, $K^+$, $Cl^-$, and $Br^-$ have been investigated in water, carbon tetrachloride, dimethyl sulfoxide, methanol, and hydrazine as solvent media (Rao and Singh, 1990, 1991a). The calculated energies are found to be within 0.3 to 3.0 kcal/mol of the available experimental values. These discrepancies have been reduced by the improvement in the computational protocol following the introduction of the so-called bond–PMF factor (Sun *et al.*, 1992). This factor accounts for the changes in free energy associated with variations in bond lengths which arise from the mutation.

Pioneering work in the determination of solvation-free energies of small molecules has been reported by Jorgensen and co-workers (1983). Monte Carlo simulations were used in place of molecular dynamics in these investigations, where the dynamic behavior of a molecule is simulated by random changes made either to internal (e.g., dihedral angle) or to Cartesian coordinates. If the energy of the resultant trial configuration is less than that of the previous configuration on which the changes were made, the former is accepted. Otherwise, algorithms commonly based on the Metropolis method (Metropolis *et al.*, 1953) are used to determine the acceptability of the new configuration (Chang *et al.*, 1988; Howard and Kollman, 1988). In general, the calculated free energies were qualitatively and semiquantitatively consistent with experimental observations (Jorgensen and Ravimohan, 1985; Jorgensen and Nguyen, 1993). For example, the free energies of hydration of some substituted benzenes (e.g., toluene, *p*-xylene, phenol, *p*-cresol, and anisole) were calculated to be within 0.5 kcal/mol of the experimental values (Jorgensen and Nguyen, 1993). Although the corresponding deviations for some other substituted benzenes (e.g., hydroquinone and benzonitrile) were higher ($>$1.0 kcal/mol), these applications of the free energy perturbation approach in a discrete solvent nevertheless demonstrate their utility in predicting relative solvation-free energies of simple uncharged molecules.

### 5. Quantum Mechanical Approaches

The advantage of the discrete solvent approach is that it accounts for specific interactions between solute and solvent molecules which are often quite important but impossible to capture using continuum approaches. On the other hand, use of a discrete solvent even in the MM framework is computationally expensive and has other shortcomings, especially its inability to account for solute polarization and to model the effect of solvent on chemical reactions. Also, this method is quite sensitive to the parameterization of both the solute and the solvent force fields.

In an attempt to develop a method that can simulate solute polarization and at the same time be used for the simulation of chemical reactions in solvent, various quantum mechanical approaches have been developed. With these methods, atomic charges are obtained naturally from the quantum mechanical wave function. In one highly parameterized semiempirical implementation, the effects of the mutual solvent–solute polarization and of the energetic effects of solute-induced solvent reorganization and dispersion forces are included in the Hamiltonian operator and the calculation iterates until self-consistency is achieved. The method relies on the AM1 semiempirical method and is called AMSOL (Cramer and Truhlar, 1991a,b; 1992a,b; Giesen *et al.*, 1995). Another semiempirical approach called SOFI (SOlvation via Field Interaction) which estimates only electrostatic contribution and uses no adjustable parameters has been shown to be surprisingly effective in reproducing experimentally determined solvation energies (Ford and Wang, 1992; Wang and Ford, 1992).

## B. Small Molecule Energetics

Quantum mechanics in conjunction with solvent continuum methods have been useful in studying individual ligands since these methods require minimal parameterization and are the only methods that account for solute polarization. However, for efficiency reasons, molecular mechanics with a continuum solvent correction have enjoyed the most widespread use in drug design.

An important test of the adequacy of various solvent models is their ability to reproduce experimentally determined conformer preferences for molecules in solution. Most force fields are parameterized to reproduce observed conformer ratios and therefore *free energies* even though these force fields do not explicitly account for entropy. The errors associated with this approximation for small molecules are normally not significant, although it has been estimated that this error may exceed 0.5 kcal/mol for larger flexible molecules (Senderowitz *et al.*, 1995).

One particularly useful set of data with which to judge the quality of solvent models is a series of simple di- and triamides which possess a relatively limited number of intramolecular hydrogen-bonding alternatives. The conformational preferences of these so-called Gellman amides in various solvents have been extensively studied using variable temperature IR and NMR techniques (Dado *et al.*, 1990; Gellman *et al.*, 1990, 1991; Liang *et al.*, 1991; Dado and Gellman, 1993).

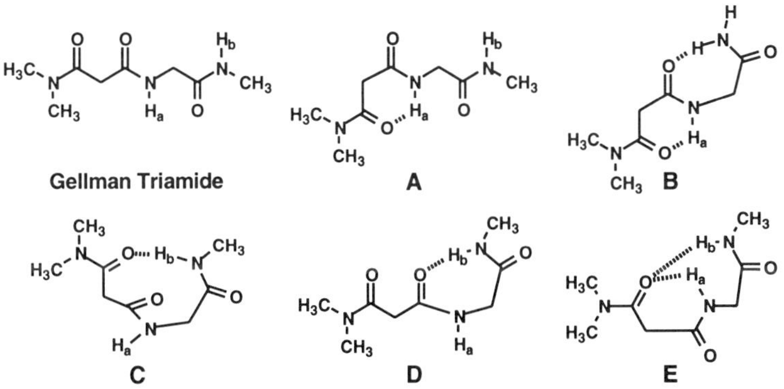

Gellman Triamide A B

C D E

At low temperatures, conformer **C** (nine-membered ring) of the Gellman triamide predominates in the relatively apolar solvent methylene chloride, whereas at increasing temperatures, conformer **A** (six-membered ring) becomes more prevalent based on IR and NMR data (Gellman *et al.*, 1990). Conformer **C** is also observed in the crystalline state (Dado *et al.*, 1990). The interpretation of these experimental results was called into question by *in vacuo* AM1 semiempirical calculations incorporating explicit solvent (Novoa and Whangbo, 1991) as well as *in vacuo* and solvent continuum AMBER molecular mechanics simulations (Smith and Vijayakumar, 1991). Both sets of calculations indicate that conformer **B**, which contains the maximum number of hydrogen bonds, is the lowest energy conformer. However, conformer **B** is predicted to have N-H···· O angles of 120° and 141° which are far from the ideal linear arrangement (Dado and Gellman, 1992). Furthermore, AM1 calculations predict an optimum in the N-H···· O angle of 130° which is at variance with *ab initio* results which show an optimum value of 150°. In addition, the N-H···· O angle potential energy surface for the AM1 calculations is much shallower compared to the *ab initio* surface. Hence it is understandable why AM1 overpredicts the relative stability of conformer **B**. As for the molecular mechanics result, the AMBER force field overpredicts the stability of the $C_7$ ($\gamma$ turn) relative to the $C_5$ conformation of dipeptides (Gellman and Dado, 1991).

The AMBER force field was reparameterized to reproduce high level *ab initio* estimates of the relative stabilities of dipeptide conformations (McDonald and Still, 1992b). The reparameterized force field (AMBER*) in conjunction with the chloroform GB/SA solvent continuum correction was then used to estimate the relative stabilities of the various conformers (McDonald and Still, 1992a). Conformer **C** was now found to be the lowest energy, with conformers **A** and **B** being 0.6 and 0.9 kcal/mol, respectively, higher in energy which is in qualitative agreement with the experimental results.

A series of dipeptides were also studied using IR and NMR methods (Gellman *et al.*, 1991). At 297°K in methylene chloride solution, the diamide **1** is predominately intramolecularily hydrogen bonded (six-membered ring) rather than extended form. None of the other amides at this temperature showed appreciable intramolecular hydrogen-bonding interactions. When the temperature of the solution was lowered to 186°K, none of the amides except **4** showed an appreciable shift in the ratio of hydrogen-bonded to nonhydrogen-bonded conformations. The temperature dependence of the conformational equilibrium of **4** was used to estimate the enthalpic and entropic energy differences between the intramolecularily hydrogen-bonded form (nine-membered ring) and the weighted average of the extended forms. These experiments showed that the hydrogen-bonded form is favored enthalpically by 1.5 ± 1 kcal/mol and disfavored entropically by 7.6 ± 0.8 e.u.

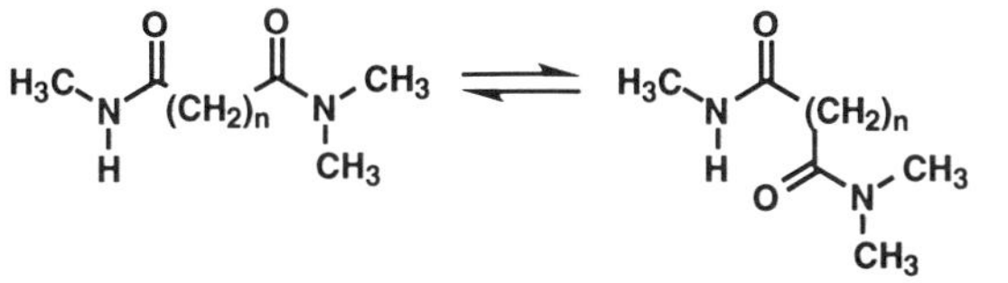

**1**, n=1
**2**, n=2
**3**, n=3
**4**, n=4
**5**, n=5
**6**, n=6

The conformational equilibria of these amides was studied using the AMBER* force field with a chloroform GB/SA solvent continuum correction (McDonald and Still, 1994). A molecular dynamics simulation at 300°K reproduced the experimental hydrogen-bonding trend ($\mathbf{1} \gg \mathbf{2} > \mathbf{3} \approx \mathbf{4} > \mathbf{5}$). Furthermore, the temperature dependence of the conformational equilibrium of **4** was also reproduced. For comparison, the conformational equilibrium of **4** was studied *in vacuo*. Whereas the calculations with the chloroform solvent continuum model closely reproduced the experimental thermodynamic parameters, the results with the *in vacuo* molecular dynamics simulation were at considerable variance with experiment. This result underscores the importance of accounting for solvation effects, even with relatively apolar solvents like methylene chloride.

## C. Macromolecular Conformation and Ligand Binding

Just as with small molecules, accounting for solvent effects is crucial to the prediction of protein conformation and dynamics. Vacuum molecular mechanics and dynamics calculations produce structures which by a variety of measures are unrealistic (Stouten *et al.*, 1993). Polar side chains on the surface of proteins are often extended to maximize their interaction with water. In contrast, *in vacuo* minimization cause these side chains to fold back onto the protein. This is partially a consequence of electrostatic attractions that are not being shielded by solvent. Another contributing factor is "van der Waals collapse" of the protein structure during *in vacuo* optimization. In the gas phase, surface atoms are only in contact with atoms on one face and therefore are "pulled" toward the center of the protein. This causes the protein structure as a whole to be denser than observed experimentally.

In order to produce satisfactory protein structures in molecular dynamics simulations, either an explicit solvent bath (Levitt and Sharon, 1988) or a solvent continuum model (Solmajer and Mehler, 1991; Arnold and Ornstein, 1994) must be used. When using atomic solvation parameters, the balance between solvation-free energy and molecular mechanics energy is critical (Schiffer *et al.*, 1992; Cregut *et al.*, 1994). If these two terms are carefully balanced, then the simulations do not seriously perturb the structure of the protein from its crystallographic starting point (Schiffer *et al.*, 1993). One drawback to this approach is that it is assumed that if a protein atom is buried within a protein, it is in a hydrophobic environment. However, there is of course a great deal of variability in the internal environments of proteins. Hence more accurate ASPs which take into account the local environment of buried protein atoms have been developed (Delarue and Koehl, 1995).

The consideration of solvent is also crucial for evaluating the energetic differences between various conformations of macromolecules. For example, molecular mechanics energy evaluations that include a solvation correction term are able to distinguish between native and incorrectly folded protein structures (Novotny *et al.*, 1988; Cregut *et al.*, 1994). The corresponding *in vacuo* calculations were unable to make this distinction. Similarly, the crystallographic conformation of protein loops could only be successfully predicted if solvation effects were included in the energy evaluation (Smith and Honig, 1994).

Successful prediction of ligand-binding affinities requires accounting for solvation effects. In free energy perturbation predictions of relative ligand affinities, explicit solvent is generally used (Section II,D) whereas empirical approaches generally rely on solvent continuum models (Section II,C,3). In an example of the latter strategy, Sharp *et al.* (1987) treated the protein–ligand system as a three-dimensional grid (with approximately 2-Å resolution), and at each grid point, the steric (van der Waals interactions) and electrostatic components of the potential were calculated using the force field (CHARMm). Solvent effects are accounted for by (a) the use of a high dielectric constant (e.g., 80 for water) and by (b) calculating electrostatic interactions by the finite difference Poisson–Boltzmann (FDPB) method which incorporates the effects of solvent ionic strength and the differing polarizabilities of protein and solvent. At grid points within the protein, a constant dielectric of 2 is employed. This technique was applied to the study of diffusion of superoxide into the electric field of superoxide dismutase and calculation of association constants as a function of ionic strength and amino acid modifications in the enzyme active site (Sharp *et al.*, 1987). The protein was treated as rigid with no conformational mobility relative to its X-ray crystallographic structure. The study demonstrated that the electric field of the enzyme enhanced association rates of the superoxide by factors exceeding 30 as evidenced by the lower association constants for the mutant enzymes in which the catalytically important arginine and lysine residues were modeled in their neutral forms. These cationic residues were thought to lower the magnitude of the negative electrostatic potential barrier around the protein which carries an overall negative charge. This observation was substantiated through simulations on mutants in which two glutamates in the vicinity of the copper site were altered to lysines, resulting in a higher association constant for the superoxide anion.

This implicit solvation method has been successfully applied to the determination of relative binding energies of substrate/protein interactions for a series of ligands for the arabinose- and sulfate-binding proteins (Shen and Quiocho, 1995) in a manner analogous to the superoxide–SOD calculations. The success of this study is attributable in part to the small structural perturbations in the ligands or proteins and that electrostatics dominate the differences in interactions between the various ligands and mutant proteins. It is not clear, however, how problems associated with conformational mobility likely to accompany larger structural variations of the protein and of the ligand will affect the calculated electrostatic and steric potentials at grid points close to the protein–solvent interface and hence the calculated binding energies.

## VI. PEPTIDOMIMETIC DESIGN—GOALS AND ACHIEVEMENTS

### A. Interplay between Modeling and Bioassay

It is clear that for any modeling strategy to be validated, it must at the very least rationalize (either qualitatively or semiquantitatively) the existing structure–activity relationships (SAR). Furthermore, it must be capable of making predictions (either extending an SAR or leading to the birth of a new SAR) that can be tested through *feasible synthetic* strategies. An essential component of the development of such a model is the availability of reliable biological assays which preferably yield consistently reproducible results and which are based on a finite number of mechanisms. Thus, inhibition of binding by a standard labeled ligand to an enzyme (typically a primary assay) would constitute an excellent example of a reliable assay which could form the basis of the pharmacophore design and evolution. Furthermore, if the structure of the macromolecular target has been solved, preferably complexed with a ligand of interest, then the SAR will provide valuable information with which to interpret the structure of the receptor and how it interacts with ligands. Regardless of the strategy followed (receptor mapping or fitting), biological assays are a critical link in the design/synthesis/testing cycle as illustrated in Fig. 6.

The success of any drug discovery effort is critically dependent on the existence of a sensitive, high-capacity primary assay which can rapidly provide activity data for large numbers of compounds. A fast turn around ensures that the assay will not become the bottleneck in the design/synthesis/testing cycle. Furthermore, a high capacity assay will provide a much more extensive SAR on which to base pharmacophore and receptor models. Finally, with the advent of large combinatorial libraries in which the amount of any one component produced is very small, it is vital that the assay be sensitive as well as rapid so that screening of these libraries is feasible.

The correlation between the activity in primary assays and the *in vivo* biological activity (e.g., in whole animals) or lack thereof normally dictates the extension of pharmacophore models or receptor structures in predicting compounds that are likely to be orally efficacious drug candidates. If a

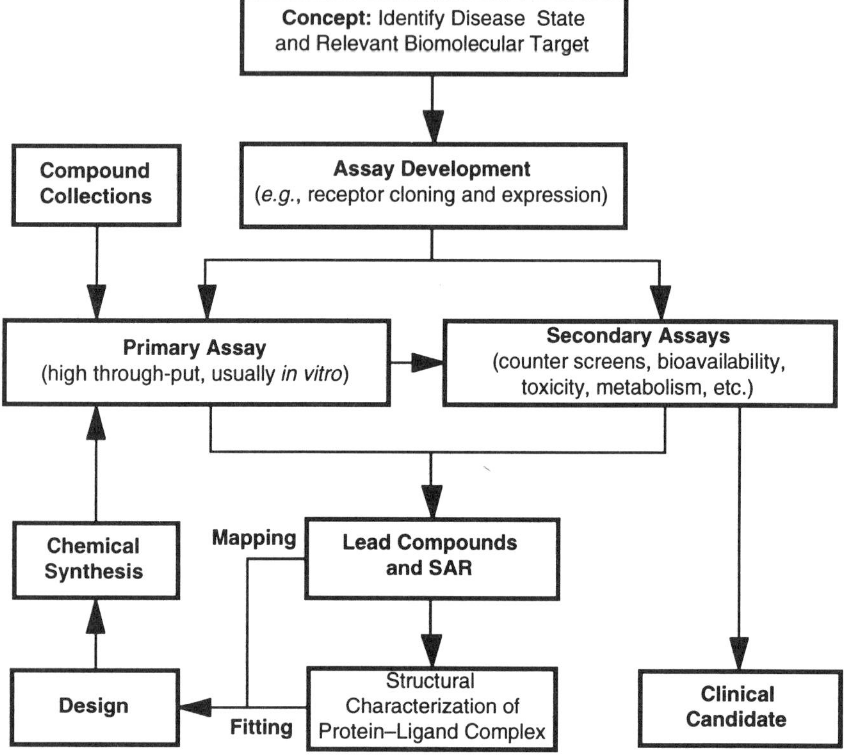


**FIGURE 6** The drug discovery cycle from concept to clinical candidate. The design process may start from either an explicit 3D structure of the macromolecular target (receptor fitting, see Section II) or from a model of the receptor based on what binds to it (receptor mapping, Section III).

suitable animal model for testing the *in vivo* efficacy of a drug is lacking, the ratio of $C_{max}$ (the maximum plasma concentration of the drug after oral administration) to $EC_{50}$ (the *in vitro* concentration that produces 50% of the maximum response) may be used as a surrogate (Kempf, 1994). Unfortunately, the degree of correlation between *in vitro* and *in vivo* potency is often low. Indeed, pharmacophore or receptor models only indicate whether a given compound is likely to bind with high affinity to a given receptor *in vitro* and have no bearing on whether the drug will reach its intended target if administered orally.

Molecular properties such as low molecular weight (<600 Da), high aqueous solubility, decreased hydrogen-bonding potential, and metabolically stable functional groups often confer good oral availability on a ligand (Kempf, 1994). For example, the anti-thrombotic agents in the benzamidine class which meet all of these requirements in general have good oral availability and consequently the correlation between *in vitro* and *in vivo* effects has been shown to be a reasonably good one. On the other hand, renin inhibitors which possess a relatively high molecular weight and are peptidic in nature constitute an example of poor correlation between the potencies in the *in vitro* and *in vivo* bioassays; this lack of correlation is attributed to poor oral bioavailability.

Currently, theoretical models cannot fully account for oral bioavailability (Nellans, 1991), metabolism (Manners *et al.*, 1988), transport to the site of action (Dearden, 1990), and related factors since methodologies rationalizing these are not yet sufficiently developed. However, one exception, log $D$, is an extensively used parameter in deriving theoretical models that relate to *in vivo* activity (Scherrer and Howard, 1977). The distribution coefficient $D$ is defined as the ratio of the concentration sum of drug in all ionization states between the two-phase octanol/water system at equilibrium whereas the partition coefficient $P$ in the ratio of unionized species between these two phases. Log $D$ and log $P$ are related by the following approximate expression

$$\log D = \log\left(\frac{[A]_{\text{octanol}}}{[A_{\text{unionized}}]_{\text{water}} + [A_{\text{ionized}}]_{\text{water}}}\right) \cong \log P \pm (\text{p}K_a - \text{pH}),$$

where the plus sign is used for acids whereas the minus sign is for bases. For neutral compounds that do not appreciably ionize at physiological pH, log $P$ = log $D$. Partition coefficients may be determined experimentally (Dearden and Bresnen, 1988) or estimated using various atomic or fragmental-based parameterizations (Viswanadhan *et al.*, 1993; Moriguchi *et al.*, 1994; Hansch and Leo, 1995). Empirical methods for estimating the acidity and basicity of organic compounds are also available (Perrin *et al.*, 1981).

It has been suggested that the optimum log $D$ that a compound should possess for transport across the blood–brain barrier is approximately 2 (Hansch *et al.*, 1987). Conversely, log $D$ should be made as low as possible without loss of efficacy in order to keep a compound out of the CNS and to minimize other side effects resulting from cytochrome P450 metabolism and excessively long clearance times. In addition, lowering log $D$ generally decreases binding to serum proteins and therefore increases the effective concentration of drug in the blood. On the other hand, increasing log $D$ has been shown to enhance oral absorption (Hamilton *et al.*, 1995).

Molecular mass also plays a crucial role in bioavailability. Most drugs currently on the market fall in the range of 250 to 600 Da (Kim *et al.*, 1995). This apparent upper limit in molecular mass for a successful orally active drug is readily understood in terms of the observed decrease in absorption and the increase in hepatic clearance as the molecular mass is increased (Hamilton *et al.*, 1995).

## B. General Design Strategies

Although combinatorial peptide libraries are capable of producing selective ligands with good *in vitro* potency, their rapid metabolism and poor oral availability make them unsuitable for use as drugs. Hence an enormous amount of effort has been expended in developing peptidomimetics with improved stability and absorption characteristics (Gante, 1994). One of the most common strategies (Fig. 7) is to replace the backbone amide bonds with various bioisosteric equivalents (e.g., ketomethylene, hydroxyethylene, ethylene, etc.).

Peptides often adopt well-defined conformational states (e.g., $\alpha$-helix, $\beta$-sheet, $\gamma$-turn) that are a crucial component of their activity. Hence an important consideration in peptidomimetic design is the retention of the bioactive conformation of the nascent peptide, hypothesized to be important in its binding to its biological receptor. Therefore, a second common strategy in developing peptidomimetics is to use rigid scaffolds which mimic these conformations (Müller *et al.*, 1993). Information concerning this conformation may be obtained by a variety of methods, including the 3D pharmacophore modeling, crystallographic structure analyses (e.g., renin and HIV-1 protease inhibitors), multidimensional NMR solution studies (Zuiderweg *et al.*, 1993), and transfer nOe techniques (Gronenborn and Clore, 1990). Database searching programs such as CAVEAT (Lauri and Bartlett, 1994) can assist in finding novel scaffolds which mimic a specific conformational state of a peptide. In addition, a number of theoretical investigations of the conformational preferences of model peptides and peptidomimetics have been reported in the literature. Although a detailed discussion of these investigations is beyond the scope of this chapter, it sufficies to say that peptidomimetics with specific conformational constraints can be designed and synthesized. For example, when the $\alpha$-carbon in an amino acid residue is tied back onto itself through a substituted cyclopropyl group, the peptide containing it adopts a $\gamma$-turn conformation (Balaji *et al.*, 1994).

One powerful driving force for binding of ligands to receptors is the hydrophobic effect. This effect is due to the release of ordered water mole-

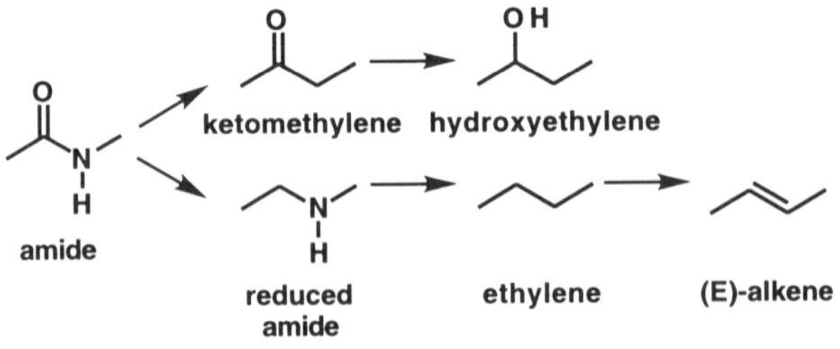

**FIGURE 7** Commonly employed bioisosteric replacements for the amide functional group (Gante, 1994).

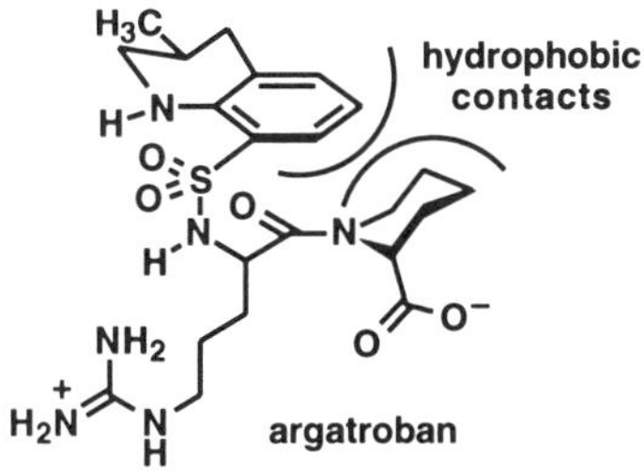

**FIGURE 8** Schematic depiction of the thrombin inhibitor argatroban in its bound conformation illustrating the hydrophobic contacts between the tetrahydroquinolyl and piperidyl rings (Banner and Hadvary, 1991). A similar conformation is observed in solution.

cules from the surface of both the receptor and of the ligand to bulk solvent (Section I,C). In an aqueous environment, hydrophobic groups located on flexible ligands frequently form intramolecular contacts in a phenomena that has been termed hydrophobic collapse (Rich, 1993). In certain instances, hydrophobic collapse may be beneficial if it locks the ligand in an active conformation. For example, the thrombin inhibitor argatroban was found to have a conformation both in solution and complexed with thrombin (Fig. 8) where the hydrophobic tetrahydroquinolyl and piperidyl groups form close contacts (Banner and Hadvary, 1991). This preorganization of the conformation of the inhibitor reduces the entropic penalty of binding and therefore boosts affinity.

More frequently, however, hydrophobic collapse can work against receptor binding. This occurs if hydrophobic groups in the ligand form intramolecular contacts that are not present in the ligand–receptor complex. In these situations, the hydrophobic contribution of these groups to binding is less than it might be since these groups are partially desolvated prior to binding. However, if these hydrophobic groups are placed on a rigid scaffold, which prohibits hydrophobic collapse from occurring, then these groups will be fully solvated in aqueous solution. Furthermore, if the scaffold locks these hydrophobic groups in a conformation that is compatible with binding, then the full advantage of the hydrophobic effect will be realized, resulting in a higher affinity ligand.

Many peptidomimetics that contain two or more hydrophobic functionalities mounted on a rigid scaffold are known. These include the 1,4-benzodiazepine ring system (e.g., diazepam, a sedative) and the A-II receptor antagonist losartan for treatment of hypertension. The high affinity of these ligands derives, in part, from their resistance to hydrophobic collapse.

These examples illustrate that consideration of hydrophobic collapse is a powerful design principle. Locking a ligand in an active conformation reduces the entropic costs of binding. This rigidification may be brought about by hydrophobic collapse itself or from the introduction of a rigid

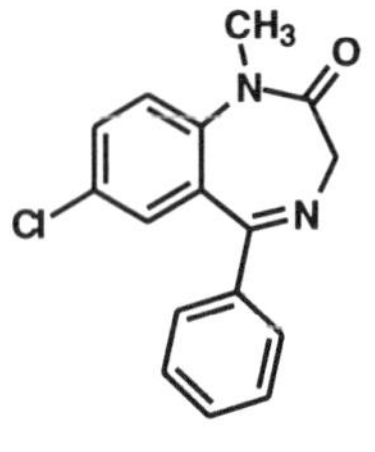

**diazepam**
**$IC_{50}$ = 8 nM**

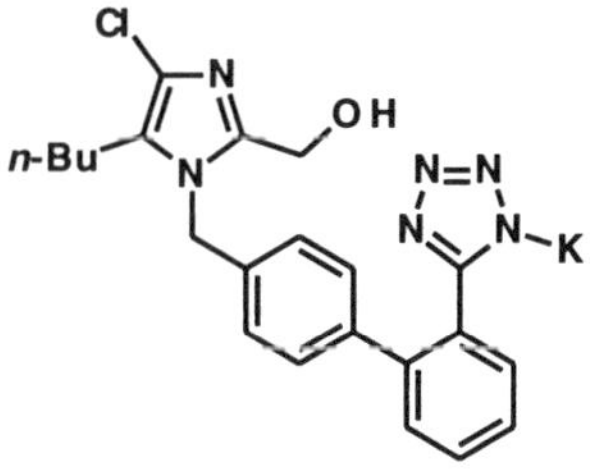

**losartan (DuP-753)**
**$IC_{50}$ = 19 nM**

scaffold which prevents hydrophobic collapse. In the latter case, a synergy results from simultaneously reducing the entropic cost and enhancing the hydrophobic contribution to binding.

Some applications where these peptidomimetic strategies have been employed with varying degrees of success include renin, HIV-1 protease and elastase inhibitors, RGDX mimics as antithrombotics and GP IIb/IIIa receptor antagonists, and cyclic enkephalins as δ receptor agonists. A few of these examples are briefly discussed in the following sections.

## C. A Sampling of Disease Targets

### 1. Aspartyl Protease Inhibitors

*a. **Human renin*** The renin–angiotensin system plays a key role in blood pressure regulation and is therefore a target for cardiovascular therapy. The first event in this cascade is the cleavage of the circulating glycoprotein angiotensinogen to the decapeptide angiotensin I (A-I) by the aspartyl protease renin. A-I is further cleaved by the angiotensin-converting enzyme (ACE) to yield the vasoconstrictor angiotensin II (A-II). Currently, ACE inhibitors are widely used for the treatment of hypertension and congestive heart failure. However, they are known to cause side effects associated with elevated levels of bradykinin (an ACE substrate). On the other hand, angiotensinogen is believed to be the only natural substrate for renin. Consequently, renin inhibitors are thought to be excellent targets for the treatment of hypertension in humans and have attracted a great deal of attention from medicinal chemists.

A number of reviews have dealt with renin inhibitors (Greenlee, 1990; Luther *et al.*, 1991; Abdel Meguid, 1993). While many highly potent and selective inhibitors have been developed, these inhibitors are beset with problems of poor oral bioavailability and pharmacokinetics. For this reason, design has focused instead on reducing the peptidic character. Some of the key strategies pursued toward this end include side chain modifications (e.g., replacement of the Phe residue by cyclohexyl alanine at the $P_1$ site) and backbone modifications (e.g., 1,2-diols as a replacement for the scissile

**ANGIOTENSIN CASCADE**

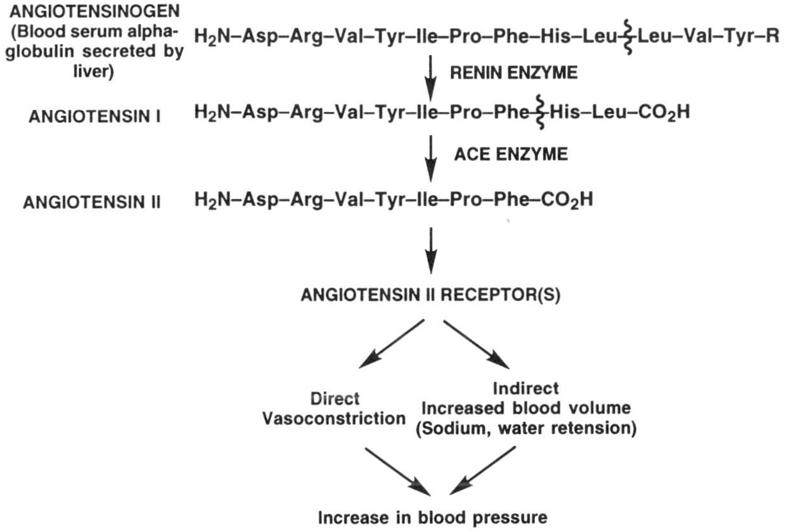

ANGIOTENSINOGEN (Blood serum alpha-globulin secreted by liver) H2N–Asp–Arg–Val–Tyr–Ile–Pro–Phe–His–Leu ~ Leu–Val–Tyr–R

↓ RENIN ENZYME

ANGIOTENSIN I H2N–Asp–Arg–Val–Tyr–Ile–Pro–Phe ~ His–Leu–CO2H

↓ ACE ENZYME

ANGIOTENSIN II H2N–Asp–Arg–Val–Tyr–Ile–Pro–Phe–CO2H

↓

ANGIOTENSIN II RECEPTOR(S)

Direct Vasoconstriction

Indirect Increased blood volume (Sodium, water retension)

Increase in blood pressure

peptide bond). Furthermore, it has been demonstrated that replacement and/or truncation of the subsites beyond the $P_3$ and $P'_2$ positions by nonpeptidic protecting groups (e.g., morpholino lactic acid) does not significantly alter their *in vitro* potencies. This has enabled the design of lower molecular weight renin inhibitors. Although these strategies have yielded renin inhibitors with subnanomolar potencies in *in vitro* assays, they continue to suffer from poor oral bioavailability. One inhibitor, A-72517, has been reported to have good oral availability in several animal models (Kleinert *et al.*, 1992; Rosenberg *et al.*, 1993).

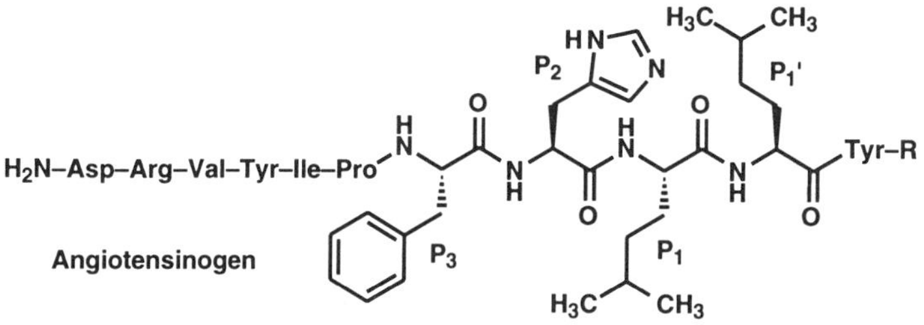

H2N–Asp–Arg–Val–Tyr–Ile–Pro– ... –Tyr–R

P3 P2 P1 P1'

Angiotensinogen

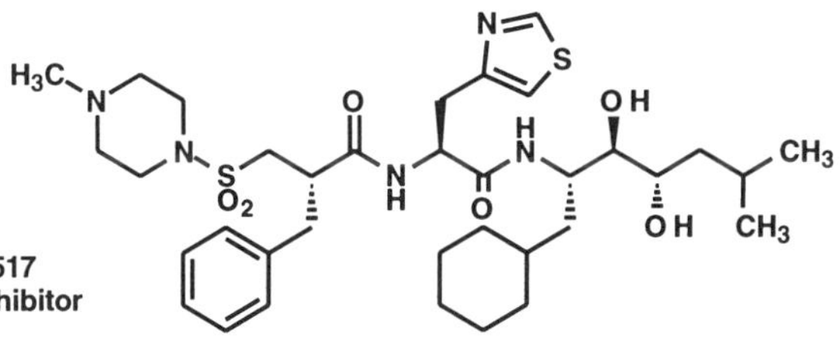

A-72517
Renin Inhibitor

When these synthetic programs started, X-ray crystallographic structures of neither human nor mouse submandibular renins were available. However, X-ray structures of several fungal aspartyl proteases that shared approximately 20% sequence homology with human renin had been solved and deposited in the Brookhaven Crystallographic Database. Consequently, extensive efforts by several research groups were undertaken to develop homology models of human renin and to apply these models to develop potent and selective inhibitors for renin. Unfortunately, most of the modeling efforts were not very successful and were primarily restricted to after the fact rationalization (Hutchins and Greer, 1991). This lack of success was attributable in part to inaccuracies in the homology models. For example, several of the models had predicted that the $S_3$ specificity pocket was larger in renin than other aspartyl proteases. This hypothesis was based on the substitution of smaller amino acid residues that line the $S_3$ pocket in human renin compared to the fungal aspartyl proteases. But subsequent X-ray crystallographic analysis revealed that while the registration of amino acid residues was correct in most of the homology models about this region, there were compensatory shifts in the position of the backbone not predicted by the models which resulted in a conservation of the size of this pocket (Dhanaraj *et al.*, 1992).

On the other hand, the models did provide useful qualitative guidance in developing new inhibitors for human renin. For example, a homology model developed at Merck predicted that the $P_2$ and $P_1'$ sidechains of bound inhibitors reside closely in space. Molecular mechanics optimization of prospective ligands in the presence of the rigidly held enzyme model suggested that 13- and 14-membered macrocycles would be tolerated by the binding site. It was hoped that the increase in rigidity afforded by the cyclization would result in an increase in potency as well as stability toward proteolytic enzymes.

**Merck Macrocyclic Renin Inhibitors**

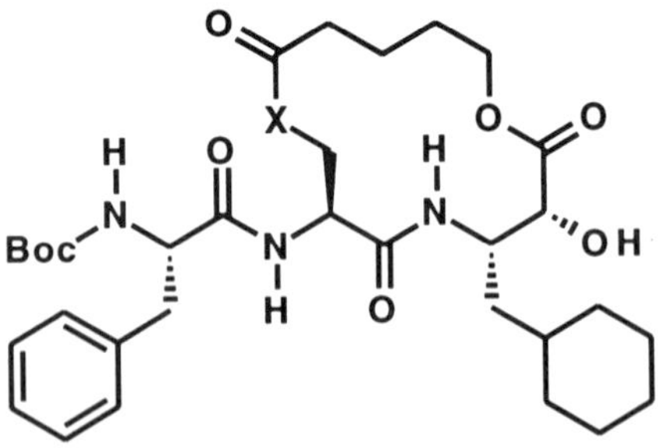

X = NH; $IC_{50}$ = 590 nM
X = O; $IC_{50}$ = 3.4 nM

Several of these suggestions were synthesized and were found to be potent inhibitors of renin (Weber *et al.*, 1992a). However, predictions of relative affinities based on the homology model and molecular mechanics interaction energies were not as successful. For example, the model predicted that replacement of a lactone with a lactam functionality in the macrocyclic ring would improve binding affinity due to a favorable hydrogen-bonding interaction with Asp-226 side chain carboxylate. However, contrary to the prediction, the *in vitro* assay demonstrated that the lactone was considerably more potent than the lactam analog (Weber *et al.*, 1992a). This discrepancy could result from errors in the homology model or the fact that solvation and entropy effects were not accounted for.

***b. HIV-1 protease*** The rapid spread of the AIDS epidemic (now a pandemic) and its putative causative agent, HIV-1 (the human immunodeficiency virus type 1), has triggered a global interest in the development of effective anti-HIV-1 therapies. The activity of the viral proteolytic enzyme HIV-1 protease, responsible for the processing of *gag* and *pol* gene products, has been demonstrated to be essential for the viral replication. Hence, selective inhibitors of this enzyme are regarded as promising drug candidates for AIDS therapy. HIV-1 protease, like human renin, is an aspartyl protease and functions as a $C_2$ symmetric homodimer, with each monomer consisting of 99 amino acids. One of its targets is the amide bond intervening the dipeptide sequence Phe/Tyr–Pro, which represents a unique cleavage sequence not shared by mammalian proteases and consequently represents an excellent template for the design of specific HIV-1 protease inhibitors. Over 100 structures of the dimer complexed with a variety of ligands have been solved using X-ray crystallographic techniques (Appelt, 1993).

In light of the similarity in the fundamental mechanism of proteolysis with human renin, many potent inhibitors of HIV-1 protease have been rapidly discovered since 1987 by applying the information obtained in the study of renin and other aspartyl proteases (Scharpe *et al.*, 1991). In particular, a number of transition state isosteres of the peptide (e.g., hydroxyethylene, hydroxyethylamine, reduced amide, $\alpha$-difluoroketone, and phosphinate) have been discovered to be potent HIV-1 protease inhibitors with varying degrees of selectivity vs mammalian aspartyl proteases.

Based on the unique symmetry and substrate preference of the enzyme, two classes of symmetrical inhibitors have been developed by workers at Abbott (Erickson *et al.*, 1990; Kempf *et al.*, 1990). At the time the design was carried out, the X-ray crystallographic structure of HIV-1 protease was unavailable. It was assumed, however, that the enzyme functioned as a homodimer because of the relatively short sequence of the enzyme, one-third as long as cellular aspartyl proteases. Subsequently, the crystal structure of the related RSV aspartyl protease confirmed that enzymes of this class function as homodimers. The inhibitor design began with the RSV protease crystal structure used as a model for HIV-1 protease (Erickson *et al.*, 1990). The structure of a reduced peptide inhibitor of rhizopuspepsin was then fitted into the active site of the RSV protease by superimposition of the catalytic sites of these two enzymes. Next, the P′ side of the docked inhibitor was truncated while the remaining portion of the inhibitor was replicated by applying the RSV protease $C_2$ crystallographic symmetry axis (see Fig. 9). The $P_1$ $\alpha$-carbon atoms were found to nearly superimpose after applying the symmetry operation. Deletion of the extra carbon atom and bond formation between the two halves led to the proposal of a diaminoalcohol structure. It should also be pointed out that the $C_2$ symmetric concept was also the starting point for the DuPont–Merck HIV protease cyclic urea inhibitor design (see Section II,C,2).

The diaminoalcohol derivative A-74702 was synthesized (Kempf *et al.*, 1990) and was shown to be a weak inhibitor of HIV-1 protease ($IC_{50}$ > 200 $\mu M$). However, it is known that substrate residues beyond the $P_1$ and $P_1'$ sites are critical for effective binding and cleavage. Therefore, an analog of A-74702 was prepared where both amino groups of the diaminoalcohol were acylated with Cbz-protected valine residues to yield A-74704 which displayed a five orders of magnitude increase in potency ($IC_{50}$ = 3 n$M$) relative to A-74702. X-ray crystallographic analysis of the A-74702/HIV-1 protease complex confirmed that the binding of the inhibitor is almost perfectly symmetrical (Erickson *et al.*, 1990).

The X-ray structure of the protease/A-74704 complex also revealed the existence of hydrogen bonds with poor geometry between the inhibitor $P_1$ and $P_1'$ amide nitrogen atoms and the enzyme carbonyl groups of Gly-27 and Gly-27′. Docking the glycol-containing $C_2$ symmetric inhibitor (Fig. 9) to HIV-1 protease suggested an improved intermolecular hydrogen-bonding geometry relative to that of the pseudosymmetric diaminoalcohol inhibitors.

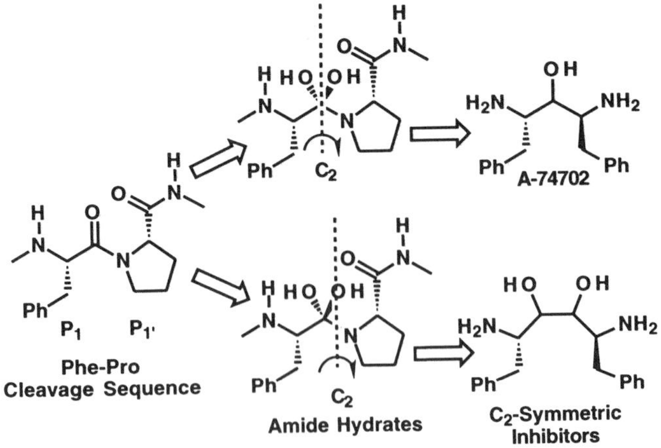

**FIGURE 9** Design of diaminoalcohol (top, A-74702) and diaminodialcohol (bottom) $C_2$ symmetric HIV-1 protease inhibitors (Erickson *et al.*, 1990; Kempf *et al.*, 1990).

These glycol inhibitor were synthesized (Kempf *et al.*, 1990) and were found to have inhibitor potencies that were approximately 10-fold greater than the corresponding diaminoalcohol inhibitors.

Both the Cbz-Val-blocked diaminoaclohol and diaminodialcohol inhibitors, however, suffer from poor solubility. Examination of the X-ray structure of the protease/A-74704 inhibitor complex revealed that the terminal phenyl rings of the Cbz groups protrude from the binding cleft to the exterior of the enzyme and therefore represent a functional group that could be modified to enhance solubility without seriously compromising potency. Based on this observation, a series of glycol-based inhibitors with modified Cbz functional groups were synthesized. One of these, A-77003, was shown to have high potency ($K_i$ = 0.15 n$M$) and good aqueous solubility ($10^4$ increase relative to A-74704) and was taken into human clinical trials (Kempf *et al.*, 1991).

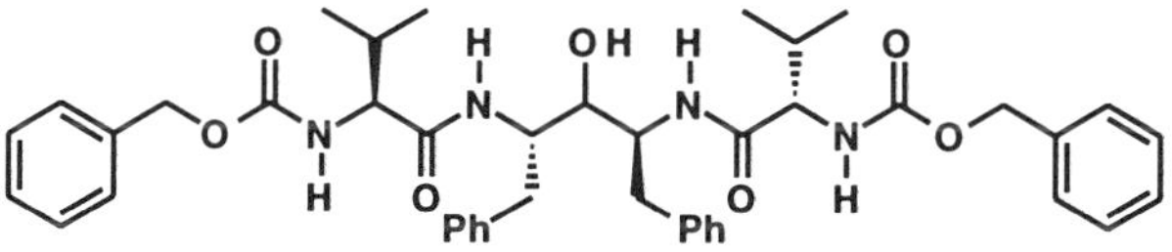

A-74704

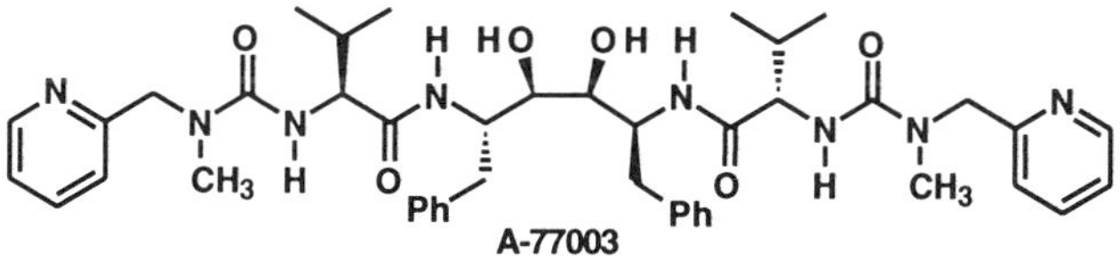

A-77003

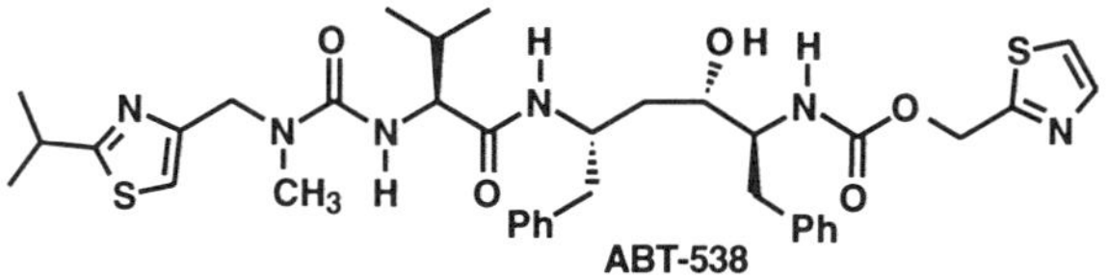

ABT-538

Potential disadvantages of A-77003 are its relatively high molecular mass (795 Da) and that it contains two peptide bonds; these features may limit its bioavailability. Indeed, the measured bioavailability of this compound in monkey is only 2.5% (Kempf, 1994). Further modification of the structure of A-77003 eventually led to the discovery of ABT-538 in which one amide and one hydroxyl group were eliminated accompanied by a reduction in the molecular mass (721 Da). Presumably because of the increased metabolic stability, the reduction in hydrogen-bonding potential, and the reduction in molecular mass, ABT-538 possesses a much improved oral bioavailability (70% in monkey) and is currently undergoing phase I/IIa human clinical trials (Kempf *et al.*, 1995). While treatment with ABT-538 results in a rapid decrease in HIV-1 plasma levels (Ho *et al.*, 1995), replacement of the wild-type virus by drug-resistant mutants occurs in a remarkably short time span of only 2 weeks (Wei *et al.*, 1995). This rapid emergence of resistance is due to the high turnover of virus and the relatively low fidelity of the HIV-1 reverse transcriptase. The molecular basis for HIV-1 resistance to ABT-538 has been shown to arise from Ile84Val and Val82Phe mutations in the viral protease (Markowitz *et al.*, 1995). Examination of a model-built drug-resistant mutant protease reveals that both mutations line the $S_1$ (and by symmetry the $S_1'$) pockets of the enzyme and therefore interact with the benzyl groups of ABT-538 (Markowitz *et al.*, 1995). The resistance conferred by the mutation at position 82 is very understandable due to a severe steric clash between the more bulky phenylalanine residue and the benzyl substituent of the inhibitor. However, mutation at position 84 where the isoleucine is replaced by a smaller valine residue actually confers more resistance. The explanation for these puzzling results may be due to movements in the backbone structure of the mutant enzyme.

### 2. Elastase Inhibitors

Human leukocyte elastase (HLE) is a serine protease that is released from leukocytes upon inflammatory stimulus. HLE is thought to aid in the migration of neutrophils to extravascular compartments through degradation of a number of structural proteins including elastin. Normally this enzyme is kept in check by indogenous inhibitors, most notably $\alpha$-1-proteinase ($\alpha$1-PI). However, $\alpha$1-PI may be damaged by cigarette smoke, or because of a genetic defect, produced in insufficient quantities. In either case, the balance between HLE and $\alpha$1-PI is upset which can lead to tissue damage which manifests itself in diseases such as emphysema, cystic fibrosis, and adult respiratory distress syndrome. Consequently there has been a concerted effort to develop low molecular weight inhibitors of HLE to treat these disease states (Edwards and Bernstein, 1994).

HLE is a trypsin-like serine protease that contains the characteristic Asp-102/His-57/Ser-195 catalytic triad (see Fig. 10). The side chain hydroxyl group of Ser-195 functions as a nucleophile which attacks the substrate amide carbonyl group. His-57 acts as a general base catalyst while the Asp-102 carboxylate anchors the His-57 in the proper orientation and tautomeric state. The reaction intermediate hemiketal alkoxide anion is stabilized through interaction with backbone NH groups of Gly-193 and Ser-195. This cavity is commonly referred to as the oxyanion hole.

HLE is a glycoprotein and because of the characteristic disorder of sugar moieties, crystallization and subsequent X-ray analysis of this enzyme have proven difficult (Bode *et al.*, 1989). Nonetheless, two crystallographic structures of HLE have been solved. The first (Brookhaven code 1PPF) is a complex between HLE and the turkey ovomucoid inhibitor (TOMI). It is believed that canonical inhibitors such as TOMI bind in a substrate-like manner. Examination of this crystal structure reveals that the oxyanion hole is occupied by the backbone carbonyl group of Leu-18I. Furthermore, the hydrophobic side chain of Leu-18I occupies what is commonly referred to

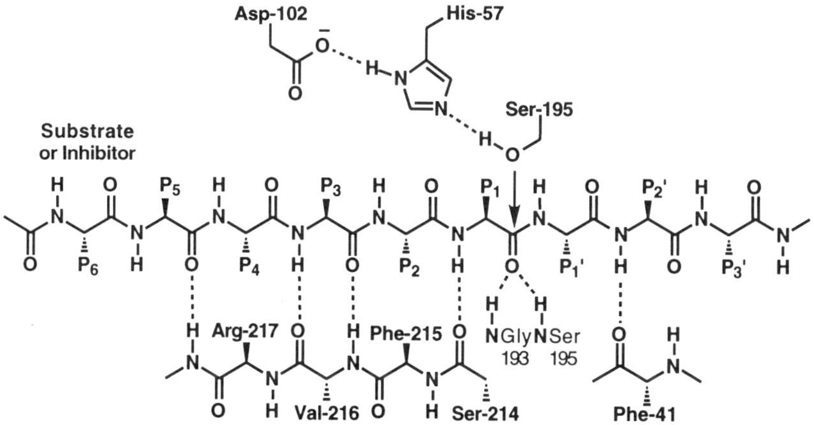

**FIGURE 10** Schematic diagram of the complex between human leukocyte elastase (HLE) with a peptide substrate or an inhibitor. The arrow represents the attack of the catalytic Ser-195 on the scissile amide bond between the $P_1$ and $P_1'$ residues of the substrate. Backbone amide NH groups from Gly-193 and Ser-195 stabilize the formation of the alkoxide anion from the carbonyl group of the amide bond undergoing cleavage. The binding of the substrate to the enzyme is stabilized in part by an antiparallel $\beta$-sheet interaction between the $P_1$–$P_5$ residues of the substrate and residues 214–218 of the enzyme.

as the $S_1$ specificity pocket consistent with the preference of HLE for substrates with a moderate-sized hydrophobic side chain in the first residue position on the carboxyl side of the scissile bone ($P_1$ position). The $P_1$–$P_5$ residues of TOMI form an antiparallel $\beta$-sheet with residues 214–218 of HLE. The second crystal structure of HLE is a complex with a covalent tetrapeptide chloromethylketone (Brookhaven code 1HNE). The peptide portion of this inhibitor binds in much the same manner as TOMI.

In contrast to HLE, porcine pancreatic elastase (PPE) readily yields crystals suitable for crystallographic analysis. The two enzymes share 40% sequence similarity and share a high degree of structural similarity, especially in the active site. Consequently, crystal structures of inhibitors complexed with PPE are often used as a surrogate for HLE.

A wide variety of inhibitor structural classes effective against other serine proteases have also been tried against HLE (Edwards and Bernstein, 1994). These are, for the most part, mechanism-based covalent inhibitors. Examples of reversible covalent inhibitors that are effective against HLE include trifluoromethylketones, boronic acids, and acyl heterocyclic inhibitors. Irreversible covalent inhibitors for HLE include chloromethyl ketones, $\beta$-lactams, and isocoumarins. These inhibitors typically contain a peptide portion that associates with the P-side of the binding cleft and a "warhead" which forms a covalent bond(s) with the catalytic Ser-195 and/or His-57. While peptidic mechanism-based inhibitors of HLE show potent *in vitro* activity, these inhibitors in general lack oral activity. Consequently there has been a great deal of effort in developing nonpeptidic inhibitors of HLE.

One particularly instructive example of the design and development of orally active nonpeptidic inhibitors of HLE where molecular modeling played an instrumental role was carried out by workers at Zeneca (Brown *et al.*, 1994). The starting point for this design was a peptidic trifluoromethylketone (TFMK) reversible covalent inhibitor ICI-200,880 (7). While this inhibitor showed potent *in vitro* activity and was active in the lung when administered as an aerosol, it was totally inactive orally. Hence an effort was initiated to find a potent and selective TFMK inhibitor of HLE with a reduced peptidic character. An additional complication with peptidic inhibitors such as ICI-200,880 is that the stereochemical center adjacent to the TFMK is readily epimerized. This leads to a mixture of two diastereomers which complicates the synthesis and pharmacokinetic evaluation of these inhibitors. An additional goal was therefore to eliminate the remainder of the stereochemical centers so that the diastereomeric mixture is replaced by a more tractable mixture of enantiomers.

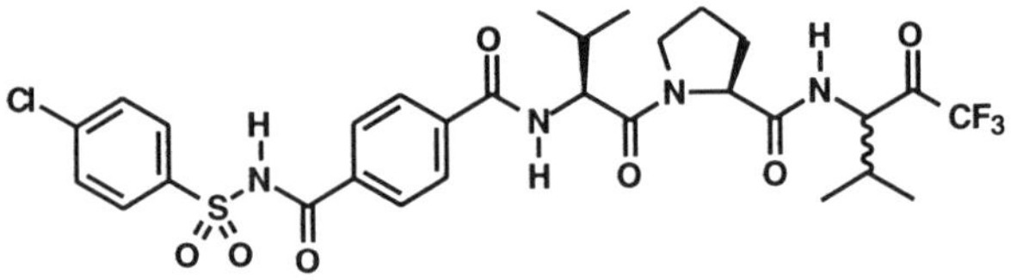

**ICI-200,880 (7)**

Since a crystallographic structure of HLE complexed with TFMK was unavailable, a model-built structure was constructed using the crystal structure of HLE complexed with TOMI (1PPF) and the structure of PPE complexed with Ac-Ala-Pro-Val-TFMK (Warner *et al.*, 1994). The structurally conserved regions of these two enzymes were overlaid and TOMI from the 1PPF structure was deleted as was PPE from the Ac-Ala-Pro-Val-TFMK complex. This left the TFMK inhibitor docked to HLE. A covalent bond was added between the catalytic Ser-195 of HLE and the electrophilic ketone of the TFMK inhibitor, and the alkoxide anion was modeled as a fluorine atom with a formal charge of zero. The resulting structure was subjected to two rounds of energy minimization. In the first round, the enzyme was held rigid while the inhibitor structure was optimized and in a second round, limited parts of the enzyme in the immediate vicinity of the enzyme were optimized.

As observed in the structure of PPE complexed with a TFMK inhibitor, a key interaction is the pair of hydrogen bonds between the $P_3$ alanine in the inhibitor with Val-216 of the enzyme as evidenced by the short intermolecular N····O distances and the low temperature factors of the atoms involved in these hydrogen-bonding interactions. In addition, these hydrogen bonds are well shielded from solvent consistent with the strength of the hydrogen bonds inferred from the crystallographic parameters. It was envisaged that bond formation between the $P_2$ and $P_3$ sidechains and bond cleavage of the $P_2$ proline moiety would lead to a pyridone structure (shown below). This transformation immediately eliminates two stereochemical centers. In order to test the compatibility of the pyridone structure with the binding cleft of HLE, the inhibitor contained within the model-built HLE/TFMK complex was mutated to the pyridone analog and the resulting structure was subjected to energy refinement. After minimization, it was found that all the key interactions between the candidate pyridone inhibitor and the enzyme were maintained with the exception of the hydrophobic contact between the $P_2$ proline side chain and the $S_2$ pocket of the enzyme. Furthermore, exhaustive conformational analysis of the inhibitor revealed that one of the low energy conformations corresponded closely with the modeled bound conformation. Hence the pyridone inhibitor appeared a viable candidate for synthesis.

8 $K_i = 210 \pm 80$ nM

9 $K_i = 2800 \pm 400$ nM

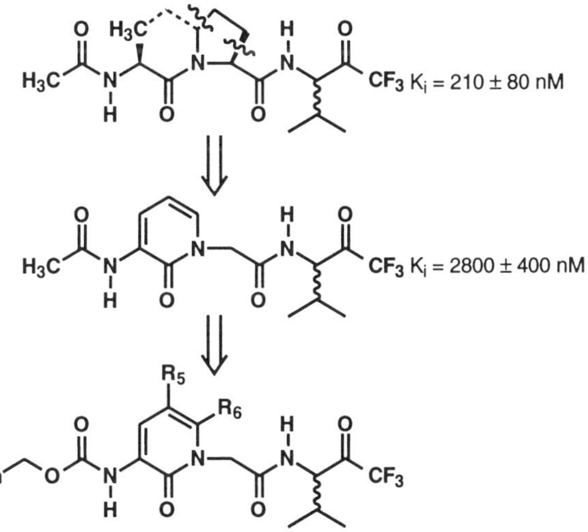

| compd | R5 | R6 | Ki (nM) |
|---|---|---|---|
| **10** | H | H | 280 ± 78 |
| **11** | $CH_2Ph$ | H | 40 ± 9 |
| **12** | H | Ph | 4.5 ± 0.8 |

Gratifyingly, it was found that the 3-N-acetate derivative **9** showed promising activity against HLE ($K_i = 2800 \pm 400$ n*M*) while the corresponding Cbz-protected derivative **10** was an order of magnitude more potent ($K_i = 280 \pm 78$ n*M*). As stated earlier, molecular modeling revealed that the pyridone ring could not engage in favorable hydrophobic interaction with the $S_2$ subsite whereas substrate kinetic evidence shows that the $P_2$ interactions contribute significantly to substrate $K_m$. Further molecular modeling revealed that either of two low energy conformations of a 5-benzyl substituent attached to the pyridone ring could interact favorably with the $S_2$ pocket of the enzyme. Similarly, an aryl group at position 6 could also interact favorably with the $S_2$ pocket. Addition of the 5-benzyl moiety (Warner *et al.*, 1994) resulted in analog **11** with a sevenfold increase in potency ($K_i$ = 40 ± 9 n*M*). The 6-phenyl analog **12** (Bernstein *et al.*, 1994) was even more potent ($K_i$ = 4.5 ± 0.8 n*M*). Subsequent crystallographic analysis (Veale *et al.*, 1995) of a complex formed between PPE and a methylsulfonamide derivative **13** confirmed the predicted binding mode (Fig. 11).

While the *in vitro* potency of the 6-phenyl analog **12** was very promising, this compound was nevertheless devoid of oral activity. This lack of oral activity may be a result of low aqueous solubility and high log *P* ($>$ 4.0) which may facilitate plasma protein binding. Consistent with this hypothesis, the *in vitro* affinity of the inhibitor for HLE was markedly decreased in the presence of human serum albumin. Hence work was initiated to synthesize more polar derivatives.

MD simulations in the presence of solvent show the Cbz group of the TFMK inhibitor to be quite mobile when bound to HLE (Bernstein *et al.*, 1994). This result suggested that the Cbz group does not contribute significantly to affinity and therefore is an excellent candidate for modification with the goal of decreasing the overall lipophilicity of the inhibitor. Further modeling suggested that the binding of the pyridone inhibitors with amide or carbamate groups at position 3 of the pyridone ring bound somewhat differently compared to peptidic inhibitors. Because of the rigidity of the pyridone ring 3-carbamate group caused by conjugation between these two functionalities, occupation of the $S_4$ pocket of the enzyme by substituents attached to position 3 of the pyridone is difficult. Instead, these substituents were directed toward a turn region consisting of Gly-218. Sulfonamides at position 3 of the pyridone ring are somewhat more flexible compared to the carbamate analogs, therefore interaction with the $S_4$ pocket becomes feasible. Furthermore, the modeling suggested that a *p*-*N*-acetylphenylsulfonamide derivative **14** would form a favorable hydrogen-bonding interaction with Gly-218. Aqueous MD simulation showed that this hydrogen bond was stable. An additional advantage of the sulfonamide compared to the carbamate is that the former is considerably more hydrophilic compared to the latter and therefore the sulfonamide analogs should be less susceptible to plasma protein binding. Experimentally, inclusion of the *p*-*N*-acetylphenylsulfonamide substituent resulted in a sixfold increase in potency relative to the Cbz analog **12**.

Unfortunately the *p*-*N*-acetylphenylsulfonamide analog still lacked significant oral activity. While the *p*-*N*-acetylphenylsulfonamide group im-

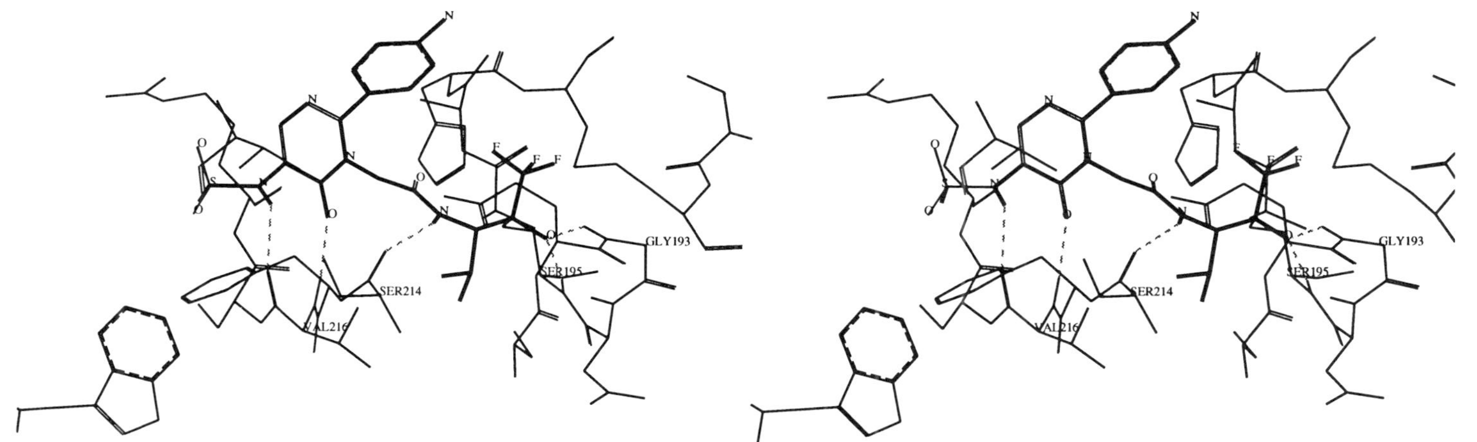


**FIGURE 11** Stereoview of the crystallographic structure [Brookhaven code 1EAT (Veale *et al.*, 1995)] of the complex between porcine pancreatic elastase (thin line) and trifluoromethylketone inhibitor **13** (thick line). Key hydrogen-bonding interactions between the enzyme and the inhibitor are represented by dashed lines. The oxyanion hole is formed by backbone amide hydrogen atoms of Gly-193 and Ser-195 while Ser-214 and Val-216 engage in a $\beta$-sheet-like interaction with the inhibitor.

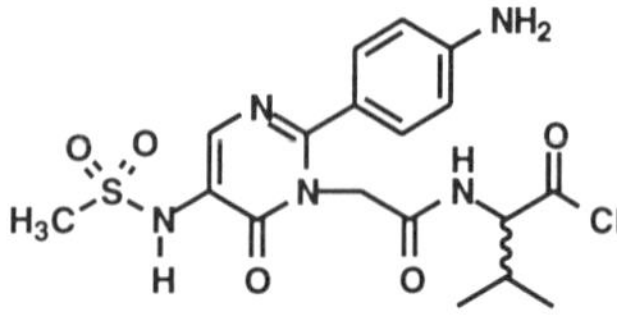

13; $K_i = 15 \pm 1.6$ nM

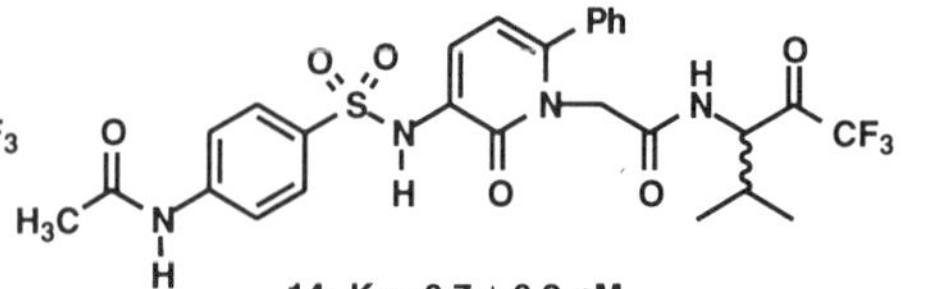

14; $K_i = 0.7 \pm 0.2$ nM

proved *in vitro* activity, it is a rather heavy substituent which increased the overall molecular mass to 593 Da. At this relatively high molecular mass, decreases in intestinal absorption and increases in hepatic clearance become significant (Hamilton *et al.*, 1995). Therefore, a number of lighter substituents at position 3 of the pyridone ring (e.g., formyl) were tried. While these lower molecular mass derivatives displayed a one to two order of magnitude reduction of *in vitro* activity, several of them showed promising *in vivo* activity. Finally, in an attempt to reduce the log *P* further, replacement of the pyridone ring with the significantly more hydrophilic pyrimidone resulted in a decrease in log *P* by greater than two orders of magnitude. Furthermore, several of the pyrimidone analogs showed excellent oral activity (Veale *et al.*, 1995). One of these, a 2-(4′-fluorophenyl)-5-aminopyrimidone derivative **15**, possessed only moderate *in vitro* activity ($K_i = 101 \pm 10$ n*M*). However, this derivative also showed the best combination of oral activity, duration of action, and selectivity against other mammalian proteases. The excellent oral activity is in part likely due to the relatively low molecular mass of this analog (400 Da).

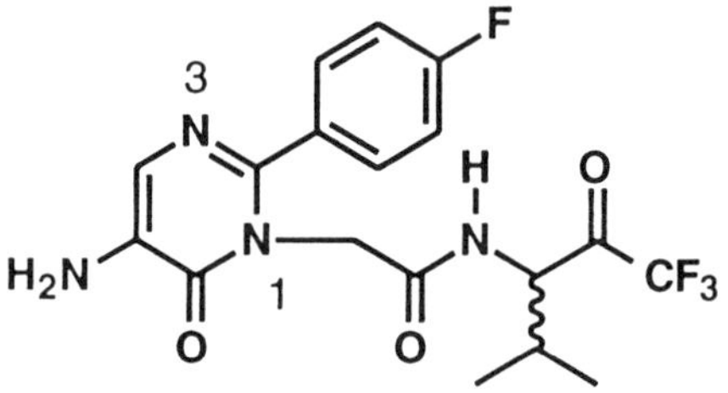

15; $K_i = 101 \pm 10$ nM

In summary, this study demonstrates the utility of model-built protein–inhibitor complexes in the design and optimization of a series of enzyme inhibitors starting with a peptidic lead which ultimately led to the discovery of potent, selective, and orally active, nonpeptidic inhibitors. *In vacuo* molecular mechanics optimization of inhibitor/enzyme complexes was used to quickly evaluate candidate inhibitors and to rule out analogs that engaged in sterically forbidden interactions with the enzyme. Molecular dynamics simulations of the surviving inhibitors in the presence of explicit water molecules then provided qualitatively correct predictions of the relative affinities of closely related analogs. While the modeling-directed optimization led to inhibitors with excellent *in vitro* activity, these derivatives lacked oral activity. In a final step, the optimized derivatives were modified to decrease molecular weight and lipophilicity. Several of the resultant analogs, while possessing inferior *in vitro* activity, nonetheless displayed excellent oral activity and duration of action.

### 3. RDG Mimics as Antithrombotic Agents

The tripeptide sequence Arg–Gly–Asp (RGD) plays a critical role in mediating interactions between adhesion proteins such as fibrinogen and their glycoprotein receptor complexes expressed on the surfaces of activated platelets. Specifically, interactions between RGD in fibrinogen and the glycoprotein complex GPIIb/IIIa are implicated in the thrombus formation and subsequent myocardial infarction. The guanidinium and the carboxylate moieties in the side chains of arginine and aspartate, respectively, are generally considered essential to biological activity. In this light, GPIIb/IIIa antagonists (based on the structures of RGD) are antithrombotic agents with considerable potential for cardiovascular therapy.

Since 1985, thousands of RGD mimetics have been synthesized and tested for their biological activity (measured in terms of *in vivo* and *in vitro* potency to block platelet aggregation when the latter are stimulated by a variety of activating factors). Generally, the most significant biological activity is seen in compounds with conformational restrictions which lead to a separation of at least 14 Å between the two ionic groups of R and D (see Table III). While RGD is conformationally very flexible due to a large number of single bonds about which hindered rotations are possible, crystal structure and NMR studies on peptides containing RGD do find distal separation of the charged groups. Furthermore, a molecular mechanics analysis of the most frequently observed families of conformations adopted by the RGD-like sequence in folded proteins reveals that structures with distal separations are energetically distinct minima from those where the two groups form an intramolecular salt bridge (Rao, 1992). Potent, semirigid nonpeptidic RDG mimics have a distance separating their charged centers that corresponds to the extended conformer family obtained from the X-ray crystallographic/molecular mechanics analysis which supports the contention that this conformation is the active one.

Replacement of guanidine (in arginine of RGD) by benzamidine and piperidine represents two of the strategies in reducing conformational flexibility while simultaneously boosting potency (Hartman *et al.*, 1992; Ku *et al.*, 1993; Eldred *et al.*, 1994; Hoekstra *et al.*, 1995; Ku *et al.*, 1995; Zablocki *et al.*, 1995b). These moieties have been incorporated into the design of numerous orally active GPIIb/IIIa antagonists along with conformationally constrained mimics for the aspartate of the RGD sequence. For example, the structure of an initial RGD-containing cyclic peptide lead (SK&F

**TABLE III** Structures, Potencies, and Minimum Distance ($D_{min}$) between Charged Carboxylate and Guanidinium Moieties of RDG Mimics[a]

| Structure | $IC_{50}$(n$M$) | $D_{min}$ (Å) |
|---|---|---|
| 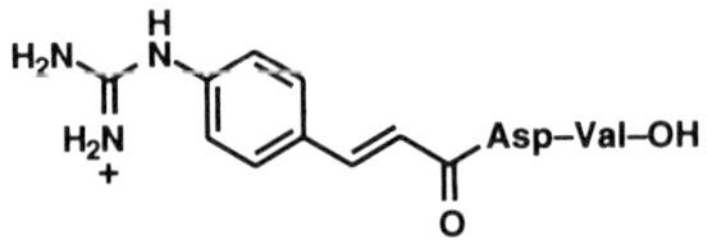 H<br>H₂N N<br>H₂N<br>+<br>Asp–Val–OH<br>O | 2000 | 8.2 |
| 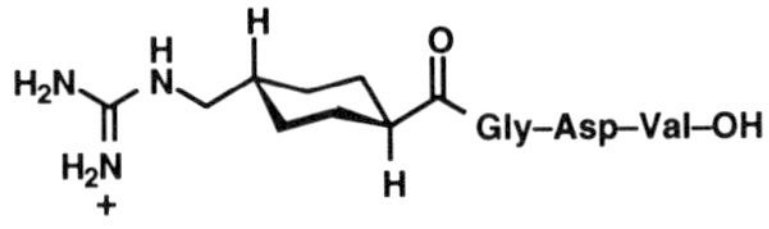 H<br>H₂N N<br>H₂N<br>+<br>H<br>O<br>Gly–Asp–Val–OH<br>H | 10 | 11.0 |
| 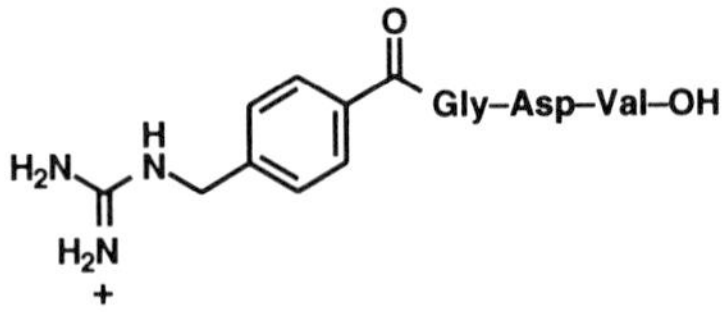 O<br>Gly–Asp–Val–OH<br>H<br>H₂N N<br>H₂N<br>+ | 9 | 10.2 |
| 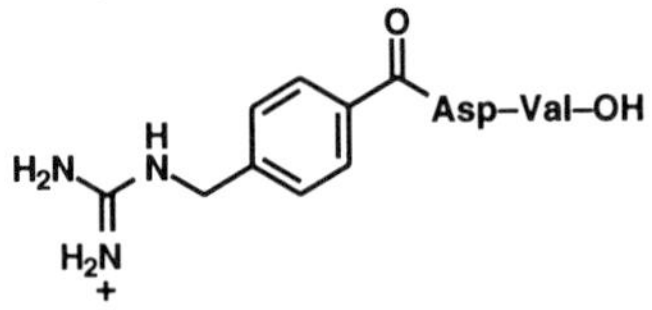 O<br>Asp–Val–OH<br>H<br>H₂N N<br>H₂N<br>+ | 900 | 4.9 |
| 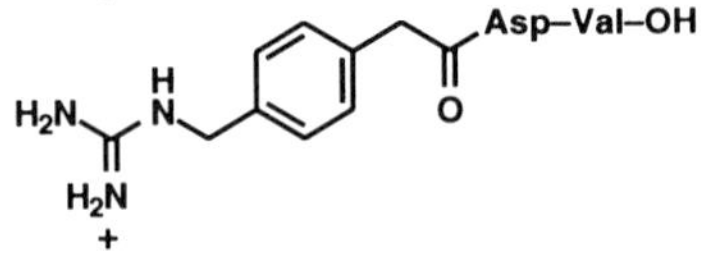 Asp–Val–OH<br>H<br>H₂N N<br>O<br>H₂N<br>+ | 2100 | 5.7 |

[a] From Rao (1992).

107260) (Ali *et al.*, 1994) and the NMR observation of a $\gamma$-turn conformation in solution in the aspartate end of the molecule led to the development of a benzodiazepine- and aminobenzamidine-containing compound (SB 207448) at SmithKline and Beecham (Ku *et al.*, 1993, 1995). While the cyclic peptide was weakly active as a platelet aggregation inhibitor in the dog PRP assay ($IC_{50} \approx 16\ \mu M$), the nonpeptidic analog with a separation of at least 14 Å between the ionic centers was considerably more potent ($IC_{50} \approx 0.15\ \mu M$). Subsequent refinement of this compound led to an orally active (dog PRP $IC_{50} \approx 0.028\ \mu M$) compound (SB 214857) that is presently undergoing clinical trials.

Database screening at Merck based on a simple pharmacophore consisting of a separation of 10–20 Å between the cationic and anionic centers led to a tyrosine-containing compound (compound 6 in Hartman *et al.*, 1992). Extensive SAR development around this lead resulted in the highly

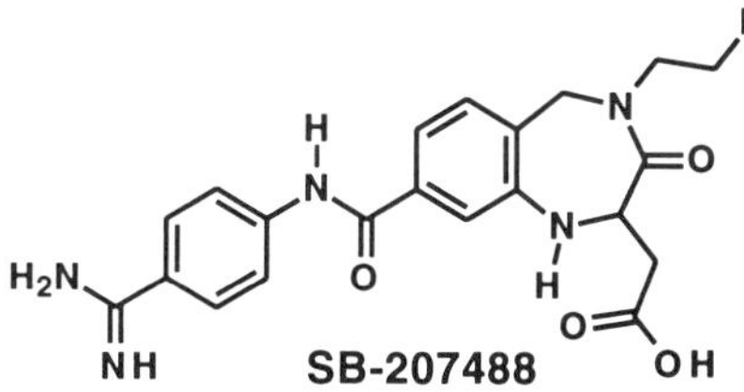

SB-207488 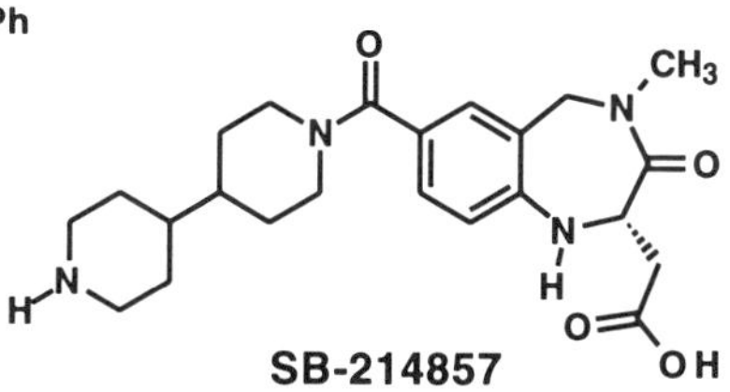 SB-214857

potent clinical candidates L-700,462 and L-703,014. More recently, nonpeptide mimics of RGD have been designed on the basis of the solution structure of the C-terminal $\gamma$-chain dodecapeptide from fibrinogen (Hoekstra *et al.*, 1995). The design was based on the type II $\beta$ turn observed for the Lys-Gln-Ala-Gly sequence at the C terminus in the solution NMR studies. The $\beta$ turn was locked in place by a conformationally constrained ring. The design of potent and orally active nonpeptide GPIIb/IIIa antagonists has clearly demonstrated in contradistinction to previously held dogma that the central glycine residue in RGD *may* be replaced by a wide variety of hydrophobic and hydrophilic spacers with retention of *in vitro* and *in vivo* activity as well as oral bioavailability (Blackburn and Gadek, 1993; Zablocki *et al.*, 1995a).

The development of RGD analogs with benzamidine and piperidine as antithrombotic agents (acting via inhibition of fibrinogen binding to GPIIb/IIIa receptors) represents one of the major success stories in the design of peptidomimetics with good oral bioavailability. This can be attributed to a number of factors, not the least of which is the design of conformationally restricted segments of the RGD peptide, which enhance *in vitro* and *in vivo* potency. It may be argued that this result is fortuitous in light of the fact that the three-dimensional structure of the GPIIb/IIIa receptor is unknown. Nevertheless, designs based on the assumption of an extended conformation were ultimately successful. Enhancement of the oral bioavailability through variations in the chemical functionalities in the carboxylate region was another strategy that contributed to the successful design of GPIIb/IIIa antagonists as antithrombotics. Whether these strategies will bear fruit in the design of RGD analogs to block the action of other integrins remains to be seen. It should be pointed out that this success story is in marked contrast to the area of peptide analgesics where the design and the development of an efficacious compound have eluded CNS medicinal chemistry efforts for several decades. The inability of peptides to cross the blood–brain barrier is considered to be one of the major stumbling blocks in the design of peptidomimetic analgesics (Morgan and Gainor, 1989).

### D. Prospects for the Future

It is clear from the numerous examples presented in this chapter that molecular modeling has finally begun to live up to its long held promise and has started to contribute in measurable ways to the drug discovery process. Many of the compounds in which molecular modeling has played a critical role in the discovery and optimization are well into human clinical trials and are nearing the market place. This is especially true in cases where explicit crystallographic or NMR structures of the drug target are available.

A number of other new technologies have also started to contribute to drug discovery. Foremost among these are synthetic combinatorial libraries (Gallop *et al.*, 1994; Gordon *et al.*, 1994). Coupled with sensitive, high capacity biological assays, these libraries promise a vastly expanded collection of structurally diverse leads. In addition, the closely related technique of rapid analog synthesis promises to greatly accelerate the optimization of these lead structures. However, discovery and optimization of novel leads are precisely the same goals as molecular modeling. Certainly synthetic combinatorial libraries are a great advance for drug discovery, but will they make molecular modeling irrelevant?

Drugs are unlikely to be found directly from screening. Leads need to be optimized and while combinatorial chemistry can help optimize a lead, design will retain an essential role in optimization. Furthermore, the types of chemistries available from combinatorial libraries will be somewhat limited for the foreseeable future. *De novo* design software is already capable of producing an astonishingly large diversity of structures approaching that of all possible structures compatible with a given receptor (Bohacek and McMartin, 1994). These computer-generated molecular repertoires greatly exceed the diversity available from synthetic combinatorial libraries.

Furthermore, the techniques of molecular modeling and combinatorial chemistry may complement one another. While combinatorial chemistry can increase the number of "needles," it also increases the size of the "haystack." Various chemometric statistical techniques may be used to design libraries that maximize diversity and hence the number of structurally diverse leads discovered (Martin *et al.*, 1995). Furthermore, these techniques may be of assistance in the creation of "targeted" libraries which are likely to have a much higher percentage of active compounds relative to a random library (Zuckermann *et al.*, 1994; Sheridan and Kearsley, 1995). In principle, automated docking coupled with an accurate scoring function should be able to identify the most potent compounds within a combinatorial library prior to synthesis. However, given the computational expense and limited accuracy of current prediction methods, computational screening is no substitute for the experimental combinatorial approach.

Clearly, computer-aided drug design has been most successful when an explicit 3D structure of the biomolecular target is available. Crystallization and structure determination of soluble proteins have become almost routine. Furthermore, sequences arising from the human genome project (Fasman *et al.*, 1994) coupled with homology modeling will provide models for many of the remaining soluble receptors. However, receptor sites are often lined by the loop regions of proteins. Since the prediction of the structures of these loops is extremely difficult, accounting for flexible receptors (in the absence of multiple experimental structures of the protein bound to different ligands) will remain an unsolved problem for the foreseeable future.

Even with an explicit receptor structure, it is still difficult to predict precisely what modifications to a ligand will enhance affinity or to predict what novel structures selected from a database will bind to the receptor with high affinity. Accurate predictions of binding affinity require an accurate account of the effects of solvation and entropy, but these remain only partially solved problems since the present approaches are computationally very expensive or possess limited accuracy. Theoretical advances and faster computers will help. Furthermore, high precision microcalorimetry devices have become widely available (Connelly, 1994). These new instruments will lead to the generation of a much larger data set of enthalpic and entropic contributions to ligand affinity which will in turn fuel theoretical advances in predicting free energies of binding.

Experimentally determined structures and, as a consequence, homology models for many important targets, especially membrane-bound receptors, will be unavailable in the foreseeable future. High throughput screening of combinatorial libraries will provide greatly expanded SAR. Furthermore, site-directed mutagenesis–SAR–modeling synergy may provide more detailed information for these receptors with unsolved structures (Strader *et al.*, 1994, 1995). This combined information may be used to construct much more accurate pharmacophore, pseudoreceptor, or homology models.

Many challenges and opportunities in drug discovery remain, and it is difficult to foresee exactly what the future holds. However, it is certain that molecular modeling will continue to play an instrumental role in this evolution.

## ACKNOWLEDGMENTS

The authors thank Gastone Gilli, Victor Matassa, and Bingze Wang for helpful discussions.

## REFERENCES

Abdel Meguid, S. S. (1993). Inhibitors of aspartyl proteinases. *Med. Res. Rev.* **13**, 731–778.

Abola, E. E., Bernstein, F. C., Bryant, S. H., Koetzle, T. F., and Weng, J. (1987). Protein Data Bank. *In* "Crystallographic Databases: Information Content, Software Systems, Scientific Applications" (F. H. Allen, G. Bergerhoff, and R. Sievers, eds.), pp. 107–132. Data Commission of the International Union of Crystallography, Bonn/Cambridge/Chester.

Abraham, D. J., Perutz, M. F., and Phillips, S. E. (1983). Physiological and X-ray studies of potential antisickling agents. *Proc. Natl. Acad. Sci. USA* **80**, 324–328.

Ali, F. E., Bennett, D. B., Calvo, R. R., Elliott, J. D., Hwang, S. M., Ku, T. W., Lago, M. A.,

Nichols, A. J., Romoff, T. T., Shah, D. H., and Samanen, J. M. (1994). Conformationally constrained peptides and semipeptides derived from RGD as potent inhibitors of the platelet fibrinogen receptor and platelet aggregation. *J. Med. Chem.* **37**, 769–780.

Allen, F. H., Bellard, S., Brice, M. D., Cartwright, B. A., Doubleday, A. Higgs, H., Hummelink, T., Hummellink-Peters, B. G., Kennard, O., Motherwell, W. D. S., Rodgers, J. R., and Watson, D. G. (1979). The Cambridge Crystal Data Centre: Computer-based search, retrieval, analysis, and display of information. *Acta Crystallogr.* **B35**, 2331–2339.

Allen, M. S., LaLoggia, A. J., Dorn, L. J., Martin, M. J., Costantino, G., Hagen, T. J., Koehler, K. F., Skolnick, P., and Cook, J. M. (1992). Predictive binding of $\beta$-carboline inverse agonists and antagonists via the CoMFA/GOLPE approach. *J. Med. Chem.* **35**, 4001–4010.

Andrews, P. R. (1993). Drug–receptor interactions. *In* "3D QSAR in Drug Design" (H. Kubinyi, ed.), pp. 13–40. ESCOM, Leiden.

Andrews, P. R., Craik, D. J., and Martin, J. L. (1984). Functional group contributions to drug–receptor interactions. *J. Med. Chem.* **27**, 1648–1657.

Appelt, K. (1993). Crystal structures of HIV-protease–inhibitor complexes. *Perspect. Drug Discov. Design* **1**, 23–48.

Appelt, K., Bacquet, R. J., Bartlett, C. A., Booth, C. L., Freer, S. T., Fuhry, M. A., Gehring, M. R., Herrmann, S. M., Howland, E. F., Janson, C. A., Jones, T. R., Kan, C.-C., Kathardekar, V., Lewis, K. K., Marzoni, G. P., Matthews, D. A., Mohr, C., Moomaw, E. W., Morse, C. A., Oatley, S. J., Ogden, R. C., Reddy, M. R., Reich, S. H., Schoettlin, W. S., Smith, W. W., Varney, M. D., Villafranca, J. E., Ward, R. W., Webber, S., Webber, S. E., Welsh, K. M., and White, J. (1991). Design of enzyme inhibitors using iteractive protein crystallographic analysis. *J. Med. Chem.* **34**, 1925–1934.

Åqvist, J., Medina, C., and Samuelsson, J.-E. (1994). A new method for predicting binding affinity in computer-aided drug design. *Prot. Engin.* **7**, 385–391.

Åqvist, J., and Mowbray, S. L. (1995). Sugar recognition by a glucose/galactose receptor: Evaluation of binding energetics from molecular dynamics simulations. *J. Biol. Chem.* **270**, 9978–9981.

Ariëns, E. J. (1987). Stereochemistry in the analysis of drug action. *Med. Res. Rev.* **7**, 367–387.

Arnold, G. E., and Ornstein, R. L. (1994). An evaluation of implicit and explicit solvent model systems for the molecular dynamics simulation of bacteriophage T4 lysozyme. *Proteins* **18**, 19–33.

Badger, J., Minor, I., Kremer, M. J., Oliveira, M. A., Smith, T. J., Griffith, J. P., Guerin, D. M., Krishnaswamy, S., Luo, M., Rossmann, M. G., McKinlay, M. A., Diana, G. D., Dutko, F. J., Fancher, M., Rueckert, R. R., and Heinz, B. A. (1988). Structural analysis of a series of antiviral agents complexed with human rhinovirus 14. *Proc. Natl. Acad. Sci. USA* **85**, 3304–3308.

Bai, Y., and Englander, S. W. (1994). Hydrogen bond strength and beta-sheet propensities: The role of a side chain blocking effect. *Proteins* **18**, 262–266.

Bairoch, A., and Boeckmann, B. (1994). The SWISS-PROT protein sequence data bank: Current status. *Nucleic Acids Res.* **22**, 3578–3580.

Bajorath, J., Stenkamp, R., and Aruffo, A. (1993). Knowledge-based model building of proteins: Concepts and examples. *Prot. Sci.* **2**, 1798–1810.

Balaji, V. N., Ramnarayan, K., Chan, M. F., and Rao, S. N. (1994). Conformational studies on model peptides with 1-aminocyclopropane 1-carboxylic acid residues. *Peptide Res.* **7**, 60–71.

Banner, D. W., and Hadvary, P. (1991). Crystallographic analysis at 3.0-Å resolution of the binding to human thrombin of four active site-directed inhibitors. *J. Biol. Chem.* **266**, 20085–20093.

Barlow, D. J., and Thornton, J. M. (1983). Ion-pairs in proteins. *J. Mol. Biol.* **168**, 867–885.

Bartlett, P. A., and Marlowe, C. K. (1983). Phosphonamidates as transition-state analogue inhibitors of thermolysin. *Biochemistry* **22,** 4618–4624.

Bartlett, P. A., and Marlowe, C. K. (1987). Evaluation of intrinsic binding energy from a hydrogen binding group in an enzyme inhibitor. *Science* **235,** 569–571.

Bartlett, P. A., Sampson, N. S., Reich, S. H., Drewry, D. H., and Lamden L. A. (1990). Interplay among enzyme mechanism, protein structure, and the design of serine protease inhibitors. *In* "Use of X-Ray Crystallography in the Design of Antiviral Agents (W. G. Lauer and G. M. Air, eds.), pp. 247–259. Academic Press, New York.

Bash, P. A., Singh, U. C., Brown, F. K., Langridge, R., and Kollman, P. A. (1987). Calculation of the relative change in binding free energy of a protein–inhibitor complex. *Science* **235,** 574–576.

Ben-Naim, A., and Mazo, R. M. (1993). Size dependence of the solvation free-energies of large solutes. *J. Phys. Chem.* **97,** 10829–10834.

Benson, D. A., Boguski, M., Lipman, D. J., and Ostell, J. (1994). GenBank. *Nucleic Acids Res.* **22,** 3441–3444.

Bernstein, F. C., Koetzle, T. F., Williams, G. J., Meyer, E. E., Jr., Brice, M. D., Rodgers, J. R., Kennard, O., Shimanouchi, T., and Tasumi, M. (1977). The Protein Data Bank: A computer-based archival file for macromolecular structures. *J. Mol. Biol.* **112,** 535–542.

Bernstein, P. R., Andisik, D., Bradley, P. K., Bryant, C. B., Ceccarelli, C., Damewood, J. R., Jr., Earley, R., Edwards, P. D., Feeney, S., Gomes, B. C., Kosmider, B. J., Steelman, G. B., Thomas, R. M., Vacek, E. P., Veale, C. A., Williams, J. C., Wolanin, D. J., and Woolson, S. A. (1994). Nonpeptidic inhibitors of human leukocyte elastase. 3. Design, synthesis, X-ray crystallographic analysis, and structure-activity relationships for a series of orally active 3-amino-6-phenylpyridin-2-one trifluoromethyl ketones. *J. Med. Chem.* **37,** 3313–3326.

Berridge, M. J. (1987). Inositol trisphosphate and diacylglycerol: Two interacting second messengers. *Annu. Rev. Biochem.* **56,** 159–193.

Bhat, T. N., Bentley, G. A., Boulot, G., Greene, M. I., Tello, D., Dall'Acqua, W., Souchon, H., Schwarz, F. P., Mariuzza, R. A., and Poljak, R. J. (1994). Bound water molecules and conformational stabilization help mediate an antigen–antibody association. *Proc. Natl. Acad. Sci. USA* **91,** 1089–1093.

Blackburn, B. K., and Gadek, T. R. (1993). Glycoprotein IIb/IIIa antagonists. *In* "Ann. Rep. Med. Chem." (D. W. Robertson, ed.), Vol. 28, pp. 79–88. Academic Press, San Diego.

Blaney, J. M., and Dixon, J. S. (1993). A good ligand is hard to find: Automated docking methods. *Perspect. Drug Discov. Design* **1,** 301–319.

Blaney, J. M., Weiner, P. K., Dearing, A., Kollman, P. A., Jorgensen, E. C., Oatley, S. J., Burridge, J. M., and Blake, C. C. F. (1984). Molecular mechanics simulation of protein–ligand interactions: Binding of thyroid hormone analogues to prealbumin. *J. Am. Chem. Soc.* **104,** 6424–6434.

Blokzijl, W., and Engberts, J. B. F. N. (1993). Hydrophobic effects. Opinions and facts. *Angew. Chem. Int. Ed. Engl.* **32,** 1545–1579.

Bode, W., Meyer, E., Jr., and Powers, J. C. (1989). Human leukocyte and porcine pancreatic elastase: X-ray crystal structures, mechanism, substrate specificity, and mechanism-based inhibitors. *Biochemistry* **28,** 1951–1963.

Bohacek, R. S., and McMartin, C. (1992). Definition and display of steric, hydrophobic, and hydrogen-bonding properties of ligand binding sites in proteins using Lee and Richards accessible surface: Validation of a high-resolution graphical tool for drug design. *J. Med. Chem.* **35,** 1671–1684.

Bohacek, R. S., and McMartin, C. (1994). Multiple highly diverse structures of complementary to enzyme binding sites: Results of extensive application of a *de novo* design method incorporating combinatorial growth. *J. Am. Chem. Soc.* **116,** 5560–5571.

Böhm, H.-J. (1994a). The development of a simple empirical scoring function to estimate the binding constant for a protein–ligand complex of known three-dimensional structure. *J. Comput.-Aided Mol. Des.* **8**, 243–256.

Böhm, H.-J. (1994b). On the use of LUDI to search the Fine Chemicals Director for ligands of proteins of known three-dimensional structure. *J. Comput.-Aided Mol. Design* **8**, 623–632.

Boobbyer, D. N., Goodford, P. J., McWhinnie, P. M., and Wade, R. C. (1989). New hydrogen-bond potentials for use in determining energetically favorable binding sites on molecules of known structure. *J. Med. Chem.* **32**, 1083–1094.

Borea, P. A., Varani, K., Guerra, L., Gilli, P., and Gilli, G. (1992). Binding thermodynamics of $A_1$ adenosine receptor ligands. *Mol. Neuropharmacol* **2**, 273–281.

Brady, K., and Abeles, R. H. (1990). Inhibition of chymotrypsin by peptidyl trifluoromethyl ketones: Determinants of slow-binding kinetics. *Biochemistry* **29**, 7608–7617.

Braxton, S., and Wells, J. A. (1991). The importance of a distal hydrogen bonding group in stabilizing the transition state in subtilisin BPN′. *J. Biol. Chem.* **266**, 11797–11800.

Brint, A. T., and Willett, P. (1987). Algorithms for the identification of three-dimensional maximal common substructures. *J. Chem. Inf. Comput. Sci.* **27**, 152–158.

Brooks, B. R., Bruccoleri, R. E., Olafson, B. D., States, D. J., Swaminathan, S., and Karplus, M. (1983). CHARMM: A program for macromolecular energy, minimization, and dynamics calculations. *J. Comput. Chem.* **4**, 187–217.

Brooks, C. L., III (1995). Methodological advances in molecular dynamics simulations of biological systems. *Curr. Opin. Struct. Biol.* **5**, 211–215.

Brooks, C. L., Karplus, M., and Pettitt, B. M. (1988). In "Proteins: A Theoretical Perspective of Dynamics, Structure, and Thermodynamics, pp. 95–111. Wiley, New York.

Brown, F. J., Andisik, D. W., Bernstein, P. R., Bryand, C. B., Ceccarelli, C., Damewood, J. R., Jr., Edwards, P. D., Earley, R. A., Feeney, S., Green, R. C., Gomes, B., Kosmider, B. J., Krell, R. D., Shaw, A., Steelman, G. B., Thomas, R. M., Vacek, E. P., Veale, C. A., Tuthill, P. A., Warner, P., Williams, J. C., Wolanin, D. J., and Woolson, S. A. (1994). Design of orally active, nonpeptidic inhibitors of human leukocyte elastase. *J. Med. Chem.* **37**, 1259–1261.

Bryan, P., Pantoliano, M. W., Quill, S. G., Hsiao, H. Y., and Poulos, T. (1986). Site-directed mutagenesis and the role of the oxyanion hole in subtilisin. *Proc. Natl. Acad. Sci. USA* **83**, 3743–3745.

Bryant, S. H., and Altschul, S. F. (1995). Statistics of sequence–structure threading. *Curr. Opin. Struct. Biol.* **5**, 236–244.

Bryant, S. H., and Lawrence, C. E. (1993). An empirical energy function for threading protein sequence through folding motif. *Proteins: Struct. Funct. Genet.* **16**, 92–112.

Buckle, A. M., Schreiber, G., and Fersht, A. R. (1994). Protein–protein recognition: Crystal structural analysis of a barnase–barstar complex at 2.0 Å resolution. *Biochemistry* **33**, 8878–8889.

Bures, M. G., Martin, Y. C., and Willett, P. (1994). Searching techniques for databases of three-dimensional chemical structures. *In* "Topics in Stereochemistry," pp. 467–511. Wiley, New York.

Burger, A. (1991). Isosterism and bioisosterism in drug design. *In* "Progress in Drug Research" (J. Ernst, ed), Vol. 37, pp. 288–371. Birkhäuser Verlag, Basel.

Burley, S. K., and Petsko, G. A. (1985). Aromatic–aromatic interation: A mechanism of protein structure stabilization. *Science* **229**, 23–28.

Casy, A. F. (1993). "The Steric Factor in Medicinal Chemistry: Dissymmetric Probes of Pharmacological Receptors." Plenum, New York.

Chakrabarti, P., and Samanta, U. (1995). CH/$\pi$ interactions in the packing of the adenine ring in protein structures. *J. Mol. Biol.* **251**, 9–14.

Chang, G., Guida, W. C., and Still, W. C. (1988). An internal coordinate Monte Carlo method for searching conformational space. *J. Am. Chem. Soc.* **111**, 4379–4386.

Chau, P.-L., and Dean, P. M. (1994). Electrostatic complementarity between proteins and ligands. *J. Comput.-Aided Mol. Design* **8**, 545–564.

Chervenak, M. C., and Toone, E. J. (1994). A direct measure of the contribution of solvent reorganization to the enthalpy of binding. *J. Am. Chem. Soc.* **116**, 10533–10539.

Chinkers, M., and Garbers, D. L. (1991). Signal transduction by guanylyl cyclases. *Annu. Rev. Biochem.* **60**, 553–575.

Chothia, C. (1992). One thousand families for the molecular biologist. *Nature* **357**, 543–544.

Cohen, J. S., and Hogan, M. E. (1994). The new genetic medicines. *Sci. Am.* **271**, 50–55.

Connelly, P. R. (1994). Acquisition and use of calorimetric data for prediction of the thermodynamics of ligand-binding and folding reactions of proteins. *Curr. Opin. Biotechnol.* **5**, 381–388.

Connelly, P. R., Aldape, R. A., Bruzzese, F. J., Chambers, S. P., Fitzgibbon, M. J., Fleming, M. A., Itoh, S., Livingston, D. J., Navia, M. A., Thomson, J. A., and Wilson, K. P. (1994). Enthalpy of hydrogen bond formation in a protein–ligand binding reaction. *Proc. Natl. Acad. Sci. USA* **91**, 1964–1968.

Connolly, M. J. (1983). Solvent-accessible surfaces of proteins and nucleic acids. *Science* **221**, 709–713.

Cook, P. D. (1991). Medicinal chemistry of antisense oligonucleotides: Future opportunities. *Anticancer Drug Des.* **6**, 585–607.

Cornell, W. D., Cieplak, P., Bayly, C. I., Gould, I. R., Merz, K. M., Jr., Ferguson, D. M., Spellmeyer, D. C., Fox, T., Caldwell, J. W., and Kollman, P. A. (1995). A second generation force field for the simulation of proteins, nucleic acids, and organic molecules. *J. Am. Chem. Soc.* **117**, 5179–5197.

Corton, J. M., Gillespie, J. G., Hawley, S. A., and Hardie, D. G. (1995). 5-Aminoimidazole-4-carboxamide ribonucleoside: A specific method for activating AMP-activated protein kinase in intact cells? *Eur. J. Biochem.* **229**, 558–565.

Cramer, C. J., and Truhlar, D. G. (1991a). General parameterized SCF model for free energies of solvation in aqueous solution. *J. Am. Chem. Soc.* **113**, 8305–8311.

Cramer, C. J., and Truhlar, D. G. (1991b). Molecular orbital theory calculations of aqueous solvation effects on chemical equilibria. *J. Am. Chem. Soc.* **113**, 8552–8554.

Cramer, C. J., and Truhlar, D. G. (1992a). AM1-SM1 and PM3-SM3 parameterized SCF solvation models for free energies in aqueous solution. *J. Comput.-Aided Mol. Design* **6**, 629–666.

Cramer, C. J., and Truhlar, D. G. (1992b). An SCF solvation model for the hydrophobic effect and absolute free energies of aqueous solvation. *Science* **256**, 213–217.

Cramer, R. D., Patterson, D. E., and Bunce, J. D. (1988). Comparative Molecular Field Analysis (CoMFA). 1. Effect of shape on binding of steroids to carrier proteins. *J. Am. Chem. Soc.* **110**, 5959–5967.

Cregut, D., Liautard, J.-P., and Chiche, L. (1994). Homology modelling of annexin I: Implicit solvation improves side-chain prediction and combination of evaluation criteria allows recognition of different types of conformational error. *Prot. Engin.* **7**, 1333–1344.

Dado, G. P., Desper, J. M., and Gellman, S. H. (1990). Hydrogen bonding in a family of triamides: Conformation-directing effects in solution vs. the solid state. *J. Am. Chem. Soc.* **112**, 8630–8632.

Dado, G. P., and Gellman, S. H. (1992). On the use of AM1 calculations for the study of intramolecular hydrogen bonding phenomena in simple amides. *J. Am. Chem. Soc.* **114**, 3138–3139.

Dado, G. P., and Gellman, S. H. (1993). Structural and thermodynamic characterization of temperature-dependent changes in the folding pattern of a synthetic triamide. *J. Am. Chem. Soc.* **115**, 4228–4245.

Dammkoehler, R. A., Karasek, S. F., Shands, E. F. B., and Marshall, G. R. (1989). Constrained search of conformational hyperspace. *J. Comput.-Aided Mol. Design* **3**, 3–21.

Dao-Pin, S., Nicholson, H., Baase, W. A., Zhang, X. J., Wozniak, J. A., and Matthews, B. W. (1991). Structural and genetic analysis of electrostatic and other interactions in bacteriophage T4 lysozyme. *In* "Protein Conformations" (D. J. Chadwick, Ed.), Vol. 161, pp. 52–62. Wiley, Chichester.

Dearden, J. C. (1990). Molecular structure and drug transport. *In* "Comprehensive Medicinal Chemistry" (C. A. Ramsden, ed.), Vol. 4, pp. 375–411. Pergamon Press, Oxford.

Dearden, J. C., and Bresnen, G. M. (1988). The measurement of partition coefficients. *Quant. Struct.-Act. Relat.* **7**, 133–144.

Delarue, M., and Koehl, P. (1995). Atomic environment energies in proteins defined from statistics of accessible and contact surface areas. *J. Mol. Biol.* **249**, 675–690.

Devereux, J., Haeberli, P., and Smithies, O. (1984). A comprehensive set of sequence analysis programs for the VAX. *Nucleic Acids Res.* **12**, 387–395.

Dhanaraj, V., Dealwis, C. G., Frazao, C., Badasso, M., Sibanda, B. L., Tickle, I. J., Cooper, J. B., Driessen, H. P., Newman, M., Aguilar, C., Wood, S. P., Blundell, T. L., Hobart, P. M., Geoghegan, K. F., Ammirati, M. J., Danley, D. E., O'Connor, B. A., and Hoover, D. J. (1992). X-ray analyses of peptide–inhibitor complexes define the structural basis of specificity for human and mouse renins. *Nature* **357**, 466–472.

Diaz-Arauzo, H., Koehler, K. F., Hagen, T. J., and Cook, J. M. (1991). Synthetic and computer assisted analysis of the pharmacophore for agonists at benzodiazepine receptors. *Life Sci.* **49**, 207–216.

Dill, K. A. (1990). Dominant forces in protein folding. *Biochemistry* **29**, 7133–7155.

Ding, J., Das, K., Moereels, H., Koymans, L., Andries, K., Janssen, P. A. J., Hughes, S. H., and Arnold, E. (1995). Structure of HIV-1 RT/TIBO R-86183 complex reveals similarity in the binding of diverse nonnucleoside inhibitors. *Nature: Struct. Biol.* **2**, 407–415.

Doig, A. J., and Williams, D. H. (1992). Binding energy of an amide–amide hydrogen bond in aqueous and nonpolar solvents. *J. Am. Chem. Soc.* **114**, 338–343.

Domenicano, A., and Hargittai, I. (eds.) (1992). "Accurate Molecular Structures: Their Determination and Importance." Oxford University Press, London.

Dunbrack, R. L., Jr., and Karplus, M. (1993). Backbone-dependent rotamer library for proteins: Application to side-chain prediction. *J. Mol. Biol.* **230**, 543–574.

Eberhardt, E. S., and Raines, R. T. (1994). Amide–amide and amide–water hydrogen bonds: Implications for protein folding and stability. *J. Am. Chem. Soc.* **116**, 2149–2150.

Edwards, P. D., and Bernstein, P. R. (1994). Synthetic inhibitors of elastase. *Med. Res. Rev.* **14**, 127–194.

Eisen, M. B., Wiley, D. C., Karplus, M., and Hubbard, R. E. (1994). HOOK: A program for finding novel molecular architectures that satisfy the chemical and steric requirements of a macromolecule binding site. *Proteins* **19**, 199–221.

Eisenberg, D., and McLachlan, A. D. (1986). Solvation energy in protein folding and binding. *Nature* **319**, 199–203.

Eisenmenger, F., Argos, P., and Abagyan, R. (1993). A method to configure protein side-chains from the main-chain trace in homology modelling. *J. Mol. Biol.* **231**, 849–860.

Eldred, C. D., Evans, B., Hindley, S., Judkins, B. D., Kelly, H. A., Kitchin, J., Lumley, P., Porter, B., Ross, B. C., Smith, K. J., Taylor, N. R., and Wheatcroft, J. R. (1994). Orally active non-peptide fibrinogen receptor (GpIIb/IIIa) antagonists: Identification of 4-[4-[4-(aminoiminomethyl)phenyl]-1-piperazinyl]-1-piperidineacetic acid as a long-acting, broad-spectrum antithrombotic agent. *J. Med. Chem.* **37**, 3882–3885.

Erhlich, P. (1913). Chemotherapeutics: Scientific principles, methods, and results. *Lancet* **2**, 445–451.

Erickson, J. Neidhart, D. J., VanDrie, J., Kempf, D. J., Wang, X. C., Norbeck, D. W., Plattner,

J. J., Rittenhouse, J. W., Turon, M., Wideburg, N., Kohlbrenner, W. E., Simmer, R., Helfrich, R., Paul, D. A., and Knigge, M. (1990). Design, activity, and 2.8 Å crystal structure of a $C_2$ symmetric inhibitor complexed to HIV-1 protease. *Science* **249**, 527–533.

Erion, M. D., Niwas, S., Rose, J. D., Ananthan, S., Allen, M., Secrist, J. A., III, Babu, Y. S., Bugg, C. E., Guida, W. C., Ealick, S. E., and Montgomery, J. A. (1993). Structure-based design of inhibitors of purine nucleoside phosphorylase. 3. 9-Arylmethyl derivatives of 9-deazaguanine substituted on the methylene group. *J. Med. Chem.* **36**, 3771–3783.

Ernst, J. A., Clubb, R. T., Zhou, H. X., Gronenborn, A. M., and Clore, G. M. (1995). Demonstration of positionally disordered water within a protein hydrophobic cavity by NMR. *Science* **267**, 1813–1817.

Fantl, W. J., Johnson, D. E., and Williams, L. T. (1993). Signalling by receptor tyrosine kinases. *Annu. Rev. Biochem.* **62**, 453–481.

Fasman, K. H., Cuticchia, A. J., and Kingsbury, D. T. (1994). The GDB Human Genome Data Base anno 1994. *Nucleic Acids Res.* **22**, 3462–3469.

Fersht, A. R. (1984). Basis of biological specificity. *Trends Biochem. Sci.* **9**, 145–147.

Fersht, A. R. (1987). The hydrogen bond in molecular recognition. *Trends Biochem. Sci.* **12**, 301–304.

Fetrow, J. S., and Bryant, S. H. (1993). New programs for protein tertiary structure prediction. *Bio/Technology* **11**, 479–484.

Fioretti, E., Angeletti, M., Coletta, M., Ascenzi, P., Bolognesi, M., Menegatti, E., Rizzi, M., and Ascoli, F. (1993). Binding of bovine basic pancreatic trypsin inhibitor (Kunitz) as well as bovine and porcine pancreatic secretory trypsin inhibitor (Kazal) to human cathepsin G: A kinetic and thermodynamic study. *J. Enzym. Inhib.* **7**, 57–64.

Fischer, E. (1894). Einfluss der configuration auf die wirkung der enzyme. *Ber. Dtsch. Chem. Ges.* **27**, 2985–2993.

Fitzgerald, P. M. D., and Springer, J. P. (1991). Structure and function of retroviral proteases. *Annu. Rev. Biophys. Biophys. Chem.* **20**, 299–320.

Fleischman, S. H., and Brooks, C. L, III (1987). Thermodynamics of aqueous solvation: Solution properties of alcohols and alkanes. *J. Chem. Phys.* **87**, 3029–3037.

Ford, G. P., and Wang, B. (1992). Incorporation of hydration effects within the semiempirical molecular orbital framework: AM1 and MNDO results for neutral molecules, cations, anions, and reacting systems. *J. Am. Chem. Soc.* **114**, 10563–10569.

Fraeyman, N., Bazan, A., and Vanscheeuwijck, P. (1994). Thermodynamic analysis of isoproterenol binding to beta-adrenoceptors in rat lung membranes. *Eur. J. Pharmacol.* **267**, 63–69.

Franks, N. P., and Lieb, W. R. (1994). Molecular and cellular mechanisms of general anaesthesia. *Nature* **367**, 607–614.

Freire, E. (1993). Structural thermodynamics: Prediction of protein stability and protein binding affinities. *Arch. Biochem. Biophys.* **303**, 181–184.

Friedman, L., and Miller, J. G. (1971). Odor incongruity and chirality. *Science* **172**, 1044–1046.

Gallop, M. A., Barrett, R. W., Dower, W. J., Fodor, S. P., and Gordon, E. M. (1994). Applications of combinatorial technologies to drug discovery. 1. Background and peptide combinatorial libraries. *J. Med. Chem.* **37**, 1233–1251.

Gante, J. (1994). Peptidomimetics: Tailored enzyme inhibitors. *Angew. Chem. Int. Ed. Engl.* **33**, 1699–1720.

Garcia, K. C., Ronco, P. M., Verroust, P. J., Brünger, A. T., and Amzel, L. M. (1992). Three-dimensional structure of an angiotensin II-FAB complex at 3 Å: Hormone recognition by an anti-idiotypic antibody. *Science* **257**, 502–507.

Gellman, S. H., Adams, B. R., and Dado, G. P. (1990). Temperature-dependent changes in the folding pattern of a simple triamide. *J. Am. Chem. Soc.* **112**, 460–461.

Gellman, S. H., and Dado, G. P. (1991). On the use of molecular mechanics calculations for the study of intramolecular hydrogen bonding phenomena. *Tet. Lett.* **32**, 7377–7380.

Gellman, S. H., Dado, G. P., Liang, G.-B., and Adams, B. R. (1991). Conformational-directing effects of a single intramolecular amide–amide hydrogen bond: Variable-temperature NMR and IR studies on a homologous diamide series. *J. Am. Chem. Soc.* **113**, 1164–1173.

George, D. G., Barker, W. C., Mewes, H. W., Pfeiffer, F., and Tsugita, A. (1994). The PIR-International Protein Sequence Database. *Nucleic Acids Res.* **22**, 3569–3573.

Ghose, A. K., Logan, M. E., Treasurywala, A. M., Wang, H., Wahl, R. C., Tomczuk, B. E., Gowravaram, M. R., Jaeger, E. P., and Wendoloski, J. J. (1995). Determination of pharmacophoric geometry for collagenase inhibitors using a novel computational method and its verification using molecular dynamics, NMR, and X-ray crystallography. *J. Am. Chem. Soc.* **117**, 4671–4682.

Giesen, D. J., Storer, J. W., Cramer, C. J., and Truhlar, D. G. (1995). General semiempirical quantum mechanical solvation model for nonpolar solvation free energies: *n*-Hexadecane. *J. Am. Chem. Soc.* **117**, 1057–1068.

Gilli, P., Ferretti, V., Gilli, G., and Borea, P. A. (1994). Enthalpy–entropy compensation in drug–receptor binding. *J. Phys. Chem.* **98**, 1515–1518.

Gilman, A. G. (1987). G proteins: Transducers of receptor-generated signals. *Annu. Rev. Biochem.* **56**, 615–649.

Gilson, M. K., and Honig, B. (1988). Calculation of the total electrostatic energy of a macromolecular system: Solvation energies, binding energies, and conformational analysis. *Proteins* **4**, 7–18.

Gilson, M. K., and Honig, B. (1991). The inclusion of electrostatic hydration energies in molecular mechanics calculations. *J. Comput.-Aided Mol. Des.* **5**, 5–20.

Glaeser, R. M., and Downing, K. H. (1993). High-resolution electron crystallography of protein molecules. *Ultramicroscopy* **52**, 478–486.

Glusker, J. P. (1994). X-ray crystallography of proteins. *Methods Biochem. Anal.* **37**, 1–72.

Golender, V. E., and Vorpagel, E. R. (1993). Computer-assisted pharmacophore identification. *In* "3D QSAR in Drug Design: Theory, Methods, and Applications" (H. Kubinyi, ed.). ESCOM Science Publishers B. V., Leiden, Netherlands.

Goodford, P. J. (1985). A computational procedure for determining energetically favorable binding sites on biologically important macromolecules. *J. Med. Chem.* **28**, 849–857.

Gordon, E. M., Barrett, R. W., Dower, W. J., Fodor, S. P., and Gallop, M. A. (1994). Applications of combinatorial technologies to drug discovery. 2. Combinatorial organic synthesis, library screening strategies, and future directions. *J. Med. Chem.* **37**, 1385–1401.

Greenlee, W. J. (1990). Renin inhibitors. *Med. Res. Rev.* **10**, 173–236.

Greer, J. (1990). Comparative modeling methods: Application to the family of the mammalian serine proteases. *Proteins* **7**, 317–334.

Greer, J., Erickson, J. W., Baldwin, J. J., and Varney, M. D. (1994). Application of the three-dimensional structures of protein target molecules in structure-based drug design. *J. Med. Chem.* **37**, 1035–1054.

Gregoret, L. M., and Cohen, F. E. (1990). Novel method for the rapid evaluation of packing in protein structures. *J. Mol. Biol.* **211**, 959–974.

Gronenborn, A. M., and Clore, G. M. (1990). Determination of ligand conformation in macromolecular complexes using the transferred nuclear Overhauser effect. *Biochem. Pharmacol.* **40**, 115–119.

Gualtieri, F. (1993). Chirality of drug–receptor interactions: A source of valuable information. *Drug Inform. J.* **27**, 1111–1120.

Guarnieri, F., and Still, W. C. (1994). A rapidly convergent simulation method: Mixed Monte Carlo/stochastic dynamics. *J. Comput. Chem.* **15**, 1302–1309.

Gussio, R., Pou, S., Chen, J. H., and Smythers, G. W. (1992). A pseudoreceptor docking study of 4,5-alpha-epoxymorphinans with a range of dielectric constants. *J. Comput.-Aided Mol. Des.* **6**, 149–158.

Hamilton, H. W., Steinbaugh, B. A., Stewart, B. H., Chan, O. H., Schmid, H. L., Schroeder, R., Ryan, M. J., Keiser, J., Taylor, M. D., Blankley, C. J., Kaltenbronn, J. S., Wright, J., and Hicks, J. (1995). Evaluation of physicochemical parameters important to the oral bioavailability of peptide-like compounds: Implications for the synthesis of renin inhibitors. *J. Med. Chem.* **38**, 1446–1455.

Hansch, C., Björkroth, J. P., and Leo, A. (1987). Hydrophobicity and central nervous system agents: On the principle of minimal hydrophobicity in drug design. *J. Pharmacol. Sci.* **76**, 663–687.

Hansch, C., and Leo, A. (1995). The hydrophobic parameter: Measurement and calculation. Calculation of octanol–water partition coefficients by fragments. *In* "Exploring QSAR: Fundamentals and Applications in Chemistry and Biology" (S. R. Heller, ed.), pp. 97–168. ACS Professional Reference Book American Chemical Society, Washington, DC.

Hansen, L. M., and Kollman, P. A. (1990). Free energy perturbation calculations on models of active sites: Applications to adenosine deaminase inhibitors. *J. Comp. Chem.* **11**, 994–1002.

Harte, W. E., Jr., and Beveridge, D. L. (1993). Predictions of the protonation state of the active site aspartyl residues in HIV-1 protease–inhibitor complexes via molecular dynamics simulation. *J. Am. Chem. Soc.* **115**, 3883–3886.

Hartman, G. D., Egbertson, M. S., Halczenko, W., Laswell, W. L., Duggan, M. E., Smith, R. L., Naylor, A. M., Manno, P. D., Lynch, R. J., Zhang, G., Chang, C., and Gould, R. (1992). Non-peptide fibrinogen receptor antagonists. 1. Discovery and design of exosite inhibitors. *J. Med. Chem.* **35**, 4640–4642.

Hayden, F., Bobo, M., Esinhar, J., and Hussey, E. (1994). "Efficacy of 4-guanidino Neu5Ac2en in experimental human influenza A virus infection." Abstracts, 34th ICAAC, Orlando, FL.

Hendsch, Z. A., and Tidor, B. (1994). Do salt bridges stabilize proteins? A continuum electrostatic analysis. *Prot. Sci.* **3**, 211–226.

Henry, D. R., McHale, P. J., Christie, B. D., and Hillman, D. (1990). Building three-dimensional structural databases: Experiences with MDDR-3D and FCD-3D. *Tetrahedron Comput. Methodol.* **3**, 531–536.

Hermans, J., Berendsen, J. J. C., Van Gunsteren, W. F., and Postam, J. P. M. (1984). GROMOS. *Biopolymers* **23**, 1513–1518.

Hibert, M. F., Gittos, M. W., Middlemiss, D. N., Mir, A. K., and Fozard, J. R. (1988). Graphics computer-aided receptor mapping as a predictive tool for drug design: Development of potent, selective, and stereospecific ligands for the 5-$HT_{1A}$ receptor. *J. Med. Chem.* **31**, 1087–1093.

Hibert, M. F., Trumpp Kallmeyer, S., Hoflack, J., and Bruinvels, A. (1993). This is not a G protein-coupled receptor. *Trends Pharmacol. Sci.* **14**, 7–12.

Hille, B. (1992). "Ionic Channels of Excitable Membranes." Sinauer Associates, Sunderland, MA.

Hitzemann, R. (1988). Thermodynamic aspects of drug–receptor interactions. *Trends Pharmacol. Sci.* **9**, 408–411.

Ho, C. M. W., and Marshall, G. R. (1995). DBMAKER: A set of programs to generate three-dimensional databases based upon user-specified criteria. *J. Comput.-Aided Mol. Design* **9**, 65–86.

Ho, D. D., Neumann, A. U., Perelson, A. S., Chen, W., Leonard, J. M., and Markowitz, M. (1995). Rapid turnover of plasma virions and CD4 lymphocytes in HIV-1 infection. *Nature* **373**, 123–126.

Hodgkin, E. E., Miller, A., and Whittaker, M. (1993). A Monte Carlo pharmacophore generation procedure: Application to the human PAF receptor. *J. Comp.-Aided Mol. Design* **7**, 515–534.

Hoekstra, W. J., Beavers, M. P., Andrade Gordon, P., Evangelisto, M. F., Keane, P. M., Press,

J. B., Tomko, K. A., Fan, F., Kloczewiak, M., Mayo, K. H., Durkin, K. A., and Liotta, D. C. (1995). Design and evaluation of nonpeptide fibrinogen gamma-chain based GPIIb/IIIa antagonists. *J. Med. Chem.* **38,** 1582–1592.

Hoflack, J., Trumpp Kallmeyer, S., and Hibert, M. (1994). Re-evaluation of bacteriorhodopsin as a model for G protein-coupled receptors. *Trends Pharmacol. Sci.* **15,** 7–9.

Holloway, M. K., Wai, J. M., Halgren, T. A., Fitzgerald, P. M. D., Vacca, J. P., Dorsey, B. D., Levin, R. B., Thompson, W. J., Chen, L. J., deSolms, S. J., Gaffin, N., Ghosh, A. K., Giuliani, E. A., Graham, S. L., Guare, J. P., Hungate, R. W., Lyle, T. A., Sanders, W. M., Tucker, T. J., Wiggins, M., Wiscount, C. M., Woltersdorf, O. W., Young, S. D., Darke, P. L., and Zugay, J. A. (1995). *A priori* prediction of activity for HIV-1 protease inhibitors employing energy minimization in the active site. *J. Med. Chem.* **38,** 305–317.

Holmes, M. A., Tronrud, D. E., and Matthews, B. W. (1983). Structural analysis of the inhibition of thermolysin by an active-site-directed irreversible inhibitor. *Biochemistry* **22,** 236–240.

Horton, N., and Lewis, M. (1992). Calculation of the free energy of association for protein complexes. *Protein Sci.* **1,** 169–181.

Howard, A. E., and Kollman, P. A. (1988). An analysis of current methodologies for conformational searching of complex molecules. *J. Med. Chem.* **31,** 1669–1675.

Hubbard, S. J., Gross, K. H., and Argos, P. (1994). Intramolecular cavities in globular proteins. *Protein Eng.* **7,** 613–626.

Humblet, C., and Dunbar, J. B. (1993). 3D database searching and docking strategies. *In* "Annual Reports in Medicinal Chemistry" (J. A. Bristol, ed.), Vol. 28, pp. 275–284. Academic Press, San Diego.

Hurst, T. (1994). Flexible 3D searching: The directed tweak technique. *J. Chem. Inf. Comp. Sci.* **34,** 190–196.

Hutchins, C., and Greer, J. (1991). Comparative modeling of proteins in the design of novel renin inhibitors. *Crit. Rev. Biochem. Mol. Biol.* **26,** 77–127.

Illman, D. L. (1994). Researchers make progress in applying virtual reality to chemistry. *Chem. Engin. News* **72** (issue 12, March 21), 22–25.

Ippolito, J. A., Alexander, R. S., and Christianson, D. W. (1990). Hydrogen bond stereochemistry in protein structure and function. *J. Mol. Biol.* **215,** 457–471.

Jackson, R. M., and Sternberg, J. E. (1995). A continuum model for protein–protein interactions: Application to the docking problem. *J. Mol. Biol.* **250,** 258–275.

Jackson, S. E., Moracci, M., el Masry, N., Johnson, C. M., and Fersht, A. R. (1993). Effect of cavity-creating mutations in the hydrophobic core of chymotrypsin inhibitor 2. *Biochemistry* **32,** 11259–11269.

Jain, A. N., Koile, K., and Chapman, D. (1994). Compass: Predicting biological activities from molecular surface properties. Performance comparisons on a steroid benchmark. *J. Med. Chem.* **37,** 2315–2327.

Janin, J. (1990). Errors in three dimensions. *Biochimie* **72,** 705–709.

Janin, J., and Chothia, C. (1990). The structure of protein–protein recognition sites. *J. Biol. Chem.* **265,** 16027–16030.

Johnson, M. S., Srinivasan, N., Sowdhamini, R., and Blundell, T. L. (1994). Knowledge-based protein modeling. *Crit. Rev. Biochem. Mol. Biol.* **29,** 1–68.

Jorgensen, W. L. (1991). Rusting of the lock and key model for protein–ligand binding. *Science* **254,** 954–955.

Jorgensen, W. L. (1992). "BOSS, version 3.2." Yale University, New Haven.

Jorgensen, W. L., Chandrasekhar, J., and Madura, J. D. (1983). Comparison of simple potential functions for simulating liquid water. *J. Chem. Phys.* **79,** 926–935.

Jorgensen, W. L., and Nguyen, T. B. (1993). Monte Carlo simulations of the hydration of substituted benzenes with OPLS potential functions. *J. Comp. Chem.* **14,** 195–205.

Jorgensen, W. L., and Ravimohan, C. (1985). Monte Carlo simulation of differences in free energy of hydration. *J. Chem. Phys.* **83**, 3050–3054.

Jorgensen, W. L., and Tirado-Rives, J. (1988). The OPLS potential functions for proteins: Energy minimizations for crystals of cyclic peptides and crambin. *J. Am. Chem. Soc.* **100**, 1657–1666.

Karuso, P., Kessler, H., and Mierke, D. F. (1990). Solution structure of FK506 from nuclear magnetic resonance and molecular dynamics. *J. Am. Chem. Soc.* **112**, 9434–9436.

Kato, Y., Inoue, A., Yamada, M., Tomioka, N., and Itai, A. (1992). Automatic superposition of drug molecules based on their common reeptor site. *J. Comput.-Aided Mol. Design* **6**, 475–486.

Kearney, P. C., Mizoue, L. S., Kumpf, R. A., Forman, J. E., McCurdy, A., and Dougherty, D. A. (1993). Molecular recognition in aqueous media: New binding studies provide further insight into the cation-$\pi$ interaction and related phenomena. *J. Am. Chem. Soc.* **115**, 9907–9919.

Kearsley, S. K., and Smith, G. M. (1990). An alternative method for the alignment of molecular structures: Maximizing electrostatic and steric overlap. *Tetrahedron Comput. Method.* **3**, 615–633.

Kearsley, S. K., Underwood, D. J., Sheridan, R. P., and Miller, M. D. (1994). Flexibases: A way to enhance the use of molecular docking methods. *J. Comput.-Aided Mol. Des.* **8**, 565–582.

Kelley, R. F., and O'Connell, M. P. (1993). Thermodynamic analysis of an antibody functional epitope. *Biochemistry* **32**, 6828–6835.

Kempf, D. J. (1994). Progress in the discovery of orally bioavailable inhibitors of HIV protease. *Perspect. Drug Discov. Design*

Kempf, D. J., Marsh, K. C., Denissen, J. F., McDonald, E., Vasavanonda, S., Flentge, C. A., Green, B. E., Fino, L., Park, C. H., Kong, X. P., Wideburg, N. E., Saldivar, A., Ruiz, L., Kati, W. M., Sham, H. L., Robins, T., Stewart, D. D., Hsu, A., Plattner, J. J., Leonard, J. M., and Norbeck, D. W. (1995). ABT-538 is a potent inhibitor of human immunodeficiency virus protease and has high oral bioavailability in humans. *Proc. Natl. Acad. Sci. USA* **92**, 2484–2488.

Kempf, D. J., Marsh, K. C., Paul, D. A., Knigge, M. F., Norbeck, D. W., Kohlbrenner, W. E., Codacovi, L., Vasavanonda, S., Bryant, P., Wang, X. C., Wideburg, N. E., Clement, J. J., Plattner, J. J., and Erickson, J. (1991). Antiviral and pharmacokinetic properties of $C_2$ symmetric inhibitors of the human immunodeficiency virus type 1 protease. *Antimicrob. Agents Chemother.* **35**, 2209–2214.

Kempf, D. J., Norbeck, D. W., Codacovi, L., Wang, X. C., Kohlbrenner, W. E., Wideburg, N. E., Paul, D. A., Knigge, M. F., Vasavanonda, S., Craig-Kennard, A., Saldivar, A., Rosenbrook, W., Jr., Clement, J. J., Plattner, J. J., and Erickson, J. (1990). Structure-based, $C_2$ symmetric inhibitors of HIV protease. *J. Med. Chem.* **33**, 2687–2689.

Kenakin, T. P. (1990). Macromolecular targets for drug action. *In* "Comprehensive Medicinal Chemistry" (C. Hansch, ed.), Vol. 1, pp. 195–208. Pergamon, New York.

Kim, E. , Baker, C. T., Dwyer, M. D., Murcko, M. A., Rao, B. G., Tung, R. D., and Navia, M. A. (1995). Crystal structure of HIV-1 protease in complex with VX-478, a potent and orally bioavailable inhibitor of the enzyme. *J. Am. Chem. Soc.* **117**, 1181–1182.

Kleanthous, C., Reilly, M., Cooper, A., Kelly, S., Price, N. C., and Coggins, J. R. (1991). Stabilization of the shikimate pathway enzyme dehydroquinase by covalently bound ligand. *J. Biol. Chem.* **266**, 10893–10898.

Klebe, G. (1994a). Structure correlation and ligand/receptor interactions. *In* "Structure Correlation" (H.-B. Bürgi and J. D. Dunitz, eds.), Vol. 2, pp. 543–603. VCH, Weinheim.

Klebe, G. (1994b). The use of composite crystal-field environments in molecular recognition and the *de novo* design of protein ligands. *J. Mol. Biol.* **237**, 212–235.

Kleinert, H. D., Rosenberg, S. H., Baker, W. R., Stein, H. H., Klinghofer, V., Barlow, J., Spina,

K., Polakowski, J., Kovar, P., Cohen, J., and Denissen, J. (1992). Discovery of a peptide-based renin inhibitor with oral bioavailability and efficacy. *Science* **257**, 1940–1943.

Knowles, J. R. (1991). Enzyme catalysis: Not different, just better. *Nature* **250**, 121–124.

Koehler, K. F., Spangler, D. P., and Snyder, J. P. (1988). Pharmacophore Identification through Molecular Similarity." 196th American Chemical Society National Meeting, September 27, American Chemical Society, Los Angeles, CA.

Kollman, P. (1993). Free energy calculations: Application to chemical and biochemical phenomena. *Chem. Rev.* **93**, 2395–2417.

Kontoyianni, M., and Lybrand, T. P. (1993). Three-dimensional models for integral membrane proteins: Possibilities and pitfalls. *Perspect. Drug Discov. Design* **1**, 291–300.

Krimm, S., and Bandekar, J. (1986). Vibrational spectroscopy and conformation of peptides, polypeptides and proteins. *Adv. Prot. Chem.* **38**, 181–364.

Krystek, S., Stouch, T., and Novotny, J. (1993). Affinity and specificity of serine endopeptidase–protein inhibitor interactions: Empirical free energy calculations based on X-ray crystallographic structures. *J. Mol. Biol.* **234**, 661–679.

Ku, T. W., Ali, F. E., Barton, L., Bean, J., Bondinell, W., Burgess, J., Callahan, J., Calvo, R. R., Chen, I., Eggleston, D., Gleason, J., Huffman, W., Hwang, S., Jakas, D., Karash, C., Keenan, R. M., Kopple, K. D., Miller, W., Newlander, K. A., Nichols, A. J., Parker, M., Peishoff, C. E., Samanen, J. M., Uzinskas, I., and Venslavsky, J. W. (1993). Direct design of a potent nonpeptide fibrinogen receptor antagonist based on the structure and conformation of a highly constrained cyclic RGD Peptide. *J. Am. Chem. Soc.* **115**, 8861–8862.

Ku, T. W., Miller, W. H., Bondinell, W. E., Erhard, K. F., Keenan, R. M., Nichols, A. J., Peishoff, C. E., Samanen, J. M., Wong, A. S., and Huffman, W. F. (1995). Potent nonpeptide fibrinogen receptor antagonists which present an alternative pharmacophore. *J. Med. Chem.* **38**, 9–12.

Kubinyi, H., ed. (1993). "3D QSAR in Drug Design: Theory, Methods, and Applications. ESCOM, Leiden.

Kuntz, I. D., Blaney, J. M., Oatley, S. J., Langridge, R., and Ferrin, T. A. (1982). Geometric approach to macromolecule–ligand interactions. *J. Mol. Biol.* **161**, 269–288.

Kuntz, I. D., Meng, E. C., and Shoichet, B. K. (1994). Structure-based molecular design. *Accts. Chem. Res.* **27**, 117–123.

Kurinov, I. V., and Harrison, R. W. (1994). Prediction of new serine proteinase inhibitors. *Nature: Struct. Biol.* **1**, 735–743.

Laduron, P. M. (1992). Towards genomic pharmacology: From membranal to nuclear receptors. *Adv. Drug Res.* **22**, 107–148.

Lam, P. Y. S., Jadhav, P. K., Eyermann, C. J., Hodge, C. N., Ru, Y., Bacheler, L. T., Meek, J. L., Otto, M. J., Rayner, M. M., Wong. Y. N., Chang, C.-H., Weber, P. C., Jackson, D. A., Sharpe, T. R., and Erickson-Viitanen, S. (1994). Rational design of potent, bioavailable, nonpeptide cyclic ureas as HIV protease inhibitors. *Science* **263**, 380–384.

Lauri, G., and Bartlett, P. A. (1994). CAVEAT: A program to facilitate the design of organic molecules. *J. Comput.-Aided Mol. Design* **8**, 51–66.

Lawrence, M. C., and Colman, P. M. (1993). Shape complementarity at protein/protein interfaces. *J. Mol. Biol.* **234**, 946–950.

Lee, B., and Richards, F. M. (1971). Interpretation of protein structure–estimation of static accessibility. *J. Mol. Biol.* **55**, 379–400.

Lee, F. S., Chu, Z. T., and Warshel, A. (1993). Microscopic and semimicroscopic calculations of electrostatic energies in proteins by the POLARIS and ENZYMIX programs. *J. Comput. Chem.* **14**, 161–185.

Lefkowitz, R. J., Cotecchia, S., Samama, P., and Costa, T. (1993). Constitutive activity of receptors coupled to guanine nucleotide regulatory proteins. *Trends Pharmacol. Sci.* **14**, 303–307.

Levitt, M., and Perutz, M. F. (1988). Aromatic rings act as hydrogen bond acceptors. *J. Mol. Biol.* **201,** 751–754.

Levitt, M., and Sharon, R. (1988). Accurate simulation of protein dynamics in solution. *Proc. Natl. Acad. Sci. USA,* **85,** 7557–7561.

Lewis, R. A., and Leach, A. R. (1994). Current methods for site-directed structure generation. *J. Comput.-Aided Mol. Design* **8,** 467–475.

Liang, G.-B., Dado, G. P., and Gellman, S. H. (1991). Anatomy of a stable intramolecularly hydrogen bonded folding pattern. *J. Am. Chem. Soc.* **113,** 3994–3995.

Lightstone, F. C., and Bruice, T. C. (1994). Geminal-dialkyl substitution intramolecular reactions, and enzyme efficiency. *J. Am. Chem. Soc.* **116,** 10789–10790.

Limbird, L. E. (1986). "Cell Surface Receptors: A Short Course on Theory and Methods." Kluwer Academic Publishers, Norwell, MA.

Linschoten, M. R., Bultsma, T., Kjzerman, A. P., and Timmerman, H. (1986). Mapping the turkey erythrocyte beta-receptor: A distance geometry approach. *J. Med. Chem.* **29,** 278–288.

Linschoten, M. R., Kranenbarg, G. W. K., De Kimpe, S. J., Wilting, J., Janssen, L. H. M., and Van Lenthe, J. H. (1990). A computer modelling study of hydrogen bonds in ligand/beta-adrenoceptor complexes: Its implications in the deduction of a receptor map. *J. Mol. Struct.* **237,** 339–354.

Lipinski, C. A. (1986). Bioisosterism in drug design. *In* "Ann. Rep. Med. Chem." (R. C. Allen, ed.), Vol. 21, pp. 283–291. Academic Press, Orlando.

Lloyd, E. J., and Andrews, P. R. (1986). A common structural model for central nervous system drugs and their receptors. *J. Med. Chem.* **29,** 453–462.

Luther, R. R., Glassman, H. N., and Boger, R. S. (1991). Renin inhibitors in hypertension. *Clin. Nephrol.* **36,** 181–186.

Lybrand, T. P. (1995). Ligand–protein docking and rational drug design. *Curr. Opin. Struct. Biol.* **5,** 224–228.

Maher, L. J., III, Wold, B., and Dervan, P. B. (1991). Oligonucleotide-directed DNA triple–helix formation: An approach to artificial repressors? *Antisense Res. Dev.* **1,** 277–281.

Manners, C. N., Payling, D. W., and Smith, D. A. (1988). Distribution coefficient, a convenient term for the relation of predictable physicochemical properties to metabolic processes. *Xenobiotica* **18,** 331–350.

Markowitz, M., Mo, H., Kempf, D. J., Norbeck, D. W., Bhat, T. N., Erickson, J. W., and Ho, D. D. (1995). Selection and analysis of human immunodeficiency virus type 1 variants with increased resistance to ABT-538, a novel protease inhibitor. *J. Virol.* **69,** 701–706.

Marquart, M., Walter, J., Deisenhoffer, J., Bode, W., and Huber, R. (1983). The geometry of the reactive site and of the peptide groups in trypsin, trypsinogen, and its complexes with inhibitors. *Acta Crystallogr. Sect. B.* **B39,** 480–490.

Martin, E. J., Blaney, J. M., Siani, M. A., Spellmeyer, D. C., Wong, A. K., and Moos, W. H. (1995). Measuring diversity: Experimental design of combinatorial libraries for drug discovery. *J. Med. Chem.* **38,** 1431–146.

Martin, Y. C. (1992). 3D database searching in drug design. *J. Med. Chem.* **35,** 2145–2154.

Martin, Y. C., Bures, M. G., Danaher, E. A., Delazzer, J., Lico, I., and Pavlik, P. A. (1993). A fast new approach to pharmacophore mapping and its application to dopaminergic and benzodiazepine agonists. *J. Comput.-Aid. Mol. Design.* **7,** 83–102.

Mattos, C., Rasmussen, B., Ding, X., Petsko, G. A., and Ringe, D. (1994). Analogous inhibitors of elastase do not always bind analogously. *Nature: Struct. Biol.* **1,** 55–58.

Mattos, C., and Ringe, D. (1993). Multiple binding modes. *In* "3D QSAR in Drug Design: Theory, Methods, and Applications" (H. Kubinyi, ed.), pp. 226–254. ESCOM, Leiden.

McDonald, D. Q. and Still, W. C. (1992a). An AMBER* study of Gellman's amides. *Tet. Lett.* **33,** 7747–7750.

McDonald, D. Q., and Still, W. C. (1992b). AMBER* torsional parameters for the peptide backbone. *Tet. Lett.* **33**, 7743–7746.
McDonald, D. Q., and Still, W. C. (1994). Conformational free energies from simulation: Stochastic dynamics/Monte Carlo simulations of a homologous series of Gellman's diamides. *J. Am. Chem. Soc.* **116**, 11550–11553.
McMartin, C., and Bohacek, R. S. (1995). Flexible matching of test ligands to a 3D pharmacophore using a molecular superposition force field: Comparison of predicted and experimental conformations of inhibitors of three enzymes. *J. Comput.-Aided Mol. Design* **9**, 237–250.
Meng, E. C., Stoichet, B. K., and Kuntz, I. D. (1992). Automated docking with grid-based energy evaluation. *J. Comp. Chem.* **13**, 505–524.
Menziani, M. C., De Benedetti, P. G., Gago, F., and Richards, W. G. (1989). The binding of benzenesulfonamides to carbonic anhydrase enzyme: A molecular mechanics study and quantitative structure–activity relationships. *J. Med. Chem.* **32**, 951–6.
Metropolis, N., Rosenbluth, A. W., Rosenbluth, M. N., Teller, A. H., and Teller, E. (1953). Equation of state calculations by fast computing machines. *J. Chem. Phys.* **21**, 1087–1092.
Miller, M. D., Kearsley, S. K., Underwood, D. J., and Sheridan, R. P. (1994). FLOG: A system to select 'quasi-flexible' ligands complementary to a receptor of known three-dimensional structure. *J. Comput.-Aided Mol. Design* **8**, 153–174.
Milne, G. W. A., Nicklaus, M. C., Driscoll, J. S., and Wang, S. (1994). National Cancer Institute drug information system 3D database. *J. Chem. Inf. Comp. Sci.* **34**, 1219–1224.
Miyamoto, S., and Kollman, P. A. (1993). Absolute and relative binding free energy calculations of the interaction of biotin and its analogs with streptavidin using molecular dynamics/free energy perturbation approaches. *Proteins* **16**, 226–245.
Mizushima, N., Spellmeyer, D., Hirono, S., Pearlman, D., and Kollman, P. (1991). Free energy perturbation calculations on binding and catalysis after mutating threonine 220 in subtilisin. *J. Biol. Chem.* **266**, 11801–11809.
Mol, J. N. M., and van der Krol, A. R., eds. (1991). Antisense Nucleic Acids and Proteins: Fundamentals and Applications." Dekker, New York.
Momany, F., Pitha, R., Klimkowski, V. J., and Venkatachalam, C. M. (1989). Drug design using a protein pseudoreceptor. *In* "Expert Systems Applications in Chemistry" (B. A. Hohne and T. H. Pierce, eds.), pp. 82–91. American Chemical Society, Washington, D.C.
Montgomery, J. A., Niwas, S., Rose, J. D., Secrist, J. A. D., Babu, Y. S., Bugg, C. E., Erion, M. D., Guida, W. C., and Ealick, S. E. (1993). Structure-based design of inhibitors of purine nucleoside phosphorylase. 1. 9-(arylmethyl) derivatives of 9-deazaguanine. *J. Med. Chem.* **36**, 55–69.
Moody, E. J., Harris, B. D., and Skolnick, P. (1994). The potential for safer anaesthesia using stereoselective anaesthetics. *Trends Pharmacol. Sci.* **15**, 387–391.
Morgan, B. A., and Gainor, J. A. (1989). Approaches to the discovery of non-peptide ligands for peptide receptors and peptidases. *In* "Ann. Rep. Med. Chem." (F. J. Vinick, ed.), Vol. 24, pp. 243–252. Academic Press, San Diego.
Moriguchi, I., Hirono, S., Nakagome, I., and Hirano, H. (1994). Comparison of reliability of log P values for drugs calculated by several methods. *Chem. Pharmacol. Bull.* **42**, 976–978.
Morris, A. L., MacArthur, M. W., Hutchinson, E. G., and Thornton, J. M. (1992). Stereochemical quality of protein structure coordinates. *Proteins* **12**, 345–64.
Morton, A., Baase, W. A., and Matthews, B. W. (1995). Energetic origins of specificity of ligand binding in an interior nonpolar cavity of T4 lysozyme. *Biochemistry* **34**, 8564–8575.
Morton, A., and Matthews, B. W. (1995). Specificity of ligand binding in a buried nonpolar cavity of T4 lysozyme: Linkage of dynamics and structural plasticity. *Biochemistry* **34**, 8576–8588.
Müller, K., Obrecht, D., Knierzinger, A., Stankovic, C., Spielger, C., Bannwarth, W., Trzeciak,

A., Englert, G., Labhardt, A. M., and Schönholzer, P. (1993). Building blocks for the induction or fixation of peptide conformation. *In* "Perspectives in Medicinal Chemistry" (B. Testa, E. Kyburz, W. Fuhrer, and R. Giger, eds.), pp. 513–531. VCH, Weinheim.

Murphy, K. P., Xie, D., Garcia, K. C., Amzel, L. M., and Freire, E. (1993). Structural energetics of peptide recognition: Angiotensin II/antibody binding. *Proteins* **15**, 113–120.

Murray-Rust, P., and Glusker, J. P. (1984). Directional hydrogen bonding to $sp^2$ and $sp^3$-hybridized oxygen atoms and its relevance to ligand–macromolecular interactions. *J. Am. Chem. Soc.* **106**, 1018–1025.

Murthy, K. H., Winborne, E. L., Minnich, M. D., Culp, J. S., and Debouck, C. (1992). The crystal structures at 2.2-Å resolution of hydroxyethylene-based inhibitors bound to human immunodeficiency virus type 1 protease show that the inhibitors are present in two distinct orientations. *J. Biol. Chem.* **267**, 22770–22778.

Needleman, S. B., and Wunsch, C. (1970). A general method applicable to the search for similarities in the amino acid sequence of two proteins. *J. Mol. Biol.* **48**, 444–453.

Nellans, H. N. (1991). Mechanisms of peptide and protein absorption. Paracellular intestinal transport: Modulation of absorption. *Adv. Drug Deliv. Rev.* **7**, 339–364.

Nicholls, A., Sharp, K. A., and Honig, B. (1991). Protein folding and association: Insights from the interfacial and thermodynamic properties of hydrocarbons. *Proteins* **11**, 281–296.

Nicklaus, M. C., Milne, G. W., and Burke, T., Jr. (1992). QSAR of conformationally flexible molecules: Comparative molecular field analysis of protein-tyrosine kinase inhibitors. *J. Comput.-Aided Mol. Des.* **6**, 487–504.

Nicklaus, M. C., Wang, W., Driscoll, J. S., and Milne, G. W. A. (1995). Conformational changes of small molecules binding to preotins. *Bioorgan. Med. Chem.* **3**, 411–428.

Novoa, J. J., and Whangbo, M.-H. (1991). The nature of intramolecular hydrogen-bonded and non-hydrogen-bonded conformations of simple di- and triamides. *J. Am. Chem. Soc.* **113**, 9017–9026.

Novotny, J., Rashin, A. A., and Bruccoleri, R. E. (1988). Criteria that discriminate between native proteins and incorrectly folded models. *Proteins* **4**, 19–30.

Ortner, M., Lackner, P., and Sippl, M. J. (1994). "PROSA: PROtein Structure Analysis." Salzburg, Techo-A Salzburg.

Pace, C. N. (1992). Contribution of hydrophobic effect to globular protein stability. *J. Mol. Biol.* **226**, 29–35.

Page, M. I. (1990a). Enzyme catalysis. *In* "Comprehensive Medicinal Chemistry" (P. G. Sammes, eds.), Vol. 2, pp. 45–60. Pergamon Press, Oxford.

Page, M. I. (1990b). Enzyme inhibition. *In* "Comprehensive Medicinal Chemistry" (P. G. Sammes, eds.), Vol. 2, pp. 61–87. Pergamon Press, Oxford.

Pascard, C. (1995). Small-molecule crystal structures as a structural basis for drug design. *Acta Crystallogr. Sect. D* **51**, 407–417.

Payne, A. W., and Glen, R. C. (1993). Molecular recognition using a binary genetic search algorithm. *J. Mol. Graph.* **11**, 74–91.

Pearlman, D. A., Case, D. A., Caldwell, J. W., Seibel, G. L., Singh, U. C., Weiner, P. K., and Kollman, P. A. (1991). "AMBER 4.0." University of California, San Francisco.

Pearlman, D. A., and Connelly, P. R. (1995). Determination of the differential effects of hydrogen bonding and water release on the binding of FK506 to native and Tyr82 → Phe82 FKBP-12 proteins using free energy simulations. *J. Mol. Biol.* **248**, 696–717.

Pearlman, R. S. (1993). 3D molecular structures: Generation and use in 3D searching. *In* "3D QSAR in Drug Design" (H. Kubinyi, ed.), pp. 41–79. ESCOM, Leiden.

Pearson, W. R. (1995). Comparison of methods for searching protein sequence databases. *Prot. Sci.* **4**, 1145–1160.

Perrin, D. D., Dempsey, B., and Serjeant, E. P. (1981). "pKa Predictions for Organic Acids and Bases." Chapman and Hall, London/New York.

Perutz, M. F., Fermi, G., Abraham, D. J., Poyart, G., and Bursaux, E. (1986). Hemoglobin as a receptor of drugs and peptides: X-ray studies of the stereochemistry of binding. *J. Am. Chem. Soc.* **108,** 1064–1078.

Pfeiffer, C. C. (1956). Optical isomerism and pharmacological action, a generalization. *Science* **124,** 29–31.

Pickersgill, R. W. (1988). A rapid method of calculating charge–charge interaction energies in proteins. *Protein Eng.* **2,** 247–248.

Ponder, J. W., and Richards, F. M. (1987). Tertiary templates for proteins: Use of packing criteria in the enumeration of allowed sequences for different structural classes. *J. Mol. Biol.* **193,** 775–793.

Pranata, J., and Jorgensen, W. L. (1991). Computational studies on FK506: Conformational search and molecular dynamics simulations in water. *J. Am. Chem. Soc.* **113,** 9483–9493.

Pranata, J., Wierschke, S. G., and Jorgensen, W. L. (1991). OPLS potential functions for nucleotide bases: Relative association constants of hydrogen-bonded base pairs in chloroform. *J. Am. Chem. Soc.* **113,** 2810–2819.

Ramasubbu, N., Parthasarathy, R., and Murray-Rust, P. (1986). Angular preferences of intermolecular forces around halogen centers: Preferred directions of approach of electrophiles and nucleophiles around the carbon–halogen bond. *J. Am. Chem. Soc.* **108,** 4308–4314.

Rao, B. G., and Murcko, M. A. (1994). Reversed stereochemical preference in binding of Ro 31-8959 to HIV-1 proteinase: A free energy perturbation analysis. *J. Comput. Chem.* **15,** 1241–1253.

Rao, B. G., and Singh, U. C. (1990). A free energy perturbation study of solvation in methanol and dimethyl sulfoxide. *J. Am. Chem. Soc.* **112,** 3803–3811.

Rao, B. G., and Singh, U. C. (1991a). A free energy perturbation study of solvation in hydrazine and carbon tetrachloride. *J. Am. Chem. Soc.* **113,** 4381–4389.

Rao, B. G., and Singh, U. C. (1991b). Studies on the binding of pepstatin and its derivatives to *Rhizopus* pepsin by quantum mechanics, molecular mechanics, and free energy perturbation methods. *J. Am. Chem. Soc.* **113,** 6735–6750.

Rao, B. G., Tilton, R. F., and Singh, U. C. (1992). Free energy perturbation studies on inhibitor binding to HIV-1 proteinase. *J. Am. Chem. Soc.* **114,** 4447–4452.

Rao, S. N. (1992). Bioactive conformation of Arg–Gly–Asp by X-ray data analyses and molecular mechanics. *Pept. Res.* **5,** 148–155.

Rao, S. N., and Kollman, P. A. (1987). Molecular mechanical simulations on double intercalation of 9-amino acridine into d(CGCGCGC).d(GCGCGCG): Analysis of the physical basis for neighbor-exclusion principle. *Proc. Natl. Acad. Sci. USA* **84,** 5735–5739.

Rao, S. N., Singh, U. C., Bash, P. A., and Kollman, P. A. (1987). Free energy perturbation calculations on binding and catalysis after mutating Asn-155 in subtilisin. *Nature* **328,** 551–554.

Reddy, M. R., Varney, M. D., Kalish, V., Viswanadhan, V. N., and Appelt, K. (1994). Calculations of relative differences in binding free energy of HIV-1 protease inhibitors: A thermodynamic perturbation approach. *J. Med. Chem.* **37,** 1145–1152.

Reithmeier, R. A. F. (1995). Characterization and modeling of membrane proteins using sequence analysis. *Curr. Opin. Struct. Biol.* **5,** 491–500.

Rich, D. H. (1993). Effect of hydrophobic collapse on enzyme–inhibitor interactions: Implications for the design of peptidomimetics. *In* "Perspectives in Medicinal Chemistry" (B. Testa, E. Kyburz, W. Fuhrer, and R. Giger, eds.), pp. 15–25. VCH, Weinheim.

Rich, D. H., Sun, C. Q., Vara-Prasad, J. V., Pathiasseril, A., Toth, M. V., Marshall, G. R., Clare, M., Mueller, R. A., and Houseman, K. (1991). Effect of hydroxyl group configuration in hydroxyethylamine dipeptide isosteres on HIV protease inhibition: Evidence for multiple binding modes. *J. Med. Chem.* **34,** 1222–1225.

Ring, C. S., Sun, E., McKerrow, J. H., Lee, G. K., Rosenthal, P. J., Kuntz, I. D., and Cohen,

F. E. (1993). Structure-based inhibitor design by using protein models for the development of antiparasitic agents. *Proc. Natl. Acad. Sci. USA* **90**, 3583–3587.

Roberts, G. C. K. (1991). Conformational flexibility and protein specificity. *In* "Host-Guest Molecular Interactions: From Chemistry to Biology (Ciba Foundation symposium 158)" (D. J. Chadwick and K. Widdows, eds.), Vol. 158, pp. 169–186. Wiley, Chichester.

Roper, D., Jacoby, E., Kruger, P. Engels, M., Grotzinger, J., Wollmer, A., and Strassburger, W. (1994). Modeling of G-protein coupled receptors with bacteriorhodopsin as a template: A novel approach based on interaction energy differences. *J. Recept. Res.* **14**, 167–186.

Rosen, M. K., Standaert, R. F., Galat, A., Nakatsuka, M., and Schreiber, S. L. (1990). Inhibition of FKBP rotamase activity by immunosuppressant FK506: Twisted amide surrogate. *Science* **248**, 863–866.

Rosenberg, S. H., Spina, K. P., Condon, S. L., Polakowski, J., Yao, Z., Kovar, P., Stein, H. H., Cohen, J., Barlow, J. L., and Klinghofer, V. (1993). Studies directed toward the design of orally active renin inhibitors. 2. Development of the efficacious, bioavailable renin inhibitor (2S)-2-benzyl-3-[[(1-methylpiperazin-4-yl)sulfonyl]propionyl]-3-thiazol-4-yl-L-alanine amide of (2S,3R,4S)-2-amino-1-cyclohexyl-3,4-dihydroxy-6-methylheptane (A-72517). *J. Med. Chem.* **36**, 460–467.

Rost, B., and Sander, C. (1994). Structure predictions of proteins: Where are we now? *Cur. Opin. Biotechnol.* **5**, 372–380.

Rutenber, E., Fauman, E. B., Keenan, R. J., Fong, S., Furth, P. S., Ortiz de Montellano, P.R., Meng, E., Kuntz, I. D., DeCamp, D. L., Salto, R., Rosè, J. R., Craik, C. S., and Stroud, R. M. (1993). Structure of a non-peptide inhibitor complexed with HIV-1 protease: Developing a cycle of structure-based drug design. *J. Biol. Chem.* **268**, 15343-15346.

Saenger, W. (1984). Structures and conformational properties of bases, furanose sugars and phosphate groups. *In* "Principles of Nucleic Acid Structure," pp. 51–104.

Scharpe, S., De-Meester, I., Hendriks, D., Vanhoof, G., van-Sande, M., and Vriend, G. (1991). Proteases and their inhibitors: Today and tomorrow. *Biochimie* **73**, 121–126.

Scherrer, R. A., and Howard, S. M. (1977). Use of distribution coefficients in quantitative structure–activity relationships. *J. Med. Chem.* **20**, 53–58.

Schiffer, C. A., Caldwell, J. W., Stroud, R. M., and Kollman, P. A. (1992). Inclusion of solvation free energy with molecular mechanics energy: Alanyl dipeptide as a test case. *Protein Sci.* **1**, 396–400.

Schiffer, C. A., Caldwell, J. W., Stroud, R. M., and Kollman, P. A. (1993). Protein structure prediction with a combined solvation free energy-molecular mechanics force field. *Mol. Simulat.* **10**, 121–149.

Schrauber, H., Eisenhaber, F., and Argos, P. (1993). Rotamers-to be or not to be: Analysis of amino acid and side-chain conformations in globular proteins. *J. Mol. Biol.* **230**, 592–612.

Secrist, J. A., III, Niwas, S., Rose, J. D., Babu, Y. S., Bugg, C. E., Erion, M. D., Guida, W. C., Ealick, S. E., and Montgomery, J. A. (1993). Structure-based design of inhibitors of purine nucleoside phosphorylase. 2. 9-Alicyclic and 9-heteroalicyclic derivatives of 9-deazaguanine. *J. Med. Chem.* **36**, 1847–1854.

Senderowitz, H., Guarnieri, F., and Still, W. C. (1995). A smart Monte Carlo technique for free energy simulations of multiconformational molecules. Direct calculations of the conformational populations of organic molecules. *J. Am. Chem. Soc.* **117**, 8211–8219.

Seri-Levy, A., West, S., and Richards, W. G. (1994). Molecular similarity, quantitative chirality, and QSAR for chiral drugs. *J. Med. Chem.* **37**, 1727–1732.

Serrano, L., Kellis, J. T., Jr., Cann, P., Matouschek, A., and Fersht, A. R. (1992). The folding of an enzyme. II. Substructure of barnase and the contribution of different interactions to protein stability. *J. Mol. Biol.* **224**, 783–804.

Shacter, E., Stadtman, E. R., Jurgensen, S. R., and P. B., C. (1988). Role of cAMP and cyclic cascade regulation. *In* "Methods in Enzymology" (J. D. Corbin and R. A. Johnson, eds.), Vol. 159, pp. 3–19. Academic Press, San Diego.

Sharp, K., Fine, R., and Honig, B. (1987). Computer simulations of the diffusion of substrate to an active site of an enzyme. *Science* **236**, 1460–1463.
Sharp, K. A., Nicholls, A., Friedman, R., and Honig, B. (1991). Extracting hydrophobic free energies from experimental data: Relationship to protein folding and theoretical models. *Biochemistry* **30**, 9686–9697.
Shen, J., and Quiocho, F. A. (1995). Calculations of binding energy differences for receptor–ligand systems using the Poisson-Boltzmann method. *J. Comput. Chem.* **16**, 445–448.
Shenkin, P. S., Yarmush, D. L., Fine, R. M., Wang, H., and Levinthal, C. (1987). Predicting antibody hypervariable loop conformations. I. Ensembles of random conformations for ringlike structures. *Biopolymers* **26**, 2053–2085.
Sheridan, R. P., and Kearsley, S. K. (1995). Using a genetic algorithm to suggest combinatorial libraries. *J. Chem. Inf. Comp. Sci.* **35**, 310–320.
Sheridan, R. P., Nilakantan, R., Dixon, J. S., and Venkataraghavan, R. (1986). The ensemble approach to distance geometry: Application to the nicotinic pharmacophore. *J. Med. Chem.* **29**, 899–906.
Shirley, B. A., Stanssens, P., Hahn, U., and Pace, C. N. (1992). Contribution of hydrogen bonding to the conformational stability of ribonuclease T1. *Biochemistry* **31**, 725–732.
Singh, U. C., Brown, F. K., Bash, P. A., and Kollman, P. A. (1987). An approach to the application of free energy perturbation methods using molecular dynamics: Applications to the transformations of $CH_3OH \rightarrow CH_3CH_3$, $H_3O^+ \rightarrow NH_4^+$, glycine → alanine, and alanine → phenylalanine in aqueous solution and to $H_3O^+(H_2O)_3 \rightarrow NH_4^+(H_2O)_3$ in the gas phase. *J. Am. Chem. Soc.* **109**, 1605–1614.
Sippl, M. J. (1993). Recognition of errors in three-dimensional structures of proteins. *Proteins* **17**, 355–362.
Sitkoff, D., Sharp, K. A., and Honig, B. (1994). Accurate calculation of hydration free energies using macroscopic solvent models. *J. Phys. Chem.* **98**, 1978–1988.
Smith, D. A., and Vijayakumar, S. (1991). Molecular modeling of intramolecular hydrogen bonding in simple oligoamides 1. In vacuo. *Tet. Lett.* **32**, 3613–3616.
Smith, K. C., and Honig, B. (1994). Evaluation of the conformational free energies of loops in proteins. *Proteins* **18**, 119–132.
Smith, P. E., and Pettitt, B. M. (1994). Modeling solvent in biomolecular systems. *J. Phys. Chem.* **98**, 9700–9711.
Snyder, J. P., Rao, S. N., Koehler, K. F., and Pellicciari, R. (1992). Drug modeling at cell membrane receptors: The concept of pseudoreceptors. *In* "Trends in Receptor Research" (P. Angel, U. Gulini, and W. Quagli, eds.), pp. 367–403. Elsevier, Amsterdam.
Snyder, J. P., Rao, S. N., Koehler, K. F., Vedani, A., and Pellicciari, R. (1993). APOLLO pharmacophores and the pseudoreceptor concept. *In* "Trends in QSAR and Molecular Modelling 92: Proceedings of the 9th European Symposium on Structure–Activity Relationships: QSAR and Molecular Modelling" (C. G. Wermuth, ed.), pp. 44–51. Elsevier, Amsterdam.
Solmajer, T., and Mehler, E. L. (1991). Electrostatic screening in molecular dynamics simulations. *Protein Eng.* **4**, 911–917.
Still, W. C., Tempczyk, A., Hawley, R. C., and Henrickson, T. (1990). Semianalytical treatment of solvation for molecular mechanics and dynamics. *J. Am. Chem. Soc.* **112**, 6127–6129.
Stinson, S. C. (1994). Chiral drugs. *Chem. Eng. News* **72** (issue 38, Sept. 19), 38–72.
Stouten, P. F. W., Frömmel, C., Nakamura, H., and Sander, C. (1993). An effective solvation term based on atomic occupancies for use in protein simulations. *Mol. Simulat.* **10**, 97–120.
Strader C. D., Fong, T. M., Graziano, M. P., and Tota, M. R. (1995). The family of G-protein-coupled receptors. *FASEB J.* **9**, 745–754.
Strader, C. D., Fong, T. M., Tota, M. R., Underwood, D., Dixon, R. A. F. (1994). Structure and function of G protein-coupled receptors. *Annu. Rev. Biochem.* **63**, 101–132.

Sufrin, J. R., Dunn, D. A., and Marshall, G. R. (1981). Steric mapping of the L-methionine binding site of ATP: L-Methionine S-adenosyltransferase. *Mol. Pharmacol.* **19,** 307–313.
Sun, Y., Spellmeyer, D., Pearlman, D. A., and Kollman, P. (1992). Simulation of the solvation free energies of methane, ethane, and propane and corresponding amino acid dipeptides: A critical test of the "bond-PMF" correction, a new set of hydrocarbon parameters, and the gas phase-water hydrophobicity scale. *J. Am. Chem. Soc.* **114,** 6798–6801.
Sussman, J. L., Harel, M., Frolow, F., Oefner, C., Goldman, A., Toker, L., and Silman, I. (1991). Atomic structure of acetylcholinesterease from *Torpedo californica:* A prototypic acetylcholine-binding protein. *Science* **253,** 872–879.
Tanokura, M. (1983). 1H-NMR study on the tautomerism of the imidazole ring of histidine residues. I. Microscopic pK values and molar ratios of tautomers in histidine-containing peptides. *Biochim. Biophys. Acta* **742,** 576–585.
Taylor, W. R., Flores, T. P., and Orengo, C. A. (1994). Multiple protein structure alignment. *Protein Sci.* **3,** 1858–1870.
Teeter, M. M., Froimowitz, M., Stec, B., and DuRand, C. J. (1994). Homology modeling of the dopamine D2 receptor and its testing by docking of agonists and tricyclic antagonists. *J. Med. Chem.* **37,** 2874–2888.
Tello, D., Goldbaum, F. A., Mariuzza, R. A., Ysern, X., Schwarz, F. P., and Poljak, R. J. (1993). Three-dimensional structure and thermodynamics of antigen binding by anti-lysozyme antibodies. *Biochem. Soc. Trans.* **21,** 943–946.
Testa, B., Jenner, P., Kilpatrick, G. J., El Tayar, N., van de Waterbeemd, H., and Marsden, C. D. (1987). Do thermodynamic studies provide information on both the binding to and the activation of dopaminergic and other receptors? *Biochem. Pharmacol.* **36,** 4041–4046.
Thanki, N., Thornton, J. M., and Goodfellow, J. M. (1988). Distributions of water around amino acid residues in proteins. *J. Mol. Biol.* **202,** 637–657.
Thompson, J. D., Higgins, D. G., and Gibson, T. J. (1994). CLUSTAL W: Improving the sensitivity of progressive multiple sequence alignment through sequence weighting, position-specific gap penalties and weight matrix choice. *Nucleic Acids Res.* **22,** 4673–4680.
Thornton, J. M., and Swindells, M. B. (1993). Modelling of related protein structures. *In* "Molecular Structures in Biology" (R. Diamond, T. F. Koetzle, K. Prout, and J. S. Richardson, eds.), pp. 82–113. Oxford University Press, Oxford.
Thorson, J. S., Chapman, E., and Schultz, P. G. (1995). Analysis of hydrogen bonding strength in proteins using unnatural amino acids. *J. Am. Chem. Soc.* **117,** 9361–9362.
Tintelnot, M., and Andrews, P. (1989). Geometries of functional group interactions in enzyme–ligands complexes: Guides for receptor modelling. *J. Comput.-Aided Mol. Des.* **3,** 67–84.
Tomasi, J., and Persico, M. (1994). Molecular interactions in solution: An overview of methods based on continuous distributions of solvent. *Chem. Rev.* **94,** 2027–2094.
Topham, C. M., McLeod, A., Eisenmenger, F., Overington, J. P., Johnson, M. S., and Blundell, T. L. (1993). Fragment ranking in modelling of protein structure: Conformationally constrained environmental amino acid substitution tables. *J. Mol. Biol.* **229,** 194–220.
Tronrud, D. E., Holden, H. M., and Matthews, B. W. (1987). Structures of two thermolysin–inhibitor complexes that differ by a single hydrogen bond. *Science* **235,** 571–574.
Tschinke, V., and Cohen, N. C. (1993). The NEWLEAD program: A new method for the design of candidate structures from pharmacophoric hypotheses. *J. Med. Chem.* **36,** 3863–3870.
Underwood, D. J., Strader, C. D., Rivero, R., Patchett, A. A., Greenlee, W., and Prendergast, K. (1994). Structural model of antagonist and agonist binding to the angiotensin II, $AT_1$ subtype, G protein coupled receptor. *Chem. Biol.* **1,** 211–221.
Vajda, S., Weng, Z., Rosenfeld, R., and DeLisi, C. (1994). Effect of conformational flexibility and solvation on receptor–ligand binding free energies. *Biochemistry* **33,** 13977–13988.
Van de Waterbeemd, H., El Tayar, N., Testa, B., Wikström, H., and Largent, B. (1987). Quantitative structure–activity relationships and eudimsic analyses of the presynaptic

dopaminergic activity and dopamine D2 and $\sigma$ receptor affinities of 3-(3-hydroxyphenyl)-piperidines and octahydrobenzo(f)quinolones. *J. Med. Chem.* **30,** 2175–2181.

van Gunsteren, W. F., and Mark, A. E. (1992). On the interpretation of biochemical data by molecular dynamics computer simulation. *Eur. J. Biochem.* **204,** 947–961.

Varghese, J. N., Epa, V. C., and Colman, P. M. (1995). Three-dimensional structure of the complex of 4-guanidino-Neu5Ac2en and influenza virus neuraminidase. *Prot. Sci.* **4,** 1081–1087.

Veale, C. A., Damewood, J. R., Jr., Steelman, G. B., Bryant, C., Gomes, B.. and Williams, J. (1995). Non-peptidic inhibitors of human leukocyte elastase. 4. Design, synthesis, and *in vitro* and *in vivo* activity of a series of $\beta$-carbolinone-containing trifluoromethyl ketones. *J. Med. Chem.* **38,** 86–97.

Vedani, A., and Dunitz, J. D. (1985). Lone-pair directionality in H-bond potential functions for molecular mechanics calculations: The inhibition of human carbonic anhydrase II by sulfonamides. *J. Am. Chem. Soc.* **107,** 7653–7658.

Vedani, A., and Huhta, D. W. (1990). An algorithm for the systematic solvation of proteins based on the directionality of hydrogen bonds. *J. Am. Chem. Soc.* **112,** 4759–4767.

Vedani, A., Zbinden, P., and Snyder, J. P. (1993). Pseudo-receptor modeling: A new concept for the three-dimensional construction of receptor binding sites. *J. Recept. Res.* **13,** 163–177.

Villar, H. O., and Kauvar, L. M. (1994). Amino acid preferences at protein binding sites. *FEBS Lett.* **349,** 125–130.

Viswanadhan, V. N., Reddy, M. R., Bacquet, R. J., and Erion, M. D. (1993). Assessment of methods used for predicting lipophilicity: Application to nucleosides and nucleoside bases. *J. Comput. Chem.* **14,** 1019–1026.

von Freyberg, B., and Braun, W. (1993). Minimization of empirical energy functions in proteins including hydrophobic surface area effects. *J. Comput. Chem.* **14,** 510–521.

von Freyberg, B., Richmond, T. J., and Braun, W. (1993). Surface area included in energy refinement of proteins: A comparative study on atomic solvation parameters. *J. Mol. Biol.* **233,** 275–292.

von Itzstein, M., Wu, W. Y., Kok. G. B., Pegg, M. S., Dyason, J. C., Jin, B., Van Phan, T., Smythe, M. L., White, H. F., Oliver, S. W., Colman, P. M., Varghese, J. N., Ryan, D. M., Woods, J. M., Bethell, R. C., Hotham, V. J., Cameron, J. M., and Penn, C. R. (1993). Rational design of potent sialidase-based inhibitors of influenza virus replication. *Nature* **363,** 418–423.

Vriend, G., Sander, C., and Stouten, P. F. W. (1994). A novel search method for protein–structure relations using property profiles. *Prot. Eng.* **7,** 23–29.

Wade, R. C., Clark, K. J., and Goodford, P. J. (1993). Further development of hydrogen bond functions for use in determining energetically favorable binding sites on molecules of known structure. 2. Ligand probe groups with the ability to form more than two hydrogen bonds. *J. Med. Chem.* **36,** 148–156.

Walters, D. E., and Hinds, R. M. (1994). Genetically evolved receptor models: A Computational approach to construction of receptor models. *J. Med. Chem.* **37,** 2527–2536.

Wang, B., and Ford, J. P. (1992). Molecular orbital theory of a solute in a continuum with an arbitrarily shaped boundary represented by finite surface elements. *J. Chem. Phys.* **97,** 4162–4169.

Warner, P., Green, R. C., Gomes, B., Strimpler, A. M. (1994). Nonpeptidic inhibitors of human leukocyte elastase. 1. The design and synthesis of pyridone-containing inhibitors. *J. Med. Chem.* **37,** 3090–3099.

Warshel, A., Sussman, F., and Hwang, J. K. (1988). Evaluation of catalytic free energies in genetically modified proteins. *J. Mol. Biol.* **201,** 139–159.

Weber, A. E., Steiner, M. G., Krieter, P. A. Colletti, A. E., Tata, J. R., Halgren, T. A., Ball, R. G., Doyle, J. J., Schorn, T. W., Stearns, R. A., Miller, R. R., Siegl, P. K. S., Greenlee,

W. J., and Patchett, A. A. (1992a). Highly potent, orally active diester macrocyclic human renin inhibitors. *J. Med. Chem.* **35**, 3755–3773.

Weber, P. C., Wendoloski, J. J., Pantoliano, M. W., and Salemme, F. R. (1992b). Crystallographic and thermodynamic comparison of natural and synthetic ligands bound to streptavidin. *J. Am. Chem. Soc.* **114**, 3197–3200.

Wei, X., Ghosh, S. K., Taylor, M. E., Johnson, V. A., Emini, E. A., Deutsch, P., Lifson, J. D., Bonhoeffer, S., Nowak, M. A., Hahn, B. H., Saag, M. S., and Shaw, G. M. (1995). Viral dynamics in human immunodeficiency virus type 1 infection. *Nature* **373**, 117–122.

Weiland, G. A., Minneman, K. P., and Molinoff, P. B. (1979). Fundamental difference between the molecular interactions of agonists and antagonists with the beta-adrenergic receptor. *Nature* **281**, 114–117.

Weiner, S. J., Kollman, P. A., Case, D. A., Singh, U. C., Ghio, C., Alagona, G., Profeta, S., and Weiner, P. (1984). A new force field for molecular mechanical simulation of nucleic acids and proteins. *J. Am. Chem. Soc.* **106**, 765–784.

Weiner, S. J., Singh, U. C., and Kollman, P. A. (1985). Simulation of formamide hydrolysis by hydroxide ion in the gas phase and in aqueous solution. *J. Am. Chem. Soc.* **107**, 2219–2229.

Wesson, L., and Eisenberg, D. (1992). Atomic solvation parameters applied to molecular dynamics of proteins in solution. *Prot. Sci.* **1**, 227–235.

Williams, D. H. (1992). The molecular basis of biological order and amide–amide hydrogen bonds: An addendum. *Aldrichim. Acta* **25**, 9.

Williams, M. A., Goodfellow, J. M., and Thornton, J. M. (1994). Buried waters and internal cavities in monomeric proteins. *Prot. Sci.* **3**, 1224-1235.

Williams, R. L., Vila, J., Perrot, G., and Scheraga, H. A. (1992). Empirical solvation models in the context of conformational energy searches: Application to bovine pancreatic trypsin inhibitor. *Proteins* **14**, 110–119.

Wilson, C., Mace, J. E., and Agard, D. A. (1991). Computational method for the design of enzymes with altered substrate specificity. *J. Mol. Biol.* **220**, 495–506.

Wilson, I. A., and Stanfield, R. L. (1993). Antibody–antigen interactions. *Curr. Opin. Struct. Biol.* **3**, 113–118.

Wodak, S. J., and Rooman, M. J. (1993). Generation and testing protein folds. *Curr. Opin. Struct. Biol.* **3**, 247–259.

Wong, C. F., and McCammon, J. A. (1986). Dynamics and design of enzyme inhibitors. *J. Am. Chem. Soc.* **108**, 3830–3832.

Yang, A.-S., and Honig, B. (1995). Free energy determinants of secondary structure formation. II. Antiparallel $\beta$-sheets. *J. Mol. Biol.* **252**, 366–376.

Zablocki, J., Rao, S. N., Baron, D. A., Flynn, D., Nicholson, N., and Feigen, L. (1995a). Fibrinogen receptor antagonists. *Curr. Pharm. Design*, in press.

Zablocki, J., Rico, J., Garland, R., Rogers, T., Williams, K., Schretzman, L., Rao, S. N., Bovy, P. R., Tjoeng, F., Lindmark, R., Toth, M., Zupec, M., McMackins, D., Adams, S. P., Miyano, M., Markos, C., Milton, M., Paulson, S., Herin, M., Jacqmin, P., Nicholson, N., Panzer-Knodle, S., Haas, N., Page, J., Szalony, J., Taite, B., Salyers, A., King, L., Campion, J., and Feigen, L. (1995b). Potent *in vitro* and *in vivo* inhibitors of platelet aggregation based upon the Arg-Gly-Asp sequence of fibrinogen (aminobenz-amidino)succinyl (ABAS) series of orally active fibrinogen receptor antagonists. *J. Med. Chem.* **38**, 2378–2394.

Zachmann, C.-D., Kast, S. M., and Brickmann, J. (1995). Quantification and visualization of molecular surface flexibility. *J. Mol. Graph.* **13**, 89–97.

Zuckermann, R. N., Martin, E. J., Spellmeyer, D. C., Stauber, G. B., Shoemaker, K. R., Kerr, J. M., Figliozzi, G. M., Goff, D. A., Siani, M. A., Simon, R. J., Banville, S. C., Brown, E. G., Wang, L., Richter, L. S., and Moos, W. H. (1994). Discovery of nanomolar ligands for 7-transmembrane G-protein-coupled receptors from a diverse N-(substituted)glycine peptoid library. *J. Med. Chem.* **37**, 2678–2685.

Zuiderweg, E. R. P., van Doren, S. R., Kurochkin, A. V., Neubig, R. R., and Majumdar, A. (1993). Modern NMR spectroscopy of proteins and peptides in solution and its relevance to drug design. *Perspect. Drug Discov. Design* **1**, 391–417.

Zwanzig, R. W. (1954). High-temperature equation of state by a perturbation method. I. Nonpolar gases. *J. Chem. Phys.* **22**, 1420–1426.

# 8

# Glossary of Terminology

**J. P. TOLLENAERE**
Department of Theoretical Medicinal Chemistry
Janssen Research Foundation
B-2340 Beerse, Belgium and
Department of Pharmaceutical Chemistry
Utrecht University
3508 TB Utrecht, The Netherlands

**Ab initio** A quantum mechanical nonparametrized molecular orbital treatment (from "first principles") for the description of chemical behavior taking into account nuclei and all electrons. In principle, it is the most accurate of the three computational methodologies: ab initio, semiempirical all-valence electron methods, and molecular mechanics.

**Active analog approach** In the absence of information regarding the receptor, a medicinal chemist may modify known active structures from which one or several pharmacophoric patterns can be deduced. Then a set of possible (low energy) conformations for each compound known to activate the receptor is calculated. For each allowed conformation, the pharmacophoric pattern is determined. The intersection of all generated pharmacophoric patterns may then yield the pharmacophore embedded in all compounds of the set of active analogs.

**Adiabatic searching** Adiabatic (Greek: not passing through) conformational searching in which no strain energy enters or leaves the molecule because at each step during the rotation around a bond all molecular strain energy is relaxed by minimizing all bond stretches and bond angles.

**All valence electron methods** In contrast to ab initio methods, the semiempirical molecular orbital methods only consider the valence electrons for the construction of the atomic orbitals. Well-known semiempirical methods are EHT, CNDO, MNDO, PCILO, AM1, and PM3. These methods are orders of magnitude faster than ab initio calculations.

**AM1** (Austin Model 1) The latest of the semiempirical quantum chemical methods originating from the M. J. S. Dewar group. The quality of the AM1 results in most cases is beyond the simpler ab initio results.

**AMBER** (Assisted Model Building with Energy Refinement) A widely used computer program to build models of molecules and to calculate their interactions using an empirical force field consisting of the usual bond stretch and bond angle deformation terms, a cosine function for the dihedrals, and where the nonbonded interactions are represented by a (6–12) Lennard–Jones potential and a Coulomb term and an optional (10–12) hydrogen bond potential.

**AMPAC** A semiempirical all valence electron program dedicated to the study of the chemical behavior of molecules and ions. Although both AMPAC and MOPAC originate from a set of subroutines developed by M. J. S. Dewar and co-workers at the University of Texas at Austin, the two programs diverged in 1985 and were called AMPAC and MOPAC. The most obvious difference between MOPAC and AMPAC is the absence of the PM3 Hamiltonian in the latter. AMPAC has a number of excellent techniques for investigating chemical reactions and in particular the location of transition states.

**Bioactive conformation** The bioactive conformation or the biologically relevant conformation can be defined either as the conformation a molecule must adopt in order to be recognized by the receptor or as the conformation of the ligand at the receptor site after binding. The dual interpretation of bioactive conformation stems from the fact that the environment of the ligand at the stage of recognition of the receptor or when it is fulfilling its biological role is not well understood.

**Born–Oppenheimer approximation** This approximation consists of separating the motion of nuclei from the electronic motion. The nuclei, being so much heavier than electrons, may then be treated as stationary as the electrons move around them. The Schrödinger equation can then be solved for the electrons alone at a definite internuclear separation. The Born–Oppenheimer approximation is quite good for the calculation of the behavior of molecules in the ground state.

**BSSE** (Basis Set Superposition Error) The BSSE in ab initio quantum chemical calculations of intermolecular interactions arises from a minor imbalance between the description given for the complex and its individual constituents. When two molecules approach each other, the description of a given molecule is energetically better within the complex than for the free monomer because orbitals of the partner molecule also become partly available leading to overestimated stabilization energies of weakly bonded complexes.

**CADD** (Computer-Aided Drug Design) In the broadest sense, CADD is the science and art of finding molecules of potential therapeutic value that satisfy a whole range of quantitative criteria such as high potency, high specificity, minimal toxic effects, and good bioavailability. CADD relies on computers, information science, statistics, mathematics, chemistry, physics, biology, and medicine. In a more narrow sense, CADD implies the use of computer graphics to visualize and manipulate chemical structures, to synthesize "in computro" new molecules, to determine their conformation, and to assess the similarities and dissimilarities between series of molecules. CADD further involves the calculation of the interaction energetics between drug molecules and hypothetical or experimentally determined macromolecular structures. CADD leads to insight in molecular recognition processes and, above all, stimulates the creativity of all those involved in drug research.

**CAMM** (computer-assisted molecular modeling); see CADD.

**CHARMM** (Chemistry at Harvard Macromolecular Mechanics) A widely used computer program that uses empirical energy functions to model macromolecular systems and ligands with various chemical functionalities. The empirical energy function is made of the usual quadratic bond stretch and bond angle deformation terms, a cosine function for the torsion angles, and a quadratic expression for the improper torsion to maintain chirality about a tetrahedral atom. The nonbonded terms are represented by the (6–12) Lennard–Jones potential, the Coulomb term, and an angle-dependent (10–12) hydrogen bond potential.

**CNDO** (Complete Neglect of Differential Overlap) One of the first semiempirical all valence electron methods formulated by J. A. Pople *et al.* in the 1960s. Because of the drastic simplifications dictated by the speed of the computers in those days, CNDO methods are superseded by more elaborate semiempirical quantum chemical calculations such as AM1 and PM3.

**CoMFA** (Comparative Molecular Field Analysis) The basic idea of CoMFA developed by R. D. Cramer *et al.* is that a suitable sampling of the steric and electrostatic field around a ligand molecule may provide all the information necessary for explaining its biological property. The steric and electrostatic contributions to the total interaction energy between the ligand and a chosen probe are calculated at regularly spaced grid points of a three-dimensional lattice encompassing the ligand.

**Computational chemistry** A branch of chemistry that can be defined as computer-assisted simulation of molecular systems and that is used to investigate the chemical behavior and properties of these systems by means of formalisms based on quantum mechanics, classical mechanics, and other mathematical techniques. Because of the ever increasing speed of computers, computational chemistry has become and will continue to be a viable alternative to chemical experimentation in cases where experiment is unfeasible, too dangerous, or too costly. According to a recent study, computational chemistry represented a $500 million market in 1991, reaching the $1 billion level in 1993 and probably $2 billion in 1996.

**CONCORD** A computer program using tables of standard bond lengths and bond angles in conjunction with expert system techniques and a simplified force field to generate a 3-D conformation from a 2-D structure representation. CONCORD is used to generate large 3-D databases containing tens or hundreds of thousands of structures which can be searched for identifying the presence of pharmacophores and finding new lead compounds. CONCORD accepts SMILES strings as input file formats.

**Conformational analysis** The study of the configuration of atoms and the relative molecular energies that result from rotation about any of the single bonds in a molecule. The possible individual arrangements of atoms in space are called conformers, conformational isomers, or rotamers. The methods of choice for the characterization of the conformation of molecules in the three aggregation states viz. solid (crystalline), dissolved, and gaseous (isolated state), are X-ray diffraction, NMR, and computational methods, respectively.

**Conformational partition function** The conformational partition function $Q_{con}$ is the summation of the Boltzmann factors $\exp[-\varepsilon(\tau_i)/kT]$ over the conformational energy levels $\varepsilon(\tau_i)$ of a molecule having $\tau_i$ torsional angles. The conformational partition function can be used when one wants to compare the conformational flexibility of a series of similar molecules.

**Conjugate gradients** A mathematical procedure to minimize a function such as a potential energy function used in molecular mechanics. The conjugate gradients is the method of choice to energy minimize large molecular systems.

**Connolly surface** The Connolly surface or solvent-accessible surface area is all the loci of the center of a solvent probe model, represented by a sphere with a given radius, free to touch but not to penetrate the van der Waals surface of the solute when the probe is rolled over the van der Waals surface of the solute.

**Constraint** A constraint in a target function such as the energy function in molecular mechanics is defined as a degree of freedom that is fixed or is not allowed to vary during the minimization of the energy.

**CORINA** The computer program CORINA originally developed to assess the influence of the spatial arrangement of atoms in a molecule on its reactivity generates 3-D models using standard bond lengths and angles. A molecule is fragmented into ring systems subdivided into small rings ($n \leq 8$), rigid and flexible macrocyclic systems, and acyclic parts. A pseudo force field is used to optimize geometries. CORINA appears to have one of the highest successful 2-D to 3-D conversion rates.

**Coulomb interaction** The Coulomb or charge–charge interaction arises from the attraction or repulsion of two charges and is inversely proportional to the distance separating the two charges. Because of this $1/r$ proportion, Coulomb interactions are long-range interactions and therefore are one of the major driving forces governing the recognition process between a ligand and its receptor. The interaction energy of two unit charges at a separation of 10 Å in a dielectric medium of $\varepsilon = 1$ amounts to about −33.2 kcal/mol.

**CPK** Corey–Pauling–Koltun or space-filling representation of a molecule in which each atom is represented by a sphere where the radius is proportional to the van der Waals radius of that atom.

**Cross-terms** Cross or off-diagonal terms in a force field account for the fact that bonds and angles in a molecule can be interdependent because the energy for a given stretch or bend depends on the actual value of neighboring bond lengths and bond angles. Cross-terms may increase the accuracy of a force field and may enhance the transferability of the diagonal terms because these are no longer contaminated by these cross-term effects.

**CSD** The Cambridge Structural Database produced by the Cambridge Crystallographic Data Centre contains bibliographic, chemical, and numerical data of crystal structures. This machine-readable file is a comprehensive compendium of molecular geometries of organic and organometallic compounds.

**Cutoff distance** In order to improve the computational efficiency in force-field calculations, nonbonded interaction energy contributions for pairs of atoms separated by distances larger than a predetermined value are neglected. As van der Waals and electrostatic interactions are significant up to 15 Å and for large systems account for more than 90% of the total computational time, a given cutoff distance is always a compromise between computational efficiency and accuracy of the calculation.

**Designer drugs** Substances of abuse that are structural analogs of substances that are subject to the provisions of the U.S. Controlled Substances Act. By "designing" compounds that produce the euphoria of the controlled substances such as narcotics, antidepressants, and stimulants but which are chemically different, laws regulating the controlled substances and the penalties that would be levied for illegally trafficking the controlled substance can be avoided.

**Diagonal terms** Diagonal terms in a force field refer to the terms representing the bond stretch and bond angle deformations, torsion angle, and out-of-plane bending contributions. Diagonal force fields do not contain cross-terms (off-diagonal terms).

**Dipole–dipole force** The dipole–dipole force, also called the Keesom force, arises from the interaction of the permanent dipoles of two interacting molecules. The interaction energy is inversely proportional to the sixth power of the distance between the two dipoles. Dipole–dipole interactions are temperature dependent as thermal motion of the molecules competes with the tendency toward favorable dipole orientations. The energy of two interacting dipoles of $\mu = 2$ Debye at a distance of 5 Å in vacuum is of the order of −0.25 kcal/mol.

**Dipole-induced dipole force** The dipole-induced dipole force, also called the induction or Debye force, arises when a permanent dipole induces a redistribution of electron density in another polarizable molecule, leading to an induced dipole. This type of interaction is inversely proportional to the sixth power of the distance between the two dipoles and is temperature independent. The average dipole-induced dipole interaction energy of a molecule of $\mu = 1$ Debye, e.g., benzene, is about −0.2 kcal/mol at a separation of 3 Å.

**Dispersion force** The dispersion or London force arises from the instantaneous transient dipoles that all molecules possess as a result of the changes in the instantaneous positions of electrons. The dispersion force, which in fact is an induced dipole-induced dipole interaction, depends on the polarizability of the interacting molecules and is inversely proportional to the sixth power of separation. For example, in the case of two $CH_4$ molecules at a separation of 3 Å, the dispersion interaction energy is of the order of −1.1 kcal/mol.

**Distance geometry** A method pioneered by G. M. Crippen that converts a set of distance bounds into a set of coordinates that are consistent with these bounds. In applying distance geometry to conformationally flexible structures, upper and lower bounds to the distance between each pair of points (atoms) are used. This approach is useful for molecular model building, conformational analysis, and has been extended to find a common pharmacophore from a set of biologically active molecules.

**Docking** An operation in which one molecule is brought into the vicinity of another while calculating the interaction energies of the many mutual orientations of the two interacting species. A docking procedure is used as a guide to identify the preferred orientation of one molecule relative to the other. In docking, the interaction energy is generally calculated by computing the van der Waals and the Coulombic energy contributions between all atoms of the two molecules.

**Drug–receptor interaction** The biomolecular reversible association of drug D and receptor R to form DR entails an unfavorable entropic consequence since both D and R each have 3° of translational freedom and 3° of rotational freedom. Those 12° of freedom are reduced to 6° of freedom for DR. Upon formation of DR, free rotation of interacting groups in D and R can be severely restricted or frozen out, leading to another unfavorable free energy change. In the case of poor structural complementarity between D and R, an enthalpic penalty has to be paid to bring D and R in their binding conformation. If the drug–receptor interaction is to take place, these three effects must be counterbalanced by factors favorable for the association of D and R. These are the free energies of interactions between polar functional groups (including the favorable effects of new vibrational modes due to new bond formation upon complexation of D and R), the entropically favorable release of water due to the hydrophobic effect, and from the fact that the molecular packing in the DR complex may be more efficient than the packing of solvated D and R.

**Dummy atom** A point in space that is treated as an atom for the purpose of a geometry definition. Well-known dummy atoms are the centroids of ring systems, the location of electron lone pairs of heteroatoms, and the endpoint of a normal of a plane.

**ECEPP** (Empirical Conformational Energy Program for Peptides) ECEPP pioneered by H. A. Scheraga *et al.* is a molecular mechanics program in which the potential energy is the sum of the electrostatic energy, the nonbonded energy, and the torsional energy. ECEPP is thus an approximation of the general force field.

**EHT** (Extended Hückel Theory) One of the first semiempirical all-valence electron methods formulated by R. Hoffmann in the early 1960s.

**EMBL data library** The main role of the European Molecular Biology Laboratory Data Library is to maintain and distribute a database of nucleotide sequences. This work is a collaborative effort with Genbank and DNA Database of Japan (DDBJ) where each participating group collects a portion of the total reported sequence data. The March 1993 release contained just under 130 million bases from over 105,000 entries. Approximately every 18 months the database doubles in size.

**Ensemble** When treating systems of interacting particles, as in a molecular dynamics simulation, it is useful to introduce the concept of ensemble, which basically means "collection." Taking a closed system with a given volume V, composition N, and temperature T and replicating it *n* times constitutes a canonical ensemble (NVT) in which all the identical closed systems are regarded as being in thermal contact with each other and having the same temperature. In the microcanonical ensemble (NVE), the condition of constant temperature is replaced by the requirement that all the systems should have the same energy. Other ensembles are the isobaric–isoenthalpic NPH ensemble and the isobaric–isothermal NPT ensemble. Depending on the molecular dynamics simulation experiment, an appropriate choice of ensemble has to be made. For example, the NVT ensemble is the appropriate choice when conformational searching of molecules is carried out in vacuum and no periodic boundary conditions are used.

**Eudismic ratio** The ratio of activity or affinity of the eutomer (enantiomeric form with higher activity or affinity) to that of the distomer (enantiomeric form with the lower activity or affinity). The eudismic ratio is a measure of stereoselectivity.

**Excluded volume** The union of volumes of a set of active ligands that is available to the ligands interacting with the receptor. Subtraction of the volume in common with the volume of the active and inactive ligands from the volume of the inactive ligand leads to the receptor essential volume, i.e., the volume required by the receptor.

**Force field** A set of equations and parameters which, when evaluated for a molecular system, yields an energy. Force fields used in molecular mechanics consider the molecular system as a collection of classical masses held together by classical forces. The contributions to the molecular energy include bond stretching, angle bending and dihedral angle deformations, van der Waals, and electrostatic interactions.

**Free energy perturbation** (FEP) A statistical mechanical method to derive the free energy difference between two states a and b from an ensemble average of a potential energy difference ($\Delta V = V_b - V_a$) that can be evaluated using molecular dynamics. In the FEP approach the free energy difference between two states of a system is computed by transforming one state into the other by changing a coupling parameter $\lambda$ in small increments such that the system is in equilibrium at all values of $\lambda$. As $\lambda$ increases from $\lambda = 0$ (state a) to $\lambda = 1$, the system is transformed into the b state. The free energy difference between the two states a and b is then calculated as the sum of free energy differences between the closely spaced $\lambda$ states.

**Frontier orbital** Frontier electron theory is based on the idea that a reaction should occur at the position of the largest electron density in the frontier orbitals. In the case of an electrophilic reaction, the frontier orbital is the HOMO, and the LUMO in the case of a nucleophilic reaction.

**G-protein-coupled receptors** Membrane-bound receptors and effector proteins can communicate via a guanine nucleotide-dependent regulatory protein, the G-protein. The ubiquitous G-protein-coupled receptor family has a common structural framework despite a remarkably wide range of structural characteristics of their activating ligands. The common architecture presumably consists of seven transmembrane $\alpha$ helices of at least 20 residues and intra- and extracellular loops of varying length. Because of their key role in a variety of physiological processes, many of these receptors are the subject of intense pharmacological and theoretical work, including computer-aided model building.

**GenBank** The National Institute of Health database of all known nucleotide and protein sequences. Entries in the database include a description of the sequence, scientific name, and taxonomy of the source organism. Collaboration with the EMBL Data Library and the DNA Database of Japan enables shared data collection and sequence information.

**Genetic algorithms** (GAs) Optimization methods based on Darwinian evolution and used for a wide range of global optimization problems having to do with high-dimensional spaces. As a conformational search method, GAs consist of successively transforming one generation of a series of conformers into the next using the operations of selection (conformers with lower energy are "fitter" than those with higher energy), crossover, and mutation. Since the selection process is biased toward conformations with lower energy, the GA method leads to a collection of low energy conformers.

**GROMOS** (Groningen Molecular Simulation) The GROMOS suite of programs uses a classical force field (bond and angle deformations, torsion, van der Waals, and Coulomb interactions) and the united atom approximation. GROMOS is widely known and is used for the simulation of the chemical and physical behavior of condensed state systems.

**Hansch analysis** A QSAR (quantitative structure–activity relationships) method based on extrathermodynamic principles which express the biological activity of a cogeneric series of molecules in terms of physical quantities, e.g., lipophilicity (log P, $\pi$), electronic (p$K$, $\sigma$), and steric effects ($E_S$ of Taft).

**HOMO** (Highest Occupied Molecular Orbital) A molecular orbital calculation yields a set of eigen values or energy levels in which all the available electrons are accommodated. The highest filled energy level is called the HOMO. The next higher energy level which is unoccupied because no more electrons are available is the LUMO or lowest unoccupied molecular orbital. On the basis of Koopman's theorem, the HOMO and LUMO of a molecule can be approximated as its ionization and electron affinity, respectively.

**Homology modeling** The art of building a protein structure knowing only its amino acid sequence and the complete three-dimensional structure of at least one other reference protein. Protein homology building is based on the fact that there are structurally conserved regions in proteins of a particular family that have nearly identical structure. In homology modeling, sequence alignment methods are used in determining which regions of the reference protein(s) and the unknown protein are conserved.

**Hydrogen bond** The stabilizing interaction, either inter- or intramolecular, between two moieties XH and Y. It is commonly assumed that for a hydrogen bond to be formed that both X and Y should be electronegative elements. Evidence is accumulating that hydrogen bonds can also be formed between CH...O and OH...$\pi$-bonded systems. Hydrogen bonds have specific geometric directionality and properties and therefore give rise to geometrically well-organized structures in biological systems such as DNA and proteins.

**Lennard–Jones potential** As two atoms approach one another there is the attraction due to London dispersion forces and eventually a van der Waals repulsion as the interatomic distance $r$ gets smaller than the equilibrium distance. A well-known potential energy function to describe this behavior is the Lennard–Jones (6–12) potential. The LJ (6–12) potential represents the attractive part as $r^{-6}$ dependent whereas the repulsive part is represented by an $r^{-12}$ term. Another often used nonbonded interaction potential is the Buckingham potential which uses a similar distance dependence for the attractive part as the LJ (6–12) potential but where the repulsive part is represented by an exponential function.

**Local density functional (LDF) theory** The LDF approach is a calculational procedure according to which all of the electronic properties of a chemical system, including the energy, can be derived from the electronic density. LDF theory, which is steadily gaining popularity in the chemical computational community, takes into account electron correlation. It requires considerably less computer time and disk space than ab initio calculations, making it feasible to deal with much larger molecular systems.

**MINDO/3** (Modified Intermediate Neglect of Differential Overlap) The MINDO/3 technique representing the third version of MINDO is a semiempirical all-valence electron self-consistent field molecular orbital approach. MINDO/3 calculations provide fairly accurate values of molecular properties on medium to large organic molecules.

**Minimization** Minimization of the energy of a molecule is a procedure used to find configurations for which the molecular energy is a minimum, i.e., finding a point in configuration space where all the forces acting on the atoms are balanced. As several points exist in large molecules where the atomic forces are balanced, finding the point of the absolute minimum energy is often not a trivial problem. Different minimization algorithms (e.g., steepest descents, conjugate gradients, Newton–Raphson) and procedures such as simulated annealing are used to find the minimum energy conformation of a molecule.

**Minimum energy conformation** (MEC) The MEC is that point in configurational space where the energy of the molecule is an absolute minimum and where all the derivatives are zero and the second derivative matrix (Hessian matrix) is positive definite.

**MNDO** (Modified Neglect of Diatomic Overlap) A semiempirical all-valence electron quantum chemical method pioneered by M. J. S. Dewar and co-workers. For the molecular properties investigated, such as heats of formation, ionization potentials, bond lengths, and dipole moments, MNDO values are quite close to the experimental ones and are superior to the MINDO/3 results, particularly for nitrogen-containing compounds. Taking into account that the computational effort for MNDO is only about 20% greater than for a MINDO/3 calculation, MNDO is considered to be a significant improvement over MINDO/3.

**Molecular dynamics** (MD) Taking the negative gradient of the potential energy as evaluated from the force field yields the force. Using this force and the mass for each atom, Newton's equation of motion (F = ma) can be numerically integrated to compute the positions of the atoms after a short time interval (typically of the order of 1 fsec, $10^{-15}$ sec). By taking successive time steps, a time-dependent trajectory of all the atomic motions can be constructed.

**Molecular electrostatic potential** (MEP) The MEP at first order is the interaction energy between a molecule and a unit positive charge. Both the molecular geometry and the charge density are considered unchanged as the positive charge approaches. The MEP may give useful information about the reactivity of a molecule such as the most probable sites of electrophilic or nucleophilic attack.

**Molecular mechanics** (MM) Molecular mechanics is an attempt to formulate a force field that can serve as a computational model for evaluating the potential energy for all degrees of freedom of a molecule. MM calculations are very popular because large structures containing many thousands of atoms can be fully energy minimized at reasonable computational costs. MM methods, however, depend heavily on the parametrization of the force field. MM is not appropriate for simulating situations where electronic effects such as orbital interactions and bond breaking are predominant.

**Molecular modeling** Molecular modeling of a molecule consists of a computer graphics visualization and representation of the geometry of a molecule. In addition, it involves the manipulation and modification of molecular structures. In combination with X-ray crystallographic or NMR data, molecular modeling implies the use of theoretical methods like ab initio, semiempirical, or molecular mechanics to evaluate and predict the minimum energy conformation and other physical and chemical properties of the molecule. Molecular modeling has become an essential tool for structural molecular biology with applications in drug design, protein engineering, and molecular recognition.

**Molecular recognition** The interaction between two molecules, e.g., ligand and receptor, is mainly dependent on the drug being able to sterically fit into the active site of the receptor and on the electrostatic complementarity between drug and receptor. The ligand and receptor are at the molecular recognition state when they are separated by more than two van der Waals radii. The contributing interactions for recognition are electrostatic, hydrogen bonding, van der Waals, and hydrophobic in nature.

**Molecular similarity** The degree of similarity between molecules, although quantitatively measurable, depends on what molecular features are used to establish the degree of similarity. One of the many comparators is the electron density of a pair of molecules. Other comparators include electrostatic potentials, reactivity indices, lipophilicity potentials, molecular geometry such as distances and angles between key atoms, and solvent accessible surface area. It is an open question as to how much or what part(s) of the molecular structure is to be compared.

**Monte Carlo** (MC) Straightforward scanning of the complete configuration space of a molecular system containing many degrees of freedom is impossible. In that case, an ensemble of configurations can be generated by the MC method which makes use of random sampling and Boltzmann factors. Given a starting configuration, a new configuration is generated by randomly displacing one or more atoms. The newly generated configuration is either accepted or rejected using an energy criterion involving the change of the potential energy ($\Delta V$) relative to the previous configuration. The current configuration is accepted only if its potential energy is lower or equal to the previous one ($\Delta V \leq 0$) or for $\Delta V > 0$ if the Boltzmann factor $\exp(-\Delta V/RT)$ is larger than a random number taken from a uniform distribution over the (0,1) interval. MC methods are in general less efficient in sampling configuration space than MD methods.

**MOPAC** A general Molecular Orbital Package based on the semiempirical all-valence electron approximation for the study of the chemical behavior of molecules and ions. In its present state of development the user can choose the level of approximation in terms of the MNDO, MINDO/3, AM1, and PM3 Hamiltonians.

**Morse potential** Often used for the bond stretching term in a force field. Instead of the quadratic dependence of a harmonic bond stretching term, a Morse potential describes the bond stretching mode as an exponential function. When a molecule is in a high energy state due to sterically overlapping atoms or at a high temperature molecular in a dynamics simulation, the Morse function may allow the bonded atoms to stretch to unrealistic bond lengths.

**Newton–Raphson** (NR) A mathematical technique used for the optimization of a function. In contrast to steepest descents and conjugate gradients methods, where the first derivative or gradient of the function is used, NR methods also use second derivative information to predict where along the gradient the function will change directions. As the second partial derivative matrix of the energy function (Hessian matrix) is calculated, the NR method is much more time-consuming than the steepests descents and conjugate gradients methods. NR minimization becomes unstable when a structure is far from the minimum where the forces are large and the second derivative (the curvature) is small. Because storage requirements scale as $3N^2$ ($N$ is the number of atoms), NR methods are not suitable for large structures such as proteins.

**NOE** (nuclear Overhauser effect) The origin of NOE is dipolar: cross relaxation between protons. Because of the $r^{-6}$ distance dependence effect, NOEs can only be measured between protons at distances shorter than 5 Å. Using an appropriate distance restraint term in a force-field energy function, based on the experimentally derived NOE data, is useful in finding theoretical conformations that are consistent with NMR-based solution structures.

**OPLS** (Optimized Potentials for Liquid Simulations) The OPLS force field addresses the classical bond stretches, angle bends, and torsions. The nonbonded interactions are represented by Coulomb and Lennard–Jones terms which are parametrized to reproduce experimental thermodynamic and structural data on organic fluids. In the OPLS model, no special functions are needed to describe hydrogen bonding nor additional interaction sites for lone pairs. The standard combination rules $A_{ij} = (A_{ii}A_{jj})^{1/2}$ and $C_{ij} = (C_{ii}\,C_{jj})^{1/2}$ are used for the parametrization of the nonbonded dispersion and repulsion interactions, respectively.

**ORTEP** (Oak Ridge Thermal Ellipsoid Program) This program is still very popular among crystallographers for drawing ball and stick-type crystal structure illustrations. The program can produce stereoscopic pairs which aid in the visualization of complex packing arrangements of atoms.

**Out-of-plane bend** The displacement of a trigonal atom above and below the molecular plane is a mode of motion distinguishable from the bond stretching, angle bending, and torsional motions. This out-of-plane coordinate is often called improper torsion because it treats the four atoms in the plane as if they were bonded in the usual way as in a proper torsional angle.

**Parametrization of force fields** The reliability of a molecular mechanics calculation depends on the potential energy equations and the numerical values of the parameters. One obstacle is the small amount of experimental data available for parametrizing and testing a force field. The energy, first and second derivatives of the energy with respect to the cartesian coordinate of a molecule obtained from high-quality ab initio calculations are used to optimize force-field parameters by adjusting the parameters to fit the energy and the energy derivatives by least-squares methods.

**Partial least squares** (PLS) A mathematical modeling technique relating physicochemical properties and one or several measurements of biological activity. The PLS results consist of two sets of computed factors which are, on the one hand, linear combinations of the chemical descriptors and, on the other hand, linear combinations of the biological activities. PLS finds many applications in chemometrics and in the CoMFA approach.

**Partition function** The partition function $Q$ is the summation of the Boltzmann factors $\exp[-\varepsilon_i/kT]$ over the energy levels $\varepsilon_i$ of a molecule. A large value of $Q$ results when the energy levels $\varepsilon_i$ are closely spaced. The partition function is a measure of the number of available translational, rotational, vibrational, and electronic energy levels. Its value depends on the molecular weight, the temperature, the molecular volume, the internuclear distances, the molecular motions, and the intermolecular forces. Although in many cases the calculation of the energy level pattern is often impossible, reasoning in terms of partition functions may provide a more concrete understanding of the free energy of drug–receptor interactions.

**Pattern recognition** (PR) A branch of artificial intelligence that provides an approach to solve the problem of recognizing an obscure property in a collection of objects from measurements made on the objects. PR techniques can be divided into display, preprocessing, supervised, and unsupervised learning. PR methods are used among others in the search for correlations between molecular structure and biological activity.

**PBC** (Periodic Boundary Conditions) The simulation of molecular systems in a periodic 3-D lattice of identical replicates of the molecular system under consideration. Using PBC allows to simulate the influence of bulk solvent in such a way as to minimize edge effects such as diffusion of a solute toward a surface or the evaporation of solvent molecules.

**PCILO** The Perturbative Configuration Interaction using Localized Orbitals method is a semiempirical all-valence electron quantum chemical method where, in addition to the ground state, singly and doubly excited configurations are taken into account. The wave function and the ground-state energy are determined by the Rayleigh–Schrödinger perturbation treatment up to the third order. Because of this summation treatment, PCILO is much faster than the self-consistent field methods such as MNDO, AM1, and PM3.

**PDB** (Protein Data Bank) Compiled at Brookhaven National Laboratory and distributed from there, the PDB contains mainly X-ray diffraction and NMR-based structural data of macromolecular structures such as proteins and nucleic acids. The PDB file is the primary source for the 3-D coordinates of macromolecular structures.

**Peptidomimetics** Compounds derived from peptides and proteins and obtained by structural modification using unnatural amino acids, conformational restraints, isosteric replacement, cyclization, etc. The peptidomimetics bridge the gap between simple peptides and the nonpeptide synthetic structures and as such may be useful in delineating pharmacophores and in helping to translate peptides into small nonpeptide compounds. Peptidomimetic is sometimes used in a broad sense to designate organic molecules mimicking some properties of peptide ligands.

**Pharmacophore** The spatial mutual orientation of atoms or groups of atoms assumed to be recognized by and to interact with a receptor or the active site of a receptor. In conjunction with the receptor concept, the notion of a pharmacophore relates directly to the lock-and-key theory proposed by Ehrlich at the beginning of this century (*Corpora non agunt nisi fixata*).

**PIR** (Protein Information Resource) These databases are maintained and distributed by an association of macromolecular sequence data collection centers. The Protein Sequence Database is a research tool for the study of protein evolution in which published protein sequences are organized by similarity and evolutionary relationship.

**PM3** (Parametrized Model 3) A reparametrized version of the AM1 method. On the whole, it seems that PM3 gives better estimates of the heat of formation than AM1.

**Potential of mean force** The thermodynamic quantity needed to estimate equilibrium constants is the $\Delta G$ between reactants and products. By sampling a reaction coordinate $r$, a potential of mean force (pmf) can be obtained. From the frequency of occurrence of different $r$ values, a distribution function $g(r)$ is calculated that is related to $w(r)$, the relative free energy or the pmf by $w(r) = -kT\ln g(r)$. By using an additional constraining or biasing potential (umbrella), a system can be forced to sample a reaction coordinate region which would be infrequently sampled in the absence of the umbrella potential because of high barriers in $w(r)$.

**Protein folding** One of the most challenging problems in structural biology is the prediction of the three-dimensional tertiary structure of a protein from its primary structure. Despite many years of experimental and theoretical studies devoted to it, the protein folding problem remains essentially unsolved. There are too many conformations that can occur in both the unfolded and the folded structure to be searched. Present-day computational approaches (because of the size of proteins this must necessarily be molecular mechanics) cannot evaluate with sufficient accuracy the relative conformational energies of the folded and unfolded states which differ from one another most probably only by a few kilocalories per mole. The problem of protein folding is further compounded by solvent and environmental effects.

**Quantitative structure–activity relationships** (QSAR) The QSAR approach pioneered by Hansch and co-workers relates biological data of congeneric structures to physical properties such as lipophilicity, electronic, and steric effects using linear regression techniques to estimate the relative importance of each of those effects contributing to the biological effect. A statistically sound QSAR regression equation can be used for lead optimization.

**Quantum chemistry program exchange** (QCPE) This initiative was undertaken by an organization committed to the promotion of the concept and practice of computational chemistry initiated by Professor H. Shull in 1962 at the Department of Chemistry at Indiana University. QCPE till 1970 financially supported by the U.S. Air Force Office of Scientific Research, receives software packages and distributes them to other theoretical chemists all over the world at a nominal fee. This organization has contributed greatly to the dissemination of a number of excellent theoretical chemistry programs to the scientific community.

**Radial distribution function** (rdf) A term that is often utilized in analyzing the results of Monte Carlo or MD calculations. The rdf, $g(r)$ gives the probability of occurrence of an atom of type a at a distance $r$ from an atom of type b. Peaks in the $g(r)$ vs $r$ plots can be associated with solvation shells or specific neighbors and can be integrated to yield coordination numbers.

**Rational drug design** The majority of drugs on the market today to treat disorders in humans, animals, and plants were discovered either by chance observation or by systematic screening of large series of synthetic and natural substances. This traditional method of drug discovery is now supplemented by methods exploiting the increasing knowledge of the molecular targets assumed to participate in some disorder, computer technology, and the physical principles underlying drug–target interactions. Rational drug design—traditional methods were or are not irrational—or better "structure-based ligand design" continues to increase in importance in the endeavor of promoting a biologically active ligand toward the status of a drug useful in human and veterinary medicine and the phytopharmaceutical world.

**Receptor** A receptor can be envisioned as a macromolecular structure such as a protein, an enzyme, or a polynucleotide being an integral part of the complex molecular structure of the cellular membrane in which it is anchored or associated with. The recognition elements or receptor sites are oriented in such a way that recognition of and interaction with ligands can take place, leading to a pharmacological effect.

**Receptor mapping** The topographical feature representation of a receptor based on the SAR and conformational aspects of active and inactive analogs of rigid and flexible molecules all putatively acting on that receptor. Inferences as to a pharmacophore on the basis of molecular interactions such as ionic and hydrogen bonding, dipolar effects, $\pi$–$\pi$ stacking interactions, and hydrophobic interactions can be used to construct a hypothetical model of the receptor in which the accessible parts of the amino acids of the receptor protein are delineated.

**Restraint** A restraint biases or forces a target function such as the energy function in molecular mechanics toward a specific value for a degree of freedom. Various restraints are in common use: torsional restraints, distance restraints, and tethering.

**Semi-ab initio method** (SAM1) The major difference between SAM1 and AM1 involves the repulsion integrals that are calculated using an STO-3G basis set and then scaled to account for electron correlation.

**SHAKE** One approach to reduce the computer time of computationally expensive MD calculations is to increase the time step $\Delta t$ used for the numerical integration of Newton's equations. For reasons of numerical stability, $\Delta t$ must be small compared to the period of the highest frequency of motions viz. bond stretching vibrations. SHAKE is an algorithm that can constrain bonds to a fixed length during a MD calculation, thereby allowing somewhat larger $\Delta t$ values.

**Simulated annealing** (SA) A technique used in locating the global minimum energy structure of polypeptides and proteins. SA uses a Monte Carlo search of conformational space starting at high temperatures where large changes in conformational energies are allowed. As the temperature is lowered with an appropriate cooling schedule, the system is (possibly) trapped into a conformation of lowest energy.

**Slow growth** The slow growth method for free energy calculations is a free energy perturbation or a thermodynamic integration approach under the assumption that the spacings $d\lambda$ of the coupling parameter $\lambda$ are so small that one needs to sample only one point at any window. This reduces the ensemble average to a single value and allows the derivative to be approximated by a finite difference.

**SMILES** (Simplified Molecular Input Line Entry System) SMILES is a chemical notation system based on the principles of molecular graph theory and denotes a molecular structure as a two-dimensional graph familiar to chemists. It allows a rigorous and unambiguous structure specification representing molecular structures by a linear string of symbols. SMILES is used for chemical structure storage, structural display, and substructure searching.

**Spacer** A chemical moiety that presumably serves to hold pharmacophoric fragments at a proper distance and in a conformation compatible with optimal binding. Spacers, however, such as double bonds, cyclohexane, and phenyl rings probably also provide additional binding at the active site.

**SPC** (Simple Point Change) In view of the importance of the water–protein interactions, it is of utmost interest to have intermolecular potential functions available for the water dimer that yield a good model for liquid water. The SPC is a three-point charge (on the hydrogen and oxygen positions) model for water with a (6–12) Lennard–Jones potential on the oxygen atom and a charge of 0.41 and −0.82 on the hydrogen and oxygen atoms, respectively.

**Steepest descents** A minimization algorithm in which the line search direction is taken as the gradient of the function to be minimized. The steepest descents method is very robust in situations where configurations are far from the minimum but converge slowly near the minimum (where the gradient approaches zero).

**STO-3G** An abbreviation employed in ab initio MO calculations indicating the basis set used. The notation STO-n G stands for Slater Type Orbital simulated by n Gaussian functions. This means that each atomic orbital consists of n Gaussian functions. STO-3G is a minimal basis set and is now hardly used any longer. More elaborate basis sets include the 6-31 G* basis set whereby six Gaussians are used for the core orbitals, three for the s-, and one for the p-valence orbitals and a single set of d-functions is indicated by the asterisk.

**Stochastic dynamics** (SD) This method is a further extension of the original MD method. A space–time trajectory of a molecular system is generated by integration of the stochastic Langevin equation which differs from the simple MD equation by the addition of a stochastic force R and a frictional force proportional to a friction coefficient $\gamma$. The SD approach is useful for the description of slow processes such as diffusion, the simulation of electrolyte solutions, and various solvent effects.

**Strain energy** Although the first strain theory is due to von Bayer in 1885, there is no generally accepted and unique definition of strain energy. The basic qualitative idea is that simple strainless molecules exist and that larger molecules are strainless if their heats of formation are equal to the summation of the bond energies and other increments from the small strainless molecules. The energy calculated by molecular mechanics is strain energy because the deformation energy occurring in a molecule is equal to the energy of minimized structure relative to the hypothetical reference structure.

**Superdelocalizability** Superdelocalizability $S_r$ is defined as the sum of the quotients of the squares of the coefficients $C_{rj}$ of the $r$th atomic orbital in the $j$th molecular orbital and the orbital energies. In the case of superdelocalizability for electrophilic attack $S_r^E$ the sum runs over all occupied orbitals and over all unoccupied orbitals for $S_r^N$ for nucleophilic attack. The superdelocalizability for radical attack is defined as $(S_r^E + S_r^N)/2$.

**SWISS-PROT** An annotated protein sequence database maintained by the Department of Medical Biochemistry of the University of Geneva and the EMBL Data Library. The SWISS-PROT database distinguishes itself from other protein databases by (i) the generous annotation information, (ii) a minimal redundancy for a given protein sequence, and (iii) the cross-reference with 12 other biomolecular databases.

**Switching function** In order to avoid discontinuities in derivatives and energies during minimization calculations, a switching function is used in conjunction with a cut-off algorithm ensuring nonbonded interactions to be smoothly reduced from full strength to zero over a predefined interatomic distance range.

**Template forcing** A type of restraint useful in the identification of possible biologically relevant conformations of a conformationally flexible molecule. By selecting atoms or groups of atoms belonging to the possible pharmacophoric pattern common to two molecules, the atoms of the flexible molecule are forced to superimpose onto the atoms of the rigid or template molecule. The energy expenditure to force the flexible molecule onto the template molecule is a measure of the similarity between the two molecules.

**Thermodynamic cycle** This approach, used to calculate relative free energies or binding constants, e.g., drug–receptor interactions, is based on the fact that the free energy is a thermodynamic function of state. Thus, as long as a system is changed reversibly, the change in free energy is independent of the path and therefore nonchemical processes (paths) can be calculated such as the conversion of one type of atom into another (computational alchemy!).

**Thermodynamic integration** (TI) An approach to free energy calculations consisting of numerically integrating the ensemble average of the derivative of the potential energy of a given configuration with respect to a coupling parameter $\lambda$. Because the free energy is evaluated directly from the ensemble average and not as the logarithm of the average of an exponential function as in the free energy perturbation (FEP), TI is not subject to certain systematic error inherent to FEP calculations.

**Three-dimensional search** Having converted the traditional 2-D databases of chemical structures to a 3-D database, 3-D searching is used to find all molecules in that database that contain a specific pharmacophore. Conformationally flexible searching addresses the problem of finding molecules with a conformation different from that which is stored in the primary 3-D database.

**Time correlation function** This function is of great value for the analysis of dynamical processes in condensed phases. A time correlation function $C(t)$ is obtained when a time-dependent quantity $A(t)$ is multiplied by itself (autocorrelation) or by another time-dependent quantity $B(t')$ evaluated at time $t'$ (cross-correlation) and the product is averaged over some equilibrium ensemble. For example, the self-diffusion coefficient can be obtained from the velocity autocorrelated function for the molecular center of mass motion.

**TIP** (Transferable Intermolecular Potential) The TIP family of potentials is used for simulating liquid water. The TIP4P potential for water involves a rigid water monomer composed of three charge centers and one Lennard–Jones center. Two charge centers ($Q = 0.52$) are placed on the hydrogen site 0.9572 Å away from the oxygen atom. The third charge center (Q = −1.04) is placed 0.15 Å away from the oxygen atom along the bisector of the HOH angle (104.52°). A Lennard–Jones center is placed on the oxygen atom. The model yields reasonable geometric and energetic results for a linear water dimer and is therefore used in simulations of aqueous solutions.

**United atom model** For the sake of speeding up an energy calculation the total number of atoms are artificially reduced by lumping together all nonpolar hydrogens into the heavy atoms (C atoms) to which they are bonded. Although this approximation may speed up the calculation severalfold, an all hydrogen atom model is preferable for accurate calculations.

**Van der Waals forces** This term denotes the short-range interactions between closed-shell molecules. Van der Waals forces include attractive forces arising from interactions between the partial electric charges and repulsive forces arising from the Pauli exclusion principle and the exclusion of electrons in overlapping orbitals. A very commonly used potential is the so-called Lennard–Jones (6–12) potential used to describe the attractive and repulsive components of van der Waals forces.

**Verlet** The Verlet algorithm is a numerical method for the integration of Newton's equations of motion used in MD calculations. Because of its simplicity, the Verlet method is easily adapted to give an algorithm with constraints on internal coordinates such as bond lengths as in the SHAKE algorithm.

**X-ray structure** Single-crystal X-ray diffraction analysis yields the three-dimensional structure of a molecule in the crystalline state. An X-ray structure is likely to be a structure in a minimum energy conformational state or close to an energy minimum. An X-ray structure, therefore, may or may not be the biologically relevant conformation. Inspection of the molecular packing arrangement may yield valuable information about intermolecular contacts and sites of intermolecular hydrogen bonds. Atomic coordinates based on X-ray diffraction data may serve as the primary input data for theoretical conformational analysis calculations.

**Z-matrix** The Z-matrix provides a description of each atom of a molecule in terms of its atomic number, bond length, bond angle, and dihedral angle, the so-called internal coordinates. The information from the Z-matrix is used to calculate the Cartesian (X, Y, Z) coordinates of the atoms.

**Zero-point energy** The residual energy of a harmonic oscillator at the lowest vibrational state. It arises from the fact that the position of a particle is uncertain and therefore its momentum and hence its kinetic energy cannot be exactly zero.

# Index